Student Solutions Manual and Study Guide

for
Serway and Jewett's

Physics for Scientists and Engineers with Modern Physics

Sixth Edition

Volume Two

John R. Gordon
James Madison University

Ralph V. McGrew
Broome Community College

Raymond A. Serway

THOMSON
™
BROOKS/COLE

Australia • Canada • Mexico • Singapore • Spain • United Kingdom • United States

Printed in the United States of America
2 3 4 5 6 7 07 06

Printer: Globus Printing

ISBN: 0-534-40856-7

For more information about our products, contact us at:
Thomson Learning Academic Resource Center
1-800-423-0563

For permission to use material from this text, contact us by:
Phone: 1-800-730-2214
Fax: 1-800-731-2215
Web: http://www.thomsonrights.com

Cover Image: Water displaced by water tanker, © Stuart Westmorland/CORBIS

Brooks/Cole—Thomson Learning
10 Davis Drive
Belmont, CA 94002-3098
USA

Asia
Thomson Learning
5 Shenton Way #01-01
UIC Building
Singapore 068808

Australia/New Zealand
Thomson Learning
102 Dodds Street
Southbank, Victoria 3006
Australia

Canada
Nelson
1120 Birchmount Road
Toronto, Ontario M1K 5G4
Canada

Europe/Middle East/South Africa
Thomson Learning
High Holborn House
50/51 Bedford Row
London WC1R 4LR
United Kingdom

Latin America
Thomson Learning
Seneca, 53
Colonia Polanco
11560 Mexico D.F.
Mexico

Spain/Portugal
Paraninfo
Calle/Magallanes, 25
28015 Madrid, Spain

Preface

This Student Solution Manual and Study Guide has been written to accompany the textbook **Physics for Scientists and Engineers**, Sixth Edition, by Raymond A. Serway and John W. Jewett, Jr. The purpose of this Student Solution Manual and Study Guide is to provide the students with a convenient review of the basic concepts and applications presented in the textbook, together with solutions to selected end-of-chapter problems from the textbook. This is not an attempt to rewrite the textbook in a condensed fashion. Rather, emphasis is placed upon clarifying typical troublesome points, and providing further drill for methods of problem solving.

Each chapter is divided into several parts, and every textbook chapter has a matching chapter in this book. Very often, reference is made to specific equations or figures in the textbook. Every feature of this Study Guide has been included to ensure that it serves as a useful supplement to the textbook. Most chapters contain the following components:

- **Equations and Concepts:** This represents a review of the chapter, with emphasis on highlighting important concepts, and describing important equations and formalisms.

- **Suggestions, Skills, and Strategies:** This offers hints and strategies for solving typical problems that the student will often encounter in the course. In some sections, suggestions are made concerning mathematical skills that are necessary in the analysis of problems.

- **Review Checklist:** This is a list of topics and techniques the student should master after reading the chapter and working the assigned problems.

- **Answers to Selected Questions:** Suggested answers are provided for approximately fifteen percent of the conceptual questions.

- **Solutions to Selected Problems:** Solutions are shown for approximately twenty percent of the problems from the text, mostly chosen to have odd numbers and to illustrate the important concepts of the chapter.

We sincerely hope that this Student Solution Manual and Study Guide will be useful to you in reviewing the material presented in the text, and in improving your ability to solve problems and score well on exams. We welcome any comments or suggestions which could help improve the content of this study guide in future editions; and we wish you success in your study.

John R. Gordon
Harrisonburg, Virginia

Ralph McGrew
Binghamton, New York

Raymond A. Serway
Leesburg, Virginia

Acknowledgments

It is a pleasure to acknowledge the excellent work of Laura and Michael Rudmin of DSC Publishing, Aleksandras Urbonas, and Richardas Preksat, whose attention to detail in the preparation of the camera-ready copy did much to enhance the quality of this Sixth Edition of the <u>Student Solutions Manual and Study Guide</u> to accompany <u>Physics for Scientists and Engineers</u>. Their graphics skills and technical expertise combined to produce illustrations for earlier editions which continue to add much to the appearance and usefulness of this volume.

Special thanks, for managing all phases of this project, go to Susan Dust Pashos, Senior Developmental Editor; Alyssa White, Assistant Editor for Physics and Chemistry; and Rebecca Heider, Associate Developmental Editor at Thomson Learning. James McLean of the State University College at Geneseo, New York, served as accuracy reviewer for this volume and made many helpful suggestions. Finally, we express our appreciation to our families for their inspiration, patience, and encouragement.

Suggestions for Study

Very often we are asked "How should I study this subject, and prepare for examinations?" There is no simple answer to this question, however, we would like to offer some suggestions which may be useful to you.

1. It is essential that you understand the basic concepts and principles before attempting to solve assigned problems. This is best accomplished through a careful reading of the textbook before attending your lecture on that material, jotting down certain points which are not clear to you, taking careful notes in class, and asking questions. You should reduce memorization of material to a minimum. Memorizing sections of a text, equations, and derivations does not necessarily mean you understand the material. Perhaps the best test of your understanding of the material will be your ability to solve the problems in the text, or those given on exams.

2. Try to solve as many problems at the end of the chapter as possible. You will be able to check the accuracy of your calculations to the odd-numbered problems, since the answers to these are given at the back of the text. Furthermore, detailed solutions to approximately half of the odd-numbered problems are provided in this study guide. Many of the worked examples in the text will serve as a basis for your study.

3. The method of solving problems should be carefully planned. First, read the problem several times until you are confident you understand what is being asked. Look for key words which will help simplify the problem, and perhaps allow you to make certain assumptions. You should also pay special attention to the information provided in the problem. In many cases a simple diagram is a good starting point; and it is always a good idea to write down the given information before proceeding with a solution. After you have decided on the method you feel is appropriate for the problem, proceed with your solution. If you are having difficulty in working problems, we suggest that you again read the text and your lecture notes. It may take several readings before you are ready to solve certain problems, though the solved problems in this Study Guide should be of value to you in this regard. However, your solution to a problem does not have to look just like the one presented here. A problem can sometimes be solved in different ways, starting from different principles. If you wonder about the validity of an alternative approach, ask your instructor.

4.	After reading a chapter, you should be able to define any new quantities that were introduced, and discuss the first principles that were used to derive fundamental formulas. A review is provided in each chapter of the Study Guide for this purpose, and the marginal notes in the textbook (or the index) will help you locate these topics. You should be able to correctly associate with each physical quantity the symbol used to represent that quantity (including vector notation if appropriate) and the SI unit in which the quantity is specified. Furthermore, you should be able to express each important formula or equation in a concise and accurate prose statement.

5.	We suggest that you use this Study Guide to review the material covered in the text, and as a guide in preparing for exams. You should also use the **Equations and Concepts** to focus in on any points which require further study. Remember that the main purpose of this Study Guide is to improve upon the efficiency and effectiveness of your study hours and your overall understanding of physical concepts. However, it should not be regarded as a substitute for your textbook or individual study and practice in problem solving.

6.	**A note concerning significant figures.** When the statement of a problem gives data to three significant figures, we state the answer to three significant figures. The last digit is uncertain; it can for example depend on the precision of the values assumed for physical constants and properties. When a calculation involves several steps, we carry out intermediate steps to many digits, but we write down only three. We 'round off' only at the end of any chain of calculations, never anywhere in the middle.

Problem Solving

Besides what you might expect to learn about physics concepts, a very valuable skill you should hope to take away from your physics course is the ability to solve complicated problems. The way physicists approach complex situations and break them down into manageable pieces is extremely useful. We have developed a general problem-solving strategy that will help guide you through the steps. To help you remember the steps of the strategy, they are called *Conceptualize, Categorize, Analyze,* and *Finalize*.

General Problem-Solving Strategy

Conceptualize

- The first thing to do when approaching a problem is to *think about* and *understand* the situation. Study carefully any diagrams, graphs, tables, or photographs that accompany the problem. Imagine a movie, running in your mind, of what happens in the problem.

- If a diagram is not provided, you should almost always make a quick drawing of the situation. Indicate any known values, perhaps in a table or directly on your sketch.

- Now focus on what algebraic or numerical information is given in the problem. Carefully read the problem statement, looking for key phrases such as "starts from rest" ($v_i = 0$), "stops" ($v_f = 0$), or "freely falls" ($a_y = -g = -9.80 \text{ m/s}^2$).

- Now focus on the expected result of solving the problem. Exactly what is the question asking? Will the final result be numerical or algebraic? If it is numerical, what units will it have? If it is algebraic, what symbols will appear in it?

- Don't forget to incorporate information from your own experiences and common sense. What should a reasonable answer look like? What should its order of magnitude be? You wouldn't expect to calculate the speed of an automobile to be 5×10^6 m/s.

Categorize

- Once you have a really good idea of what the problem is about, you need to *simplify* the problem. Remove the details that are not important to the solution. For example, model a moving object as a particle. If appropriate, ignore air resistance or friction between a sliding object and a surface.

- Once the problem is simplified, it is important to *categorize* the problem. How does it fit into a framework of ideas that you construct to understand the world? Is it a simple *plug-in problem*, such that numbers can be simply substituted into a definition? If so, the problem is likely to be finished when this substitution is done. If not, you face what we can call an *analysis problem* — the situation must be analyzed more deeply to reach a solution.

- If it is an analysis problem, it needs to be categorized further. Have you seen this type of problem before? Does it fall into the growing list of types of problems that you have solved previously? Being able to classify a problem can make it much easier to lay out a plan to solve it. For example, if your simplification shows that the problem can be treated as a particle moving under constant acceleration and you have already solved such a problem (such as the examples in Section 2.5), the solution to the present problem follows a similar pattern.

Analyze

- Now, we need to analyze the problem and strive for a mathematical solution. Because you have already categorized the problem, it should not be too difficult to select relevant equations that apply to the type of situation in the problem. For example, if your categorization shows that the problem involves a particle moving under constant acceleration, Equations 2.9 to 2.13 are relevant.

- Use algebra (and calculus, if necessary) to solve symbolically for the unknown variable in terms of what is given. Substitute in the appropriate numbers, calculate the result, and round it to the proper number of significant figures.

Finalize

- This is the most important part. Examine your numerical answer. Does it have the correct units? Does it meet your expectations from your conceptualization of the problem? What about the algebraic form of the result—before you substituted numerical values? Does it make sense? Try looking at the variables in it to see whether the answer would change in a physically meaningful way if they were drastically increased or decreased or even became zero. Looking at limiting cases to see whether they yield expected values is a very useful way to make sure that you are obtaining reasonable results.

- Think about how this problem compares with others you have done. How was it similar? In what critical ways did it differ? Why was this problem assigned? You should have learned something by doing it. Can you figure out what? Can you use your solution to expand, strengthen, or otherwise improve your framework of ideas? If it is a new category of problem, be sure you understand it so that you can use it as a model for solving future problems in the same category.

When solving complex problems, you may need to identify a series of subproblems and apply the problem-solving strategy to each. For very simple problems, you probably don't need this strategy at all. But when you are looking at a problem and you don't know what to do next, remember steps in the strategy and use them as a guide.

Table of Contents

Chapter 23
ELECTRIC FIELDS

EQUATIONS AND CONCEPTS

The **magnitude of the electrostatic force** between two stationary point charges, q_1 and q_2, separated by a distance r is given by Coulomb's law. *The term point charge is used to indicate a particle of zero size that carries an electric charge.*

$$F_e = k_e \frac{|q_1||q_2|}{r^2} \qquad (23.1)$$

The **Coulomb constant** is stated in SI units in Equation 23.2; the constant can also be expressed in terms of the permittivity constant, ϵ_0.

$$k_e = 8.987\,5 \times 10^9 \text{ N·m}^2/\text{C}^2 \qquad (23.2)$$

$$k_e = \frac{1}{4\pi\epsilon_0} \qquad (23.3)$$

$$\epsilon_0 = 8.854\,2 \times 10^{-12} \text{ C}^2/\text{N·m}^2 \qquad (23.4)$$

An **approximate value** for k_e may be used in solving end-of-chapter problems.

$$k_e = 8.99 \times 10^9 \text{ N·m}^2/\text{C}^2$$

The **charge on an electron** $(-e)$ or a proton $(+e)$ is the smallest unit of charge known in nature.

$$e = 1.602\,19 \times 10^{-19} \text{ C} \qquad (23.5)$$

The **vector form of Coulomb's law** includes a unit vector. $\mathbf{F}_{12}$ is the force on q_2 due to q_1. The unit vector $\hat{\mathbf{r}}$ is directed from q_1 to q_2. *Coulomb's law applies exactly only to point charges. Regardless of the relative magnitudes of the two charges, $\mathbf{F}_{21} = -\mathbf{F}_{12}$; this follows from Newton's third law.*

$$\mathbf{F}_{12} = k_e \frac{q_1 q_2}{r^2}\hat{\mathbf{r}} \qquad (23.6)$$

1

The **relative direction of the electrostatic force** on a charge is determined from the experimental observation that like sign charges experience forces of mutual repulsion and unlike sign charges attract each other.

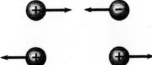

When **more than two point charges** are present, the resultant electrostatic force exerted on any one of them equals the vector sum of the forces exerted on that charge by the others individually. *The principle of superposition applies.*

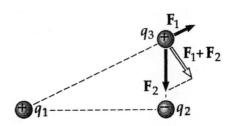

The **electric field vector** at any point in space is defined as the electric force per unit charge exerted on a small positive test charge placed at the point where the field is to be determined. *It is important to note that E is the electric field produced by a source charge (charge or distribution of charges) separate from the test charge; E is not the field due to the test charge. In the figure at right, Q is the source charge and gives rise to the electric field; q_0 is the test charge. The direction of the electric field at any point is the direction of force on a positive test charge placed at the point.*

$$E \equiv \frac{F_e}{q_0} \qquad (23.7)$$

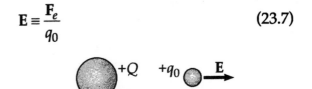

The **electric field a distance r from a point charge** is given by Equation 23.9. The unit vector $\hat{r}$ is directed away from q and toward the point where the field is to be calculated. *The direction of the electric field is radially outward from a positive point charge and radially inward toward a negative point charge.*

$$E = k_e \frac{q}{r^2}\hat{r} \qquad (23.9)$$

The **electric field due to a group** of source charges equals the vector sum of the electric fields of all the charges. *First use Equation 23.9 to calculate the electric field vector due to each individual charge and then add them vectorially.*

$$\mathbf{E} = k_e \sum_i \frac{q_i}{r_i^2} \hat{\mathbf{r}}_i \quad \text{(vector sum)} \qquad (23.10)$$

Electric field lines are a convenient graphical representation of the electric field in the vicinity of a group of charges or a charge distribution. Lines are drawn so that:

- Lines begin on a positive charge (or at infinity) and terminate on a negative charge (or at infinity).

- The number of lines leaving a positive charge (or approaching a negative charge) is proportional to the magnitude of the charge.

- No two field lines can cross.

Electric field lines are related to the electric field in the following manner:

- The electric field vector is tangent to an electric field line at each point.

- The number of lines per unit area through a surface perpendicular to the field lines is proportional to the magnitude of the electric field in that region.

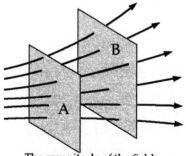

The magnitude of the field at A is greater than the magnitude of the field at B.

The **electric field of a continuous charge distribution** is found by integrating over the entire region that contains the charge. This is a vector operation and can usually be carried out easily when the charge is distributed along a line, over a surface or throughout a volume.

$$E = k_e \int \frac{dq}{r^2} \, \hat{r} \qquad (23.11)$$

The concept of **charge density** is utilized to perform the integration described above. It is convenient to represent a charge increment dq as the product of an element of length, area, or volume and the charge density over that region.

For an element
$$\text{of length } dx, \qquad dq = \lambda \, dx$$
$$\text{of area } dA, \qquad dq = \sigma \, dA$$
$$\text{of volume } dV, \qquad dq = \rho \, dV$$

For **uniform charge distributions** the volume charge density (ρ), the surface charge density (σ), and the linear charge density (λ) can be calculated based on the total charge and the geometric size (length, area, or volume) of the charge distribution.

$$\rho \equiv \frac{Q}{V} \qquad \sigma \equiv \frac{Q}{A} \qquad \lambda \equiv \frac{Q}{\ell}$$

For a nonuniform distribution, the densities λ, σ, and ρ must be stated as functions of position.

The **acceleration of a charged particle** in an electric field has a magnitude that is proportional to the magnitude of the field and has a direction that depends on the sign of the charge. *Positively charged particles accelerate along the direction of the field and negatively charged particles accelerate opposite the direction of the field.*

$$a = \frac{qE}{m} \qquad (23.12)$$

4

SUGGESTIONS, SKILLS, AND STRATEGIES

PROBLEM-SOLVING STRATEGY FOR ELECTRIC FORCES AND FIELDS

- **Units:** When performing calculations that involve the use of the Coulomb constant k_e that appears in Coulomb's law, charges must be in coulombs and distances in meters. If they are given in other units, you must convert them to SI.

- **Applying Coulomb's law to point charges:** It is important to use the superposition principle properly when dealing with a collection of interacting point charges. When several charges are present, the resultant force on any one of them is found by finding the individual force that every other charge exerts on it and then finding the vector sum of all these forces. The magnitude of the force that any charged object exerts on another is given by Coulomb's law, and the direction of the force is found by noting that the forces are repulsive between like charges and attractive between unlike charges.

- **Calculating the electric field of point charges:** Remember that the superposition principle can be applied to electric fields, which are vector quantities. To find the total electric field at a given point, first calculate the electric field at the point due to each individual charge. The resultant field at the point is the vector sum of the fields due to the individual charges. If charge 2 is between charge 1 and the field point, blocking the "line of sight" from charge 1 to the given point, its particular location makes no difference. Both charges contribute to the field according to Equation 23.9, wherever they are.

- **Evaluating the electric field of a continuous charge distribution:** To evaluate the electric field of a continuous charge distribution, it is convenient to employ the concept of charge density. Depending on the particular charge distribution, charge density can be written as: charge per unit volume, ρ; charge per unit area, σ; or charge per unit length, λ. The total charge distribution is then subdivided into small elements of volume dV, area dA, or length dx. Each element contains an increment of charge dq (equal to ρdV, σdA, or λdx). If the charge is nonuniformly distributed over the region, then the charge densities must be written as functions of position. For example, if the charge density along a line or long bar is proportional to the distance from one end of the bar, then the linear charge density could be written as $\lambda = bx$, where b is a constant of proportionality, and the charge increment dq becomes $dq = (bx)dx$.

- **Symmetry:** When dealing with either a distribution of point charges or a continuous charge distribution, take advantage of any symmetry in the system to simplify your calculations.

REVIEW CHECKLIST

You should be able to:

▷ Describe the fundamental properties of electric charge and the nature of electrostatic forces between charged bodies. (Section 23.1)

▷ Use Coulomb's law to determine the net electrostatic force (magnitude and direction) on a point electric charge due to a known distribution of a finite number of point charges. (Section 23.3)

▷ Calculate the electric field E (magnitude and direction) at a specified location in the vicinity of a group of point charges. (Section 23.4)

▷ Calculate the electric field due to a continuous charge distribution. The charge may be distributed uniformly or nonuniformly along a line, over a surface, or throughout a volume. (Section 23.5)

▷ Correctly use the rules to draw and interpret electric field lines in the vicinity of point charges. (Section 23.6)

▷ Describe quantitatively the motion of a charged particle in a uniform electric field. (Section 23.7)

ANSWERS TO SELECTED QUESTIONS

9. A balloon is negatively charged by rubbing and then clings to a wall. Does this mean that the wall is positively charged? Why does the balloon eventually fall?

Answer No. The balloon induces polarization of the molecules in the wall, so that a layer of positive charge exists near the balloon. This is just like the situation in Figure 23.5a, except that the signs of the charges are reversed. The attraction between these charges and the negative charges on the balloon is stronger than the repulsion between the negative charges on the balloon and the negative charges in the polarized molecules (because they are farther from the balloon), so that there is a net attractive force toward the wall. Ionization processes in the air surrounding the balloon provide ions to which excess electrons in the balloon can transfer, reducing the charge on the balloon and eventually causing the attractive force to be insufficient to support the weight of the balloon.

□   □   □   □

25. Would life be different if the electron were positively charged and the proton were negatively charged? Does the choice of signs have any bearing on physical and chemical interactions? Explain.

Answer   No, life would not be different. The character and effect of electric forces is defined by (1) the fact that there are only two types of electric charge — positive and negative, and (2) the fact that opposite charges attract, while like charges repel. The choice of signs is completely arbitrary.

As a related exercise, you might consider what would happen in a world where there were three types of electric charge, or in a world where opposite charges repelled, and like charges attracted.

□   □   □   □

SOLUTIONS TO SELECTED PROBLEMS

3.   The Nobel laureate Richard Feynman once said that if two persons stood at arm's length from each other and each person had 1% more electrons than protons, the force of repulsion between them would be enough to lift a "weight" equal to that of the entire Earth. Carry out an order-of-magnitude calculation to substantiate this assertion.

Solution   Suppose each person has mass 70 kg. In terms of elementary charges, each person consists of precisely equal numbers of protons and electrons and a nearly equal number of neutrons. The electrons comprise very little of the mass, so we find the number of protons-and-neutrons in each person:

$$(70 \text{ kg})\left(\frac{1 \text{ u}}{1.66 \times 10^{-27} \text{ kg}}\right) = 4 \times 10^{28} \text{ u}$$

Of these, nearly one half, 2×10^{28}, are protons, and 1% of this is 2×10^{26},

constituting a charge of $(2 \times 10^{26})(1.60 \times 10^{-19} \text{ C}) = 3 \times 10^{7} \text{ C}$

Thus, Feynman's force is $F = \dfrac{k_e q_1 q_2}{r^2} = \dfrac{(8.99 \times 10^{9} \text{ N} \cdot \text{m}^2 / \text{C}^2)(3 \times 10^{7} \text{ C})^2}{(0.5 \text{ m})^2} \sim 10^{26} \text{ N}$

where we have used a half-meter arm's length. According to the particle in a gravitational field model, if the Earth were in an externally-produced uniform gravitational field of magnitude 9.80 m/s^2,

it would weigh $F_g = mg = (6 \times 10^{24} \text{ kg})(10 \text{ m} / \text{s}^2) \sim 10^{26} \text{ N}$

Thus, the forces are of the same order of magnitude.   ◊

7. Three point charges are located at the corners of an equilateral triangle as shown in Figure P23.7. Calculate the resultant electric force on the 7.00-μC charge.

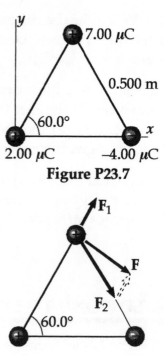

Figure P23.7

Solution The 7.00-μC charge experiences a repulsive force $\mathbf{F}_1$ due to the 2.00-μC charge, and an attractive force $\mathbf{F}_2$ due to the −4.00-μC charge, where $F_2 = 2F_1$. If we sketch vectors representing $\mathbf{F}_1$ and $\mathbf{F}_2$ and the resultant, $\mathbf{F}$ (see figure at lower right) we find that the resultant appears to be about the same magnitude as F_2 and is directed to the right about 30.0° below the horizontal.

We can find the net electric force by adding the two separate forces acting on the 7.00-μC charge. These individual forces can be found by applying Coulomb's law to each pair of charges.

The force on the 7.00-μC charge by the 2.00-μC charge is

$$\mathbf{F}_1 = k_e \frac{q_1 q_2}{r^2} \hat{\mathbf{r}} = \frac{\left(8.99 \times 10^9 \ \text{N} \cdot \text{m}^2/\text{C}^2\right)\left(7.00 \times 10^{-6} \ \text{C}\right)\left(2.00 \times 10^{-6} \ \text{C}\right)}{(0.500 \ \text{m})^2} \left(\cos 60°\hat{\mathbf{i}} + \sin 60°\hat{\mathbf{j}}\right)$$

$$\mathbf{F}_1 = \left(0.252\hat{\mathbf{i}} + 0.436\hat{\mathbf{j}}\right) \ \text{N}$$

Similarly, the force on the 7.00-μC charge by the −4.00-μC charge is

$$\mathbf{F}_2 = k_e \frac{q_1 q_3}{r^2} \hat{\mathbf{r}} = -\frac{\left(8.99 \times 10^9 \ \text{N} \cdot \text{m}^2/\text{C}^2\right)\left(7.00 \times 10^{-6} \ \text{C}\right)\left(-4.00 \times 10^{-6} \ \text{C}\right)}{(0.500 \ \text{m})^2} \left(\cos 60°\hat{\mathbf{i}} - \sin 60°\hat{\mathbf{j}}\right)$$

$$\mathbf{F}_2 = \left(0.503\hat{\mathbf{i}} - 0.872\hat{\mathbf{j}}\right) \ \text{N}$$

Thus, the total force on the 7.00-μC charge, expressed as a set of components, is

$$\mathbf{F} = \mathbf{F}_1 + \mathbf{F}_2 = \left(0.755\hat{\mathbf{i}} - 0.436\hat{\mathbf{j}}\right) \ \text{N} \qquad \diamond$$

We can also write the total force as:

$$\mathbf{F} = \sqrt{(0.755 \ \text{N})^2 + (0.436 \ \text{N})^2} \quad \text{at} \quad \tan^{-1}\left(\frac{0.436 \ \text{N}}{0.755 \ \text{N}}\right) \ \text{below the} +x \text{ axis}$$

$$\mathbf{F} = 0.872 \ \text{N} \ \text{at} \ 30.0° \ \text{below the} +x \text{ axis} \qquad \diamond$$

Our calculated answer agrees with our initial estimate. An equivalent approach to this problem would be to find the net electric field due to the two lower charges and apply $\mathbf{F} = q\mathbf{E}$ to find the force on the upper charge in this electric field.

13. What are the magnitude and direction of the electric field that will balance the weight of (a) an electron and (b) a proton? (Use the data in Table 23.1.)

Solution   If the forces balance, then $mg + q\mathbf{E} = 0$.

(a)   For an electron:

$$\mathbf{E} = -\frac{mg}{q} = \frac{-(9.11\times10^{-31}\text{ kg})(9.80\text{ m}/\text{s}^2)(-\hat{\jmath})}{-1.602\times10^{-19}\text{ C}} = (-5.57\times10^{-11}\,\hat{\jmath})\text{ N}/\text{C}$$

(b)   For a proton:

$$\mathbf{E} = -\frac{mg}{q} = \frac{-(1.67\times10^{-27}\text{ kg})(9.80\text{ m}/\text{s}^2)(-\hat{\jmath})}{+1.602\times10^{-19}\text{ C}} = (1.02\times10^{-7}\,\hat{\jmath})\text{ N}/\text{C}$$

27.   A uniformly charged ring of radius 10.0 cm has a total charge of 75.0 μC. Find the electric field on the axis of the ring at (a) 1.00 cm, (b) 5.00 cm, (c) 30.0 cm, and (d) 100 cm from the center of the ring.

Solution   Using the result of Example 23.8:

$$E = \frac{k_e x Q}{\left(x^2+a^2\right)^{3/2}} = \frac{(8.99\times10^9\text{ N}\cdot\text{m}^2/\text{C}^2)(75.0\times10^{-6}\text{ C})x}{\left(x^2+(0.100\text{ m})^2\right)^{3/2}} = \frac{(6.74\times10^5\text{ N}\cdot\text{m}^2/\text{C})x}{\left(x^2+(0.100\text{ m})^2\right)^{3/2}}.$$

Now, using your calculator and taking the x direction pointing away from the ring along its axis,

(a)   At $x = 0.0100$ m, $\mathbf{E} = 6.64\times10^6\,\hat{\imath}$ N / C

(b)   At $x = 0.0500$ m, $\mathbf{E} = 2.41\times10^7\,\hat{\imath}$ N / C

(c)   At $x = 0.300$ m, $\mathbf{E} = 6.40\times10^6\,\hat{\imath}$ N / C

(d)   At $x = 1.00$ m, $\mathbf{E} = 6.64\times10^5\,\hat{\imath}$ N / C

33.   A uniformly charged insulating rod of length 14.0 cm is bent into the shape of a semicircle as shown in Figure P23.33. The rod has a total charge of –7.50 μC. Find the magnitude and direction of the electric field at O, the center of the semicircle.

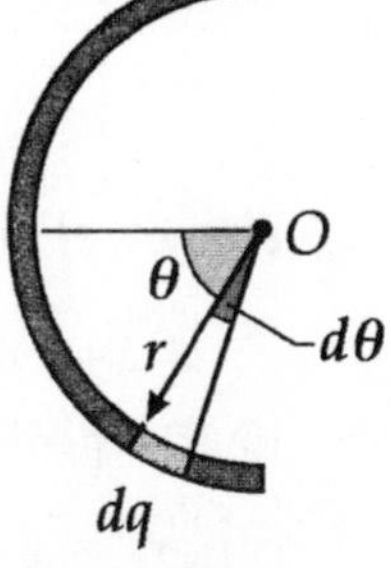

Solution   Let λ be the charge per unit length.

Then, $dq = \lambda ds = \lambda r\,d\theta$ and $dE = \frac{k_e dq}{r^2}$

In component form, $E_y = 0$ (from symmetry) $dE_x = dE\cos\theta$

Figure P23.33 (modified)

9

Integrating, $$E_x = \int dE_x = \int \frac{k_e \lambda r \cos\theta}{r^2} d\theta = \frac{k_e \lambda}{r} \int_{-\pi/2}^{\pi/2} \cos\theta \, d\theta = \frac{2k_e \lambda}{r}$$

But $$Q_{total} = \lambda \ell, \quad \text{where} \quad \ell = 0.140 \text{ m}, \quad \text{and} \quad r = \ell / \pi$$

Thus, $$E_x = \frac{2\pi k_e Q}{\ell^2} = \frac{2\pi (8.99 \times 10^9 \text{ N} \cdot \text{m}^2 / \text{C}^2)(-7.50 \times 10^{-6} \text{ C})}{(0.140 \text{ m})^2}$$

$$\mathbf{E} = (-2.16 \times 10^7 \text{ N} / \text{C})\hat{\mathbf{i}} \qquad \Diamond$$

---

35. A thin rod of length ℓ and uniform charge per unit length λ lies along the x axis, as shown in Figure P23.35. (a) Show that the electric field at P, a distance y from the rod along its perpendicular bisector, has no x component and is given by $E = 2k_e \lambda \sin\theta_0 / y$. (b) **What if?** Using your result to part (a), show that the field of a rod of infinite length is $E = 2k_e \lambda / y$. (*Suggestion:* First calculate the field at P due to an element of length dx, which has a charge λdx. Then change variables from x to θ, using the relationships $x = y \tan\theta$ and $dx = y \sec^2\theta \, d\theta$, and integrate over θ.)

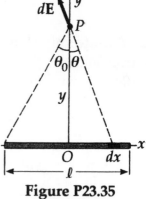

Figure P23.35
(modified)

Solution

(a) The segment of rod from x to $(x + dx)$ has a charge of λdx, and creates an electric field upward along the line from dq to P:

$$d\mathbf{E} = \frac{k_e dq}{r^2} \hat{\mathbf{r}} = \frac{k_e \lambda dx}{y^2 + x^2} \hat{\mathbf{r}}$$

This bit of field has an x component $\quad dE_x = \frac{k_e \lambda \, dx}{y^2 + x^2}(-\sin\theta) = \frac{-k_e \lambda x \, dx}{(y^2 + x^2)^{3/2}}$

and a y component $\quad dE_y = \frac{k_e \lambda \, dx}{y^2 + x^2}(\cos\theta) = \frac{k_e \lambda y \, dx}{(y^2 + x^2)^{3/2}}$

The total field has an x component $\quad E_x = \int_{\text{All } q} dE_x = \int_{-\ell/2}^{\ell/2} -\frac{k_e \lambda x \, dx}{(y^2 + x^2)^{3/2}}$

To integrate, make a change of variables to θ, such that $\quad x = y \tan\theta$

When $\quad x = -\ell / 2, \quad \theta = -\theta_0;$ $\qquad$ and when $\qquad x = \ell / 2, \quad \theta = \theta_0.$

Further, $(y^2 + x^2)^{3/2} = (y^2 + y^2 \tan^2\theta)^{3/2} = y^3 \sec^3\theta$ $\quad$ and $\quad dx = y \sec^2\theta \, d\theta$

Thus, $\quad E_x = \int_{-\theta_0}^{\theta_0} -\dfrac{k_e \lambda y (\tan\theta) y \left(\sec^2\theta\right) d\theta}{y^3 \sec^3\theta} = -\dfrac{k_e\lambda}{y}\int_{-\theta_0}^{\theta_0} \sin\theta\, d\theta$

$$E_x = +\dfrac{k_e\lambda}{y}\cos\theta\Big]_{-\theta_0}^{+\theta_0} = \dfrac{k_e\lambda}{y}\left(\cos\theta_0 - \cos(-\theta_0)\right) = \dfrac{k_e\lambda}{y}\left(\cos\theta_0 - \cos\theta_0\right) = 0$$

This answer has to be zero because each segment of rod on the left produces a field whose contribution cancels out that of the corresponding segment of rod on the right. But every incremental bit of charge produces at P a contribution to the field with upward y component:

$$E_y = \int_{\text{All } q} dE_y = \int_{-\ell/2}^{\ell/2} \dfrac{k_e\lambda y\, dx}{(y^2 + x^2)^{3/2}}$$

Think of k_e, λ, and y as known constants. Now E_y is the unknown and x is the variable of integration, which we again change to θ, with $x = y\tan\theta$:

$$E_y = \int_{-\theta_0}^{\theta_0} \dfrac{k_e\lambda y\,(y\sec^2\theta)\, d\theta}{y^3 \sec^3\theta} = \dfrac{k_e\lambda}{y}\int_{-\theta_0}^{\theta_0}\cos\theta\, d\theta = \dfrac{k_e\lambda}{y}\sin\theta\Big]_{-\theta_0}^{\theta_0}$$

$$E_y = \dfrac{k_e\lambda}{y}\left(\sin\theta_0 - \sin(-\theta_0)\right) = \dfrac{k_e\lambda}{y}\left(\sin\theta_0 + \sin\theta_0\right) = \dfrac{2k_e\lambda\sin\theta_0}{y} \qquad \Diamond$$

(b) As ℓ goes to infinity, θ_0 goes to $90°$ and $\sin\theta_0$ becomes 1. Then the infinite amount of charge produces a finite field at P:

$$\mathbf{E} = 0\hat{\mathbf{i}} + \dfrac{2k_e\lambda}{y}\hat{\mathbf{j}} \qquad \Diamond$$

39. A negatively charged rod of finite length carries charge with a uniform charge per unit length. Sketch the electric field lines in a plane containing the rod.

Solution

Since the rod has negative charge, field lines point inwards. Any field line points nearly toward the center of the rod at large distances, where the rod would look like just a point charge.

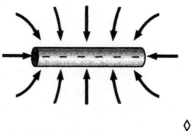

The lines curve to reach the rod perpendicular to its surface, where they end at equally-spaced points. $\qquad \Diamond$

41. Three equal positive charges q are at the corners of an equilateral triangle of side a as shown in Figure P23.41. (a) Assume that the three charges together create an electric field. Sketch the field lines in the plane of the charges. Find the location of a point (other than ∞) where the electric field is zero. (b) What are the magnitude and direction of the electric field at P due to the two charges at the base?

Figure P23.41

Solution

(a) **Conceptualize:** The electric field has the general appearance shown by the black arrows in the figure to the right. This drawing indicates that $E = 0$ at the center of the triangle, since a small positive charge placed at the center of this triangle will be pushed away from each corner equally strongly. This fact could be verified by vector addition as in part (b) below.

The electric field at point P should be directed upwards and about twice the magnitude of the electric field due to just one of the lower charges as shown in Figure P23.17. For part (b), we must ignore the effect of the charge at point P, because a charge cannot exert a force on itself.

Categorize: The electric field at point P can be found by adding the electric field vectors due to each of the two lower point charges: $\mathbf{E} = \mathbf{E}_1 + \mathbf{E}_2$

(b) **Analyze:** The electric field from a point charge is $\mathbf{E} = k_e \dfrac{q}{r^2}\,\hat{\mathbf{r}}$

As shown in the figure to the right, $\mathbf{E}_1$ and $\mathbf{E}_2$ both point $60°$ above the horizontal,

$\mathbf{E}_1 = k_e \dfrac{q}{a^2}$ to the right and upward, $\qquad$ $\mathbf{E}_2 = k_e \dfrac{q}{a^2}$ to the left and upward

$$\mathbf{E} = \mathbf{E}_1 + \mathbf{E}_2 = k_e \frac{q}{a^2}\left[\left(\cos 60°\hat{\mathbf{i}} + \sin 60°\hat{\mathbf{j}}\right) + \left(-\cos 60°\hat{\mathbf{i}} + \sin 60°\hat{\mathbf{j}}\right)\right]$$

$$\mathbf{E} = k_e \frac{q}{a^2}\left[2\left(\sin 60°\hat{\mathbf{j}}\right)\right] = 1.73 k_e \frac{q}{a^2}\hat{\mathbf{j}}$$

Finalize: The net electric field at point P is indeed nearly twice the magnitude due to a single charge and is entirely vertical as expected from the symmetry of the configuration. In addition to the center of the triangle, the gray electric field lines in the figure above indicate three other points near the middle of each leg of the triangle where $E = 0$, but they are more difficult to find mathematically.

43. A proton accelerates from rest in a uniform electric field of 640 N/C. At some later time, its speed is 1.20×10^6 m/s (nonrelativistic, since v is much less than the speed of light). (a) Find the acceleration of the proton. (b) How long does it take the proton to reach this speed? (c) How far has it moved in this time? (d) What is its kinetic energy at this time?

Solution

(a) We use the particle in an electric field model and the particle under net force model.

$$a = \frac{F}{m} = \frac{qE}{m} = \frac{\left(1.602 \times 10^{-19} \text{ C}\right)\left(640 \text{ N}/\text{C}\right)}{1.67 \times 10^{-27} \text{ kg}} = 6.14 \times 10^{10} \text{ m}/\text{s}^2 \qquad \diamond$$

(b) We use the particle under constant acceleration model.

$$\Delta t = \frac{\Delta v}{a} = \frac{1.20 \times 10^6 \text{ m}/\text{s}}{6.14 \times 10^{10} \text{ m}/\text{s}^2} = 19.5 \text{ } \mu s \qquad \diamond$$

(c) $\quad \Delta x = v_i t + \frac{1}{2}at^2 = 0 + \frac{1}{2}\left(6.14 \times 10^{10} \text{ m}/\text{s}^2\right)\left(19.5 \times 10^{-6} \text{ s}\right)^2 = 11.7 \text{ m} \qquad \diamond$

(d) $\quad K = \frac{1}{2}mv^2 = \frac{1}{2}\left(1.67 \times 10^{-27} \text{ kg}\right)\left(1.20 \times 10^6 \text{ m}/\text{s}\right)^2 = 1.20 \times 10^{-15} \text{ J} \qquad \diamond$

---

45. The electrons in a particle beam each have a kinetic energy K. What are the magnitude and direction of the electric field that will stop these electrons in a distance d?

Solution

Conceptualize: We should expect that a larger electric field would be required to stop electrons with greater kinetic energy. Likewise, E must be greater for a shorter stopping distance, d. The electric field should be in the same direction as the motion of the negatively charged electrons in order to exert an opposing force that will slow them down.

Categorize: The electrons will experience an electrostatic force $\mathbf{F} = q\mathbf{E}$. Therefore, the work done by the electric field can be equated with the initial kinetic energy since energy should be conserved.

Analyze: The work done on the charge is $\qquad$ $W = \mathbf{F} \cdot \mathbf{d} = q\mathbf{E} \cdot \mathbf{d}$

and $\qquad$ $K_i + W = K_f = 0$

Assuming $\mathbf{v}$ is in the $+x$ direction, $\qquad$ $K + (-e)\mathbf{E} \cdot d\hat{\mathbf{i}} = 0$

$$e\mathbf{E} \cdot (d\hat{\mathbf{i}}) = K$$

E is therefore in the direction of the electron beam: $\qquad$ $\mathbf{E} = \dfrac{K}{ed}\hat{\mathbf{i}}$ $\qquad$ ◊

Finalize: As expected, the electric field is proportional to K, and inversely proportional to d. The direction of the electric field is important; if it were otherwise the electron would speed up instead of slowing down! If the particles were protons instead of electrons, the electric field would need to be directed opposite to $\mathbf{v}$ in order for the particles to slow down.

---

47. A proton moves at 4.50×10^5 m/s in the horizontal direction. It enters a uniform vertical electric field with a magnitude of 9.60×10^3 N/C. Ignoring any gravitational effects, find (a) the time interval required for the proton to travel 5.00 cm horizontally, (b) its vertical displacement during the time interval in which it travels 5.00 cm horizontally, and (c) the horizontal and vertical components of its velocity after it has traveled 5.00 cm horizontally.

Solution

E is directed along the y direction; therefore, $a_x = 0$ and $x = v_{xi}t$.

(a) $\quad t = \dfrac{x}{v_{xi}} = \dfrac{0.0500 \text{ m}}{4.50 \times 10^5 \text{ m}/\text{s}} = 1.11 \times 10^{-7}$ s $\qquad$ ◊

(b) $\quad a_y = \dfrac{qE_y}{m} = \dfrac{(1.60 \times 10^{-19} \text{ C})(9.60 \times 10^3 \text{ N}/\text{C})}{1.67 \times 10^{-27} \text{ kg}} = 9.20 \times 10^{11} \text{ m}/\text{s}^2$

$\quad y = v_{yi}t + \dfrac{1}{2}a_y t^2 = \left(\dfrac{1}{2}\right)(9.20 \times 10^{11} \text{ m}/\text{s}^2)(1.11 \times 10^{-7} \text{ s})^2 = 5.68$ mm $\qquad$ ◊

(c) $\quad v_{xf} = v_{xi} = 4.50 \times 10^5 \text{ m}/\text{s}$ $\qquad$ ◊

$\quad v_{yf} = v_{yi} + a_y t = 0 + (9.20 \times 10^{11} \text{ m}/\text{s}^2)(1.11 \times 10^{-7} \text{ s}) = 1.02 \times 10^5 \text{ m}/\text{s}$ $\qquad$ ◊

---

55. A charged cork ball of mass 1.00 g is suspended on a light string in the presence of a uniform electric field as shown in Figure P23.55. When $E = (3.00\hat{i} + 5.00\hat{j}) \times 10^5$ N/C, the ball is in equilibrium at $\theta = 37.0°$. Find (a) the charge on the ball and (b) the tension in the string.

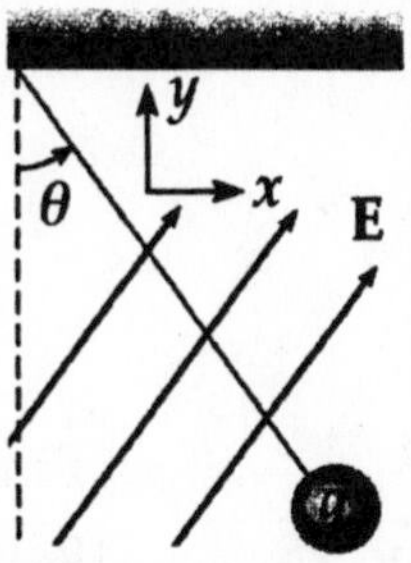

Figure P23.55

56. A charged cork ball of mass m is suspended on a light string in the presence of a uniform electric field as shown in Figure P23.55. When $E = (A\hat{i} + B\hat{j})$ N/C, where A and B are positive numbers, the ball is in equilibrium at the angle θ. Find (a) the charge on the ball and (b) the tension in the string.

Solution

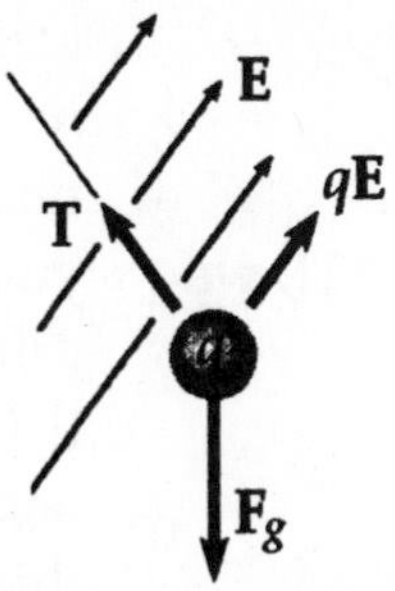

(a) **Conceptualize:** Since the electric force must be in the same direction as **E**, the ball must be positively charged.

If we examine the free body diagram that shows the three forces acting on the ball, the sum of which must be zero, we can see that the tension is about half the magnitude of the weight.

Categorize: The tension can be found from applying Newton's second law to this statics problem (electrostatics, in this case!). Since the force vectors are in two dimensions, we must apply $\Sigma F = ma$ to both the x and y directions.

Analyze: $\qquad \sum F = T + qE + F_g = 0$

We are given $\quad E_x = 3.00 \times 10^5$ N/C

and $\qquad\qquad E_y = 5.00 \times 10^5$ N/C

Applying Newton's Second Law in the x and y directions,

$$\sum F_x = qE_x - T\sin 37.0° = 0 \qquad [1]$$

$$\sum F_y = qE_y + T\cos 37.0° - mg = 0 \qquad [2]$$

Solution

Conceptualize: This is the general version of the preceding problem. The known quantities are A, B, m, g, and θ. The unknowns are q and T.

Categorize: The approach to this problem should be the same as for the last problem, but without numbers to substitute for the variables. Likewise, we can use the free body diagram given in the solution to Problem 55.

Analyze:

Again, Newton's second law gives us

$$-T\sin\theta + qA = 0 \qquad [1]$$

and $\quad +T\cos\theta + qB - mg = 0 \qquad [2]$

(a) Substituting $T = qA/\sin\theta$ into Eq. [2],

$$\frac{qA\cos\theta}{\sin\theta} + qB = mg$$

Isolating q on the left,

$$q = \frac{mg}{(A\cot\theta + B)} \qquad \Diamond$$

Substitute T from Eq. [1] into Eq. [2]:

$$q = \frac{mg}{E_y + \dfrac{E_x}{\tan 37.0°}}$$

$$q = \frac{(1.00 \times 10^{-3} \text{ kg})(9.80 \text{ m / s}^2)}{5.00 \times 10^5 \text{ N / C} + \dfrac{3.00 \times 10^5 \text{ N / C}}{\tan 37.0°}}$$

$$q = 1.09 \times 10^{-8} \text{ C} \qquad \diamond$$

(b) Using this result for q in Equation [1], we find that the tension is

$$T = \frac{qE_x}{\sin 37.0°} = 5.44 \times 10^{-3} \text{ N} \qquad \diamond$$

Finalize: The tension is slightly more than half the weight of the ball ($F_g = 9.80 \times 10^{-3}$ N), so our result seems reasonable based on our initial prediction.

(b) Substituting this value into Eq. [1],

$$T = \frac{mgA}{(A\cos\theta + B\sin\theta)} \qquad \diamond$$

Finalize: If we had solved this general problem first, we would only need to substitute the appropriate values in the equations for q and T to find the numerical results needed for Problem 55. If you find this problem more difficult than Problem 55, the little list at the Conceptualize step is useful. It shows what symbols to think of as known data, and what to consider unknown. The list is a guide for deciding what to solve for in the Analyze step, and for recognizing when we have an answer.

---

65. Two small spheres of mass m are suspended from strings of length ℓ that are connected at a common point. One sphere has charge Q; the other has charge $2Q$. The strings make angles θ_1 and θ_2 with the vertical. (a) How are θ_1 and θ_2 related? (b) Assume θ_1 and θ_2 are small. Show that the distance r between the spheres is given by

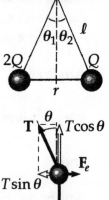

$$r \approx \left(\frac{4k_e Q^2 \ell}{mg} \right)^{1/3}$$

Solution We use the particle in equilibrium model.

(a) The spheres have different charges, but each exerts an equal force on the other, given by $F_e = k_e(Q)(2Q)/r^2$, where r is the distance between them. Since their masses are equal, $\qquad \theta_2 = \theta_1 \qquad \diamond$

(b) For equilibrium, $\sum F_y = 0$: $\quad T\cos\theta - mg = 0$, $\quad$ thus $\quad T = \dfrac{mg}{\cos\theta}$

$$\sum F_x = 0: \quad F_e - T\sin\theta = 0$$

- Substituting for T, $$F_e = \frac{mg\sin\theta}{\cos\theta} = mg\tan\theta$$

  For small angles, $$\tan\theta \approx \sin\theta = \frac{r}{2\ell}$$

  Therefore, $$F_e \approx mg\frac{r}{2\ell}$$

  The force F_e is $$\frac{k_eQ(2Q)}{r^2} \approx mg\frac{r}{2\ell}$$

  so that $$4k_eQ^2\ell \approx mgr^3 \qquad \text{and} \qquad r \approx \left(\frac{4k_eQ^2\ell}{mg}\right)^{1/3} \qquad \Diamond$$

---

71. Review problem. A negatively charged particle $-q$ is placed at the center of a uniformly charged ring, where the ring has a total positive charge Q as shown in Example 23.8. The particle, confined to move along the x axis, is displaced a small distance x along the axis (where $x \ll a$) and released. Show that the particle oscillates in simple harmonic motion with a frequency given by

$$f = \frac{1}{2\pi}\left(\frac{k_eqQ}{ma^3}\right)^{1/2}$$

Solution From Example 23.8, the electric field at points along the x axis is

$$\mathbf{E} = \frac{k_exQ\,\hat{\mathbf{i}}}{(x^2+a^2)^{3/2}}$$

The field is zero at $x=0$, so the negative charge is in equilibrium at this point. When it is displaced by an amount x that is small compared to a,

$$\sum\mathbf{F} = m\frac{d^2\mathbf{x}}{dt^2}: \qquad (-q)\frac{k_exQ\,\hat{\mathbf{i}}}{a^3} = m\frac{d^2\mathbf{x}}{dt^2}$$

The particle's acceleration is proportional to its distance (x) from the equilibrium position and is oppositely directed, so it moves in simple harmonic motion. $\qquad \Diamond$

Since $\dfrac{d^2x}{dt^2} = -\omega^2x$, $\qquad\qquad \omega = \left(\dfrac{k_eqQ}{ma^3}\right)^{1/2} = 2\pi f$

and $$f = \frac{1}{2\pi}\left(\frac{k_eqQ}{ma^3}\right)^{1/2} \qquad \Diamond$$

Chapter 24
GAUSS'S LAW

EQUATIONS AND CONCEPTS

Electric flux is a measure of the number of electric field lines that penetrate a surface. The flux equals the product of the magnitude of the field and the projection of the area onto a plane perpendicular to the direction of the field. Equation 24.2 gives the value of the flux through a plane area when a uniform field makes an angle θ with the normal to the surface. *Electric flux is a scalar quantity and has SI units of $N \cdot m^2/C$.*

$$\Phi_E = EA\cos\theta \qquad (24.2)$$

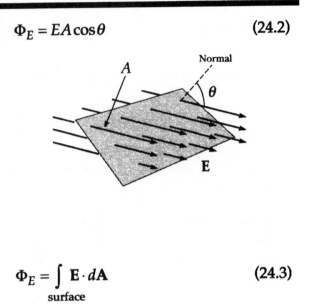

For a general surface in a nonuniform field, the flux is calculated by integrating the normal component of the field over the surface in question. *The integrand is the "dot" product of two vectors and the integral must be evaluated over the entire surface in question.*

$$\Phi_E = \int_{\text{surface}} \mathbf{E} \cdot d\mathbf{A} \qquad (24.3)$$

The **net flux through a closed surface** is proportional to the net number of lines leaving the surface (the number of lines leaving the surface minus the number entering the surface). E_n *represents the component of the electric field normal to the surface; and the integral must be evaluated over a closed surface.*

$$\Phi_E = \oint E_n dA \qquad (24.4)$$

Gauss's law states that the net flux through any closed surface surrounding a net charge q equals the net charge enclosed by the surface divided by the constant ϵ_0. The net flux is independent of the shape and size of the surface. *The surface is called a "gaussian" surface.*

$$\Phi_E = \oint \mathbf{E} \cdot d\mathbf{A} = \frac{q_{in}}{\epsilon_0} \qquad (24.6)$$

18

The **electric field due to symmetric charge distributions** can be determined by evaluating Equation 24.6. Some examples are the magnitude of the electric field at:

a distance r from a point charge q

$$E = k_e \frac{q}{r^2}$$

a point exterior to a uniformly charged insulating sphere of radius a and charge Q

$$E = k_e \frac{Q}{r^2} \qquad \text{for } r > a$$

a point interior to a uniformly charged insulating sphere with total charge Q

$$E = k_e \left(\frac{Q}{a^3} \right) r \qquad \text{for } r < a$$

a point outside a thin uniformly charged spherical shell of radius a and charge Q

$$E = k_e \frac{Q}{r^2} \qquad \text{for } r > a$$

a point inside a thin spherical shell of radius a and total charge Q

$$E = 0 \qquad \text{(Why?)}$$

a distance r from an infinitely long uniform line of charge with linear charge density λ

$$E = 2k_e \frac{\lambda}{r} \qquad (24.7)$$

any distance from **an infinite plane of charge** with surface charge density σ

$$E = \frac{\sigma}{2\epsilon_0} \qquad (24.8)$$

a point just outside the surface of a charged conductor in equilibrium with surface charge density σ

$$E_n = \frac{\sigma}{\epsilon_0} \qquad (24.9)$$

SUGGESTIONS, SKILLS, AND STRATEGIES

Gauss's law is a very powerful theorem which relates any charge distribution to the resulting electric field at any point in the vicinity of the charge. In this chapter you should learn how to apply Gauss's law to those cases in which the charge distribution has a sufficiently high degree of symmetry. As you review the examples presented in Section 24.3 of the text, observe how each of the following steps have been included in the application of the equation $\oint \mathbf{E} \cdot d\mathbf{A} = q / \epsilon_0$ to that particular situation.

- The gaussian surface should be chosen to have the **same symmetry as the charge distribution.**

- The dimensions of the surface should be such that **the surface includes the point where the electric field is to be calculated.**

- From the symmetry of the charge distribution, you should be able to correctly describe the direction of the electric field vector, **E**, relative to the direction of an element of surface area vector, $d\mathbf{A}$, which **points outward from each region of the gaussian surface.**

- Write $\mathbf{E} \cdot d\mathbf{A}$ as $E\, dA \cos \theta$ and separate the gaussian surface into regions such that over each region the electric field has a constant value and can therefore be removed from the integral. Each of the separate regions should satisfy one or more one of the following:

    $\mathbf{E} \perp d\mathbf{A}$ so that $E\, dA \cos \theta = 0$ (as is the case over each end of the cylindrical gaussian surface in the figure).

    $\mathbf{E} \parallel d\mathbf{A}$ so that $E\, dA \cos \theta = E\, dA$ (as is the case over the curved portion of the cylindrical gaussian surface in the figure).

    E and $d\mathbf{A}$ are oppositely directed so that $E\, dA \cos \theta = -E\, dA$.

    $\mathbf{E} = 0$ (as is the case over a surface inside a conductor).

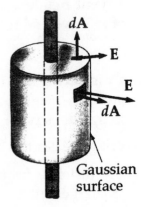

Gaussian surface

- If the gaussian surface has been chosen and subdivided so that the magnitude of **E** is constant over those regions where $\mathbf{E} \cdot d\mathbf{A} = E\, dA$, then over each of those regions

$$\int \mathbf{E} \cdot d\mathbf{A} = E \int dA = EA$$

- The total charge enclosed by the gaussian surface is $q = \int dq$. It is often convenient to represent the charge distribution in terms of the charge density ($dq = \lambda\, dx$ for a line of charge, $dq = \sigma\, dA$ for a surface of charge, or $dq = \rho\, dV$ for a volume of charge). The integral of dq is then evaluated only over that length, area, or volume which **includes that portion of the charge inside the gaussian surface**.

- After the left and right sides of Gauss's law have been evaluated, you can calculate the electric field on the gaussian surface, assuming the charge distribution is given in the problem. Conversely, if the electric field is known, you can determine the charge distribution that produces the field.

REVIEW CHECKLIST

You should be able to:

▷＊ Calculate the electric flux through a surface; in particular, find the net electric flux through a closed surface. (Section 24.1)

▷ Construct a gaussian surface to match charge distributions which have a high degree of symmetry (spherical, cylindrical, or planar); and use Gauss's law to evaluate the electric field at points interior or exterior to the charge distributions. (Sections 24.2 and 24.3)

▷ Describe the properties which characterize an electrical conductor in electrostatic equilibrium. (Section 24.4)

ANSWERS TO SELECTED QUESTIONS

5. If the total charge inside a closed surface is known but the distribution of the charge is unspecified, can you use Gauss's law to find the electric field? Explain.

Answer No. If we wish to use Gauss's law to find the electric field, we must be able to bring the electric field, E, out of the integral. This can be done, in some cases — when the field is constant, for example. However, since we do not know the charge distribution, we cannot claim that the field is constant, and thus cannot find the electric field.

To illustrate this point, consider a sphere that contained a net charge of 100 μC. The charges could be located near the center, or they could all be grouped at the northernmost point within the sphere. In either case, the net electric flux would be the same, but the electric field would vary greatly.

□   □   □   □

10. A person is placed in a large hollow metallic sphere that is insulated from ground. If a large charge is placed on the sphere, will the person be harmed upon touching the inside of the sphere? Explain what will happen if the person also has an initial charge whose sign is opposite that of the charge on the sphere.

Answer The metallic sphere is a good conductor, so any excess charge on the sphere will reside on the outside of the sphere. From Gauss's law, we know that the field inside the sphere will then be zero. As a result, when the person touches the inside of the sphere, no charge will be exchanged between the person and the sphere, and the person will not be harmed.

What happens, then, if the person has an initial charge? Regardless of the sign of the person's initial charge, the charges in the conducting surface will redistribute themselves to maintain a net zero charge within the **conducting metal**. Thus, if the person has a 5.00 μC charge on his skin, exactly –5.00 μC will gather on the inner surface of the sphere; so the electric field within the metallic material will be zero. When the person touches the metallic sphere then, he will receive a shock due to the charge on his own skin.

□   □   □   □

12. A common demonstration involves charging a rubber balloon, which is an insulator, by rubbing it on your hair, and touching the balloon to a ceiling or wall, which is also an insulator. The electrical attraction between the charged balloon and the neutral wall results in the balloon sticking to the wall. Imagine now that we have two infinitely large flat sheets of insulating material. One is charged and the other is neutral. If these are brought into contact, will an attractive force exist between them, as there was for the balloon and the wall?

Answer There will not be an attractive force. There are two factors to consider in the attractive force between a balloon and a wall, or between any pair of charged and neutral objects. The first factor is that the molecules in the wall will orient themselves with their negative ends toward the balloon, and their positive ends pointing away from the balloon. The second factor to consider is that the balloon is of finite curved dimensions, and thus the molecules in the wall are in a nonuniform electric field. Therefore the nearby "negative ends" of the molecules in the wall will experience an attractive electrostatic force that will be greater in magnitude than the repulsive force exerted on the more distant "positive ends" of the molecules. The net result is an overall force of attraction.

Now consider the infinite sheets brought into contact. The polarization of the molecules in the neutral sheet will indeed occur, as in the wall. But the electric field from the charged sheet is **uniform**, and therefore is independent of the distance from the sheet. Thus, both the negative and positive charges in the neutral sheet will experience the same electric field and the same magnitude of electric force. The attractive force on the negative charges will cancel with the repulsive force on the positive charges, and there will be no net force.

□   □   □   □

SOLUTIONS TO SELECTED PROBLEMS

3. A 40.0-cm-diameter loop is rotated in a uniform electric field until the position of maximum electric flux is found. The flux in this position is measured to be 5.20×10^5 N·m^2/C. What is the magnitude of the electric field?

Solution

We calculate the flux as $\qquad\qquad\qquad\qquad\qquad\qquad\qquad\qquad\qquad \Phi = \mathbf{E} \cdot \mathbf{A} = EA\cos\theta$

The maximum value of the flux occurs when $\qquad\qquad\qquad\qquad\qquad\qquad \theta = 0$

Therefore, we can calculate the field strength at this point as $\qquad E = \dfrac{\Phi_{max}}{A} = \dfrac{\Phi_{max}}{\pi r^2}$

$$E = \frac{5.20 \times 10^5}{\pi (0.200 \text{ m})^2} = 4.14 \times 10^6 \text{ N} / \text{C} \qquad\qquad\qquad \diamondsuit$$

---

9. The following charges are located inside a submarine: 5.00 μC, –9.00 μC, 27.0 μC, and –84.0 μC. (a) Calculate the net electric flux through the hull of the submarine. (b) Is the number of electric field lines leaving the submarine greater than, equal to, or less than the number entering it?

Solution

The total charge within the closed surface is

$$5.00 \ \mu C - 9.00 \ \mu C + 27.0 \ \mu C - 84.0 \ \mu C = -61.0 \ \mu C$$

so the total electric flux is

$$\Phi_E = \frac{q}{\epsilon_0} = \frac{-61.0 \times 10^{-6} \text{ C}}{(8.85 \times 10^{-12} \text{ C}^2 / \text{N} \cdot \text{m}^2)} = -6.89 \times 10^6 \text{ N} \cdot \text{m}^2 / \text{C} \qquad\qquad \diamondsuit$$

The minus sign means that more lines enter the surface than leave it.

---

15. A point charge Q is located just above the center of the flat face of a hemisphere of radius R as shown in Figure P24.15. What is the electric flux (a) through the curved surface and (b) through the flat face?

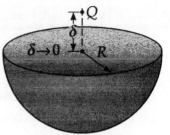

Figure P24.15

Solution

Conceptualize: From Gauss's law, the flux through a sphere with a point charge in it should be Q/ϵ_0, so we should expect the electric flux through a hemisphere to be half this value:

$$\Phi_{curved} = Q/2\epsilon_0$$

Since the flat section appears like an infinite plane to a point just above its surface so that half of all the field lines from the point charge are intercepted by the flat surface, the flux through this section should also equal $Q/2\epsilon_0$.

Categorize: We can apply the definition of electric flux directly for part (a) and then use Gauss's law to find the flux for part (b).

Analyze:

(a) With δ very small, all points on the hemisphere are nearly at distance R from the charge, so the field everywhere on the curved surface is $k_e Q/R^2$ radially outward (normal to the surface). Therefore, the flux is this field strength times the area of half a sphere:

$$\Phi_{curved} = \int \mathbf{E} \cdot d\mathbf{A} = E_{local} A_{hemisphere}$$

$$\Phi_{curved} = \left(k_e \frac{Q}{R^2} \right) \left(\frac{1}{2} \right) \left(4\pi R^2 \right) = \frac{1}{4\pi\epsilon_0} Q(2\pi) = \frac{Q}{2\epsilon_0} \qquad \Diamond$$

(b) The closed surface encloses zero charge so Gauss's law gives

$$\Phi_{curved} + \Phi_{flat} = 0 \qquad \text{or} \qquad \Phi_{flat} = -\Phi_{curved} = \frac{-Q}{2\epsilon_0} \qquad \Diamond$$

Finalize: The direct calculations of the electric flux agree with our predictions, except for the negative sign in part (b), which comes from the fact that the area unit vector is defined as pointing outward from an enclosed surface, and in this case, the electric field has a component in the opposite direction (down).

29. Consider a long cylindrical charge distribution of radius R with a uniform charge density ρ. Find the electric field at distance r from the axis where $r < R$.

Solution   If ρ is positive, the field must everywhere be radially outward. Choose as the gaussian surface a cylinder of length L and radius r, contained inside the charged rod. Its volume is $\pi r^2 L$ and it encloses charge $\rho \pi r^2 L$. The circular end caps have no electric flux through them; there $\mathbf{E} \cdot d\mathbf{A} = E\, dA \cos 90.0° = 0$. The curved surface has $\mathbf{E} \cdot d\mathbf{A} = E\, dA \cos 0°$, and E must be the same strength everywhere over the curved surface.

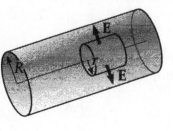

Then
$$\oint \mathbf{E} \cdot d\mathbf{A} = \frac{q}{\epsilon_0}$$
becomes
$$E \int\limits_{\substack{\text{Curved}\\\text{Surface}}} dA = \frac{\rho \pi r^2 L}{\epsilon_0}$$

Noting that $2\pi r L$ is the lateral surface area of the cylinder, $E(2\pi r)L = \dfrac{\rho \pi r^2 L}{\epsilon_0}$

Thus, $\mathbf{E} = \dfrac{\rho r}{2\epsilon_0}$ radially away from the axis   ◊

---

31. Consider a thin spherical shell of radius 14.0 cm with a total charge of 32.0 μC distributed uniformly on its surface. Find the electric field (a) 10.0 cm and (b) 20.0 cm from the center of the charge distribution.

Solution

(a)   A gaussian sphere, radius 10.0 cm, encloses 0 charge: $\mathbf{E} = 0$ ◊

(b)   For a gaussian sphere of radius 20.0 cm, $\oint \mathbf{E} \cdot d\mathbf{A} = \dfrac{q_{in}}{\epsilon_0}$

The field is radially outward, and $E(4\pi r^2) = \dfrac{q}{\epsilon_0}$

$$E = \frac{k_e q}{r^2} = \frac{(8.99 \times 10^9 \ \text{N} \cdot \text{m}^2/\text{C}^2)(32.0 \times 10^{-6} \ \text{C})}{(0.200 \ \text{m})^2} = 7.19 \times 10^6 \ \text{N/C}$$

so $\mathbf{E} = (7.19 \times 10^6 \ \text{N/C})\hat{\mathbf{r}}$ ◊

35. A uniformly charged, straight filament 7.00 m in length has a total positive charge of 2.00 μC. An uncharged cardboard cylinder 2.00 cm in length and 10.0 cm in radius surrounds the filament at its center, with the filament as the axis of the cylinder. Using reasonable approximations, find (a) the electric field at the surface of the cylinder and (b) the total electric flux through the cylinder.

Solution   The approximation in this case is that the filament length is so large when compared to the cylinder length that the "infinite line" of charge can be assumed.

(a) $E = \dfrac{2k_e\lambda}{r}$ where $\lambda = \dfrac{2.00\times10^{-6}\ C}{7.00\ m} = 2.86\times10^{-7}\ C/m$

so $E = \dfrac{(2)(8.99\times10^9\ N\cdot m^2/C)(2.86\times10^{-7}\ C/m)}{0.100\ m} = 5.14\times10^4\ N/C$ ◊

(b) $\Phi_E = 2\pi rLE = 2\pi rL\left(\dfrac{2k_e\lambda}{r}\right) = 4\pi k_e\lambda L$

so $\Phi_E = 4\pi(8.99\times10^9\ N\cdot m^2/C^2)(2.86\times10^{-7}\ C/m)(0.0200\ m) = 6.46\times10^2\ N\cdot m^2/C$ ◊

---

37. A large flat horizontal sheet of charge has a charge per unit area of 9.00 μC/m². Find the electric field just above the middle of the sheet.

Solution   For a large insulating sheet, **E** will be perpendicular to the sheet, and will have a magnitude of

$$E = \frac{\sigma}{2\epsilon_0} = 2\pi k_e\sigma = (2\pi)(8.99\times10^9\ N\cdot m^2/C^2)(9.00\times10^{-6}\ C/m^2)$$

so $\mathbf{E} = 5.08\times10^5\ N/C\,\hat{\jmath}$ ◊

---

39. A long, straight metal rod has a radius of 5.00 cm and a charge per unit length of 30.0 nC/m. Find the electric field (a) 3.00 cm, (b) 10.0 cm, and (c) 100 cm from the axis of the rod, where distances are measured perpendicular to the rod.

Solution   Outside the conductor, $E = 2k_e\lambda/r$

(a)   Inside the conductor, $E = 0$ ◊

(b)   At $r = 0.100$ m,

$E = 2\dfrac{k_e\lambda}{r} = 2\dfrac{(8.99\times10^9\ N\cdot m^2/C^2)(30.0\times10^{-9}\ C/m)}{0.100\ m} = 5.39\times10^3\ N/C$ ◊

(c)   At $r = 1.00$ m,

$$E = 2\frac{k_e\lambda}{r} = 2\frac{(8.99 \times 10^9 \text{ N} \cdot \text{m}^2/\text{C}^2)(30.0 \times 10^{-9} \text{ C}/\text{m})}{1.00 \text{ m}} = 539 \text{ N}/\text{C} \qquad \Diamond$$

47.   A long, straight wire is surrounded by a hollow metal cylinder whose axis coincides with that of the wire. The wire has a charge per unit length of λ, and the cylinder has a net charge per unit length of 2λ. From this information, use Gauss's law to find (a) the charge per unit length on the inner and outer surfaces of the cylinder and (b) the electric field outside the cylinder, a distance r from the axis.

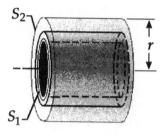

Solution

(a)   Use a cylindrical gaussian surface S_1 within the conducting cylinder; $E = 0$.

Thus, $\oint E_n dA = \left(\dfrac{1}{\epsilon_0}\right) q_{in} = 0$ and $\lambda_{inner} = -\lambda$ $\qquad \Diamond$

Also, $\lambda_{inner} + \lambda_{outer} = 2\lambda$ and $\lambda_{outer} = 3\lambda$ $\qquad \Diamond$

(b)   For a gaussian surface S_2 outside the conducting cylinder,

$$\oint E_n dA = \left(\frac{1}{\epsilon_0}\right) q_{in} \qquad \text{or} \qquad E(2\pi rL) = \frac{1}{\epsilon_0}(\lambda - \lambda + 3\lambda)L: \qquad E = \frac{3\lambda}{2\pi\epsilon_0 r} \quad \Diamond$$

49.   A thin square conducting plate 50.0 cm on a side lies in the xy plane. A total charge of 4.00×10^{-8} C is placed on the plate. Find (a) the charge density on the plate, (b) the electric field just above the plate, and (c) the electric field just below the plate. You may assume that the charge density is uniform.

Solution   In this problem ignore "edge" effects and assume that the total charge distributes uniformly over each side of the plate (one half the total charge on each side).

(a) $$\sigma = \frac{q}{A} = \frac{4.00 \times 10^{-8} \text{ C}}{2(0.500 \text{ m})^2} = 8.00 \times 10^{-8} \text{ C}/\text{m}^2 \qquad \Diamond$$

(b)   Just above the plate, $E = \dfrac{\sigma}{\epsilon_0} = \dfrac{8.00 \times 10^{-8} \text{ C}/\text{m}^2}{8.85 \times 10^{-12} \text{ C}^2/\text{N} \cdot \text{m}^2} = 9.04 \times 10^3 \text{ N}/\text{C}$ upward $\Diamond$

(c)   Just below the plate, $E = \dfrac{\sigma}{\epsilon_0} = 9.04 \times 10^3 \text{ N}/\text{C}$ downward $\qquad \Diamond$

53. A sphere of radius R surrounds a point charge Q, located at its center. (a) Show that the electric flux through a circular cap of half-angle θ (Fig. P24.53) is

$$\Phi_E = \frac{Q}{2\epsilon_0}(1 - \cos\theta)$$

What is the flux for (b) $\theta = 90°$ and (c) $\theta = 180°$?

Solution

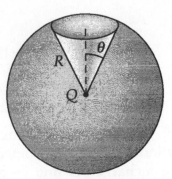

Figure P24.53

(a) The electric field of the point charge has constant strength k_eQ/R^2 over the cap and points radially outward. To find the area of the curved cap, we think of it as formed of rings, each of radius $r = R\sin\phi$, where ϕ ranges from 0 to θ. The width of each ring is $ds = R\,d\phi$, so its area is the product of its two perpendicular dimensions,

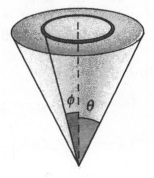

$$dA = (2\pi r)\,ds = 2\pi(R\sin\phi)(R\,d\phi)$$

The whole cap has an area of:

$$A = \int dA = \int_0^\theta 2\pi R^2 \sin\phi\,d\phi = 2\pi R^2(-\cos\phi)\Big|_0^\theta = 2\pi R^2(-\cos\theta + 1)$$

The flux through it is

$$\Phi_E = \int \mathbf{E}\cdot d\mathbf{A} = \int E\,dA\cos 0° = E\int dA = EA$$

$$\Phi_E = \frac{k_eQ}{R^2}2\pi R^2(1-\cos\theta) = \left(\frac{1}{4\pi\epsilon_0}\right)(2\pi Q)(1-\cos\theta) = \frac{Q}{2\epsilon_0}(1-\cos\theta) \qquad \Diamond$$

(b) For $\theta = 90°$, the cap is a hemisphere and intercepts half the flux from the charge:

$$\Phi_E = \frac{Q}{2\epsilon_0}(1-\cos 90°) = \frac{Q}{2\epsilon_0} \qquad \Diamond$$

(c) For $\theta = 180°$, the cap is a full sphere and all the field lines go through it:

$$\Phi_E = \frac{Q}{2\epsilon_0}(1-\cos 180°) = \frac{Q}{\epsilon_0} \qquad \Diamond$$

57. A solid, insulating sphere of radius a has a uniform charge density ρ and a total charge Q. Concentric with this sphere is an uncharged, conducting hollow sphere whose inner and outer radii are b and c, as shown in Figure P24.57. (a) Find the magnitude of the electric field in the regions $r<a$, $a<r<b$, $b<r<c$, and $r>c$. (b) Determine the induced charge per unit area on the inner and outer surfaces of the hollow sphere.

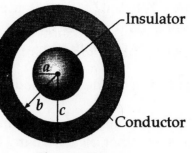

Figure P24.57

Solution

(a) Choose as the gaussian surface a concentric sphere of radius r. The electric field will be perpendicular to its surface, and will be uniform in strength over its surface.

The sphere of radius $r<a$ encloses charge

$$\rho\left(\frac{4}{3}\pi r^3\right)$$

so $\Phi=q/\epsilon_0$ becomes

$$E\left(4\pi r^2\right)=\frac{\rho\left(\frac{4}{3}\pi r^3\right)}{\epsilon_0} \quad \text{and} \quad E=\frac{\rho r}{3\epsilon_0} \qquad \diamond$$

For $a<r<b$, we have

$$E\left(4\pi r^2\right)=\frac{4}{3}\rho\frac{\pi a^3}{\epsilon_0}=\frac{Q}{\epsilon_0} \quad \text{and} \quad E=\frac{\rho a^3}{3\epsilon_0 r^2}=\frac{Q}{4\pi\epsilon_0 r^2} \qquad \diamond$$

For $b<r<c$, we must have

$$E=0 \qquad \diamond$$

because any nonzero field would be moving charges in the metal. Free charges did move in the metal to deposit charge $-Q_b$ on its inner surface, at radius b, leaving charge $+Q_c$ on its outer surface, at radius c.

Since the shell as a whole is neutral,

$$Q_c-Q_b=0$$

For $r>c$, $\Phi=q/\epsilon_0$ becomes

$$E\left(4\pi r^2\right)=\frac{Q+Q_c-Q_b}{\epsilon_0} \quad \text{and} \quad E=\frac{Q}{4\pi\epsilon_0 r^2} \qquad \diamond$$

(b) For a gaussian surface of radius $b<r<c$, we have

$$0=\frac{Q-Q_b}{\epsilon_0}$$

So $Q_b=Q$ and the charge density on the inner surface is

$$\frac{-Q_b}{A}=\frac{-Q}{4\pi b^2} \qquad \diamond$$

Then $Q_c=Q_b=Q$ and the charge density on the outer surface is

$$+\frac{Q}{4\pi c^2} \qquad \diamond$$

63. What if? Repeat the calculations for Problem 62 when both sheets have *positive* uniform surface charge densities of value σ.

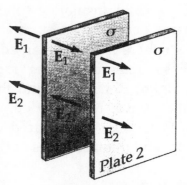

Solution The new, modified problem statement reads:

"Two infinite, nonconducting sheets of charge are parallel to each other, as shown in Figure P24.62. The sheet on the left has a uniform surface charge density σ, and the one on the right has a uniform charge density σ. Calculate the electric field at points (a) to the left of, (b) in between, and (c) to the right of the two sheets."

**Figure P24.62
(modified)**

Conceptualize: When both sheets have the same charge density, a positive test charge at a point midway between them will experience the same force in opposite directions from each sheet. Therefore, the electric field here will be zero. (We should ask: can we also conclude that the test charge will experience equal and oppositely directed forces *everywhere* in the region between the plates?)

Outside the sheets the electric field will point away and should be twice the strength due to one sheet of charge, so $E = \sigma/\epsilon_0$ in these regions.

Categorize: The principle of superposition can be applied to add the electric field vectors due to each sheet of charge.

Analyze: For each sheet, the electric field at any point is $|\mathbf{E}| = \sigma/2\epsilon_0$ directed away from the sheet.

(a) At a point to the left of the two parallel sheets $\quad \mathbf{E} = E_1(-\hat{\mathbf{i}}) + E_2(-\hat{\mathbf{i}}) = 2E(-\hat{\mathbf{i}}) = -\dfrac{\sigma}{\epsilon_0}\hat{\mathbf{i}}$ ◊

(b) At a point between the two sheets $\quad\quad\quad\quad \mathbf{E} = E_1\hat{\mathbf{i}} + E_2(-\hat{\mathbf{i}}) = 0 \quad\quad\quad\quad\quad$ ◊

(c) At a point to the right of the two parallel sheets $\mathbf{E} = E_1\hat{\mathbf{i}} + E_2\hat{\mathbf{i}} = 2E\hat{\mathbf{i}} = \dfrac{\sigma}{\epsilon_0}\hat{\mathbf{i}} \quad\quad$ ◊

Finalize: We essentially solved this problem in the Conceptualize step, so it is no surprise that these results are what we expected. A better check is to confirm that the results are complementary to the case where the plates are oppositely charged (Problem 62).

67. A solid insulating sphere of radius R has a nonuniform charge density that varies with r according to the expression $\rho = Ar^2$, where A is a constant and $r < R$ is measured from the center of the sphere. (a) Show that the magnitude of the electric field outside $(r > R)$ the sphere is $E = AR^5/5\epsilon_0 r^2$. (b) Show that the magnitude of the electric field inside $(r < R)$ the sphere is $E = Ar^3/5\epsilon_0$. (*Suggestion:* The total charge Q on the sphere is equal to the integral of ρdV, where r extends from 0 to R; also the charge q within a radius $r < R$ is less than Q. To evaluate the integrals, note that the volume element dV for a spherical shell of radius r and thickness dr is equal to $4\pi r^2 dr$.)

Solution

(a) We call the constant A', reserving A to denote area. The whole charge of the ball is

$$Q = \int\limits_{\text{ball}} dQ = \int\limits_{\text{ball}} \rho dV = \int\limits_{r=0}^{R} A'r^2 4\pi r^2 dr = 4\pi A' \frac{r^5}{5} \bigg|_0^R = \frac{4\pi A' R^5}{5}$$

To find the electric field, consider as gaussian surface a concentric sphere of radius r outside the ball of charge:

In this case, $\qquad\qquad \int \mathbf{E} \cdot d\mathbf{A} = \dfrac{Q}{\epsilon_0} \qquad$ reads $\qquad EA\cos 0° = \dfrac{Q}{\epsilon_0}$

Solving, $\qquad\qquad\qquad E(4\pi r^2) = \dfrac{4\pi A' R^5}{5\epsilon_0}$

Thus, the electric field is $\qquad E = \dfrac{A'R^5}{5\epsilon_0 r^2}$ $\qquad\qquad\qquad\qquad$ ◊

(b) Let the gaussian sphere lie inside the ball of charge: $\qquad \displaystyle\int\limits_{\substack{\text{sphere,}\\\text{radius } r}} \mathbf{E}\cdot d\mathbf{A} = \int\limits_{\substack{\text{sphere,}\\\text{radius } r}} dQ/\epsilon_0$

Now the integral becomes $\quad E(\cos 0)\displaystyle\int dA = \int \dfrac{\rho dV}{\epsilon_0} \quad$ or $\quad EA = \displaystyle\int_0^r \dfrac{A'r^2(4\pi r^2)dr}{\epsilon_0}$

Performing the integration, $\quad E(4\pi r^2) = \left(\dfrac{A'4\pi}{\epsilon_0}\right)\left(\dfrac{r^5}{5}\right)\bigg|_0^r = \dfrac{A'4\pi r^5}{5\epsilon_0}$

and $\qquad\qquad\qquad\qquad E = \dfrac{A'r^3}{5\epsilon_0}$ $\qquad\qquad\qquad\qquad$ ◊

71. Review problem. A slab of insulating material (infinite in two of its three dimensions) has a uniform positive charge density ρ. An edge view of the slab is shown in Figure P24.71. (a) Show that the magnitude of the electric field a distance x from its center and inside the slab is $E = \rho x/\epsilon_0$. (b) **What if?** Suppose an electron of charge $-e$ and mass m_e can move freely within the slab. It is released from rest at a distance x from the center. Show that the electron exhibits simple harmonic motion with a frequency

$$f = \frac{1}{2\pi}\sqrt{\frac{\rho e}{m_e \epsilon_0}}$$

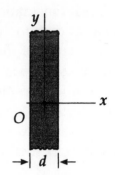

Figure P24.71

Solution

(a) The slab has left-to-right symmetry, so its field must be equal in strength at x and at $-x$. It points everywhere away from the central plane. Take as gaussian surface a rectangular box of thickness $2x$ and height and width L, centered on the $x=0$ plane. The charge it contains is $\rho V = \rho(2xL^2)$. The total flux leaving it is EL^2 through the right face, EL^2 through the left face, and zero through each of the other four sides.

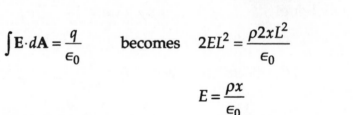

Thus Gauss's law $\qquad \int \mathbf{E}\cdot d\mathbf{A} = \dfrac{q}{\epsilon_0} \qquad$ becomes $\qquad 2EL^2 = \dfrac{\rho 2xL^2}{\epsilon_0}$

so $\qquad\qquad\qquad\qquad\qquad\qquad\qquad\qquad E = \dfrac{\rho x}{\epsilon_0} \qquad\qquad\qquad\qquad\qquad \Diamond$

(b) The electron experiences a force opposite to **E**. When displaced to $x>0$, it experiences a restoring force to the left.

For it, $\sum \mathbf{F} = m_e\mathbf{a}$ reads $\qquad q\mathbf{E} = m_e\mathbf{a} \qquad$ or $\qquad \dfrac{-e\rho x\hat{\mathbf{i}}}{\epsilon_0} = m_e\mathbf{a}$

Solving for the acceleration, $\quad \mathbf{a} = -\left(\dfrac{e\rho}{m_e\epsilon_0}\right)x\hat{\mathbf{i}} \quad$ or $\qquad \mathbf{a} = -\omega^2 x\hat{\mathbf{i}}$

That is, its acceleration is proportional to its displacement and oppositely directed, as is required for simple harmonic motion.

Solving for the frequency, $\qquad \omega^2 = \dfrac{e\rho}{m_e\epsilon_0} \qquad$ and $\qquad f = \dfrac{\omega}{2\pi} = \dfrac{1}{2\pi}\sqrt{\dfrac{e\rho}{m_e\epsilon_0}} \qquad \Diamond$

Chapter 25
ELECTRIC POTENTIAL

EQUATIONS AND CONCEPTS

The **potential difference** between two points, $\Delta V = V_B - V_A$, is defined as the change in the potential energy of a charge-field system when a test charge, q_0, is moved from point A to point B divided by the test charge q_0. ΔV can be evaluated by integrating $\mathbf{E} \cdot d\mathbf{s}$ along any path from A to B. *Electric potential is a scalar quantity characteristic of an electric field, independent of any charges that may be placed in the field. The SI unit of electric potential is the volt (V); 1 V = 1 J/C.*

$$\Delta V = \frac{\Delta U}{q_0} = -\int_A^B \mathbf{E} \cdot d\mathbf{s} \qquad (25.3)$$

The **electron volt** (eV) is a unit of energy which is defined as the energy that a charge-field system gains or loses when a charge of magnitude e is moved through a potential difference of 1 V.

$$1\,\text{eV} = 1.60 \times 10^{-19}\ \text{J} \qquad (25.5)$$

In a **uniform electric field**, the potential difference between two points depends on the displacement d along the direction parallel to $\mathbf{E}$ and on the magnitude of E. *Electric field lines always point in the direction of decreasing electric field. In the figure, the potential at point B is lower than that at point A.*

$$\Delta V = -E \int_A^B ds = -Ed \qquad (25.6)$$

$$(\text{when } \mathbf{d} \parallel \mathbf{E})$$

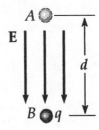

The **change in potential energy**, ΔU, of a charge-field system as a charge moves from point A to point B in an electric field depends on the sign and magnitude of the charge as well as on the change in potential, ΔV. *The electric potential energy of a charge-field system decreases when a positive charge moves in the direction of the field. See figure above.*

$$\Delta U = q_0 \Delta V = -q_0 E d \qquad (25.7)$$

$$\text{(for } \mathbf{d} \parallel \mathbf{E} \text{ in a uniform field)}$$

The **electric potential due to a single point charge**, at a distance r from the charge, depends inversely on the distance from the charge. The potential is given by Equation 25.11 when the zero reference level for potential is taken to be at infinity.

$$V = k_e \frac{q}{r} \qquad (25.11)$$

The **electric potential in the vicinity of several point charges** is the scalar sum of the potentials due to the individual charges.

$$V = k_e \sum_i \frac{q_i}{r_i} \qquad (25.12)$$

The **potential energy of a system of two charges** separated by a distance r_{12} represents the work required to assemble the charges from an infinite separation. Hence, the negative of the potential energy equals the minimum work required to separate them to an infinite distance with no final kinetic energy. *The electric potential energy associated with a system of two charges is positive if the two charges have the same sign, and negative if they are of opposite sign.*

$$U = k_e \frac{q_1 q_2}{r_{12}} \qquad (25.13)$$

The **total potential energy of a group of charges** is found by calculating U for each pair of charges and summing the terms algebraically. *One way to do this sum is to do a double sum over the group, and then divide by 2 because of double counting (e.g. the double sum includes both a "1,2" and a "2,1" term).*

$$U = \frac{k_e}{2} \sum_{i=1}^{N} \sum_{j=1}^{N} \frac{q_i q_j}{r_{ij}} \qquad \text{when } i \neq j$$

The **components of the electric field** are the negative partial derivatives of the electric potential. The electric field components in Cartesian coordinates are given in Equation 25.18.

$$E_x = -\frac{\partial V}{\partial x}$$

$$E_y = -\frac{\partial V}{\partial y}$$

$$E_z = -\frac{\partial V}{\partial z}$$

(25.18)

The **vector expression for the electric field** can be evaluated at any point $P(x, y, z)$ within the region if the electric potential over the region is known.

$$\mathbf{E} = -\left(\frac{\partial}{\partial x}\hat{\mathbf{i}} + \frac{\partial}{\partial y}\hat{\mathbf{j}} + \frac{\partial}{\partial z}\hat{\mathbf{k}} \right) V$$

The **potential** (relative to zero at infinity) **due to a continuous charge distribution** can be calculated by integrating the contribution due to each charge element dq over the line, surface, or volume which contains all the charge. *Recall, if the electric field is known (e.g. from Gauss's law) the potential can be calculated by using Equation 25.3.*

$$V = k_e \int \frac{dq}{r}$$

(25.20)

SUGGESTIONS, SKILLS, AND STRATEGIES

The vector expressions giving the electric field **E** over a region can be obtained from the scalar function which describes the electric potential, V, over the region by using a vector differential operator called the gradient operator, ∇:

$$\nabla = \hat{\mathbf{i}}\frac{\partial}{\partial x} + \hat{\mathbf{j}}\frac{\partial}{\partial y} + \hat{\mathbf{k}}\frac{\partial}{\partial z}$$

$$\mathbf{E} = -\nabla V$$

This is equivalent to

$$\mathbf{E} = -\hat{\mathbf{i}}\frac{\partial V}{\partial x} - \hat{\mathbf{j}}\frac{\partial V}{\partial y} - \hat{\mathbf{k}}\frac{\partial V}{\partial z}$$

The derivatives in the above expression are called *partial derivatives*. This means that when the derivative is taken with respect to any one coordinate, any other coordinates which appear in the expression for the potential function are treated as constants.

Since the electrostatic force is a conservative force, the work done by the electrostatic force in moving a charge q from an initial point A to a final point B depends only on the location of the two points and is independent of the path taken between A and B. When calculating potential differences using the equation

$$\Delta V = -\int_A^B \mathbf{E} \cdot d\mathbf{s} \qquad (25.3)$$

any path between A and B may be chosen to evaluate the integral; therefore you should select a path for which the evaluation of the "line integral" in Equation 25.3 will be as convenient as possible.

For example; if **E** is in the form
$$\mathbf{E} = Lx\hat{\mathbf{i}} + My\hat{\mathbf{j}} + Nz\hat{\mathbf{k}}$$

where L, M, and N are constants,

the potential is integrated as
$$\int_A^B \mathbf{E} \cdot d\mathbf{s} = \int_A^B (Lx\,dx + My\,dy + Nz\,dz).$$

This integral can be most easily evaluated over a path which moves first parallel to the x-axis from x_A to x_B, then parallel to the y-axis, then parallel to the z-axis. Along the first leg of this path, for instance, $dy = dz = 0$.

Thus Equation 25.3 becomes
$$V_B - V_A = -\left[L\int_{x_A}^{x_B} x\,dx + M\int_{y_A}^{y_B} y\,dy + N\int_{z_A}^{z_B} z\,dz \right]$$

$$V_B - V_A = \frac{L}{2}(x_A^2 - x_B^2) + \frac{M}{2}(y_A^2 - y_B^2) + \frac{N}{2}(z_A^2 - z_B^2)$$

PROBLEM-SOLVING STRATEGY

- When working problems involving electric potential, remember that potential is a *scalar quantity* (rather than a vector quantity like the electric field), so there are no components to worry about. Therefore, when using the superposition principle to evaluate the electric potential at a point due to a system of point charges, you simply take the algebraic sum of the potentials due to each charge. However, you must keep track of signs. The potential ($V = k_e q / r$) due to each positive charge is positive, while the potential due to each negative charge is negative.

- Only *changes* in electric potential are significant, hence the point where you choose the potential to be zero is arbitrary. When dealing with point charges or a finite-sized charge distribution, we usually define $V = 0$ to be at a point infinitely far from the charges. However, if the charge distribution itself extends to infinity, some other nearby point must be selected as the reference point.

- The electric potential at some point P due to a continuous distribution of charge can be evaluated by dividing the charge distribution into infinitesimal elements of charge dq located at a distance r from the point P. You then treat this element as a point charge, so that the potential at P due to the element is $dV = k_e \, dq / r$. The total potential at P is obtained by integrating dV over the entire charge distribution. In performing the integration for most problems, it is necessary to express dq and r in terms of a single variable. In order to simplify the integration, it is important to give careful consideration of the geometry involved in the problem.

- Another method that can be used to obtain the potential due to a finite continuous charge distribution is to start with the definition of the potential difference given by Equation 25.3. If E is known or can be obtained easily (say from Gauss's law), then the line integral of $\mathbf{E} \cdot d\mathbf{s}$ can be evaluated. An example of this method is given in Example 25.8 in the textbook.

- When you know the electric potential in a region around a point, it is possible to obtain the electric field at that point by remembering that

$$E_x = -\frac{\partial V}{\partial x}, \quad E_y = -\frac{\partial V}{\partial y}, \text{ and } E_z = -\frac{\partial V}{\partial z}$$

- In this chapter the symbol V is used to represent the electric potential at some point and ΔV is used to represent the potential difference between two points. For example, the expression $\Delta V = V_B - V_A$ is read *"the potential difference between points A and B is the potential at point B minus the potential at point A."*

In practice, a variety of phrases are used to describe the potential difference between two points, the most common being "voltage". A voltage applied to a device or across a device has the same meaning as the potential difference across the device. For example, if we say that the voltage across a certain capacitor is 12 volts, we mean that the potential difference between the capacitor's plates is 12 volts.

REVIEW CHECKLIST

You should be able to:

▷ Calculate the change in electric potential energy (or work done by an external force) when a charge q moves between any two points in an electric field. (Section 25.1)

▷ Calculate the electric potential difference between any two points in a uniform electric field, and the electric potential difference between any two points in the vicinity of a group of point charges. (Sections 25.1 and 25.2)

▷ Calculate the electric potential energy associated with a group of point charges. (Section 25.3)

▷ Obtain an expression for the electric field (a vector quantity) over a region of space if the scalar electric potential function for the region is known. (Section 25.4)

▷ Calculate the electric potential due to continuous charge distributions of reasonable symmetry — such as a charged ring, sphere, line, disk or infinite plane. (Section 25.5)

ANSWERS TO SELECTED QUESTIONS

3. Give a physical explanation of the fact that the potential energy of a pair of charges with the same sign is positive whereas the potential energy of a pair of charges with opposite signs is negative.

Answer You may remember from the chapter on gravitational potential energy that potential energy of a system is defined to be positive when positive work must have been performed by an external agent to change the system from an initial configuration, to which we assign a zero value of potential energy, to a final configuration. For example the system of a flag and the Earth has positive potential energy when the flag is raised if we define zero potential energy as the configuration with the flag at the ground, since positive work must be done by an external force in order to raise it from the ground to the top of the pole.

When assembling like charges from an infinite separation, a configuration for which we have defined the potential energy as having a value of zero, it takes work to move them closer together to some distance r; therefore energy is being stored in the system of charges, and the potential energy is positive.

When assembling unlike charges from an infinite separation, the charges tend to accelerate toward each other, and thus energy is released as they approach a separation of distance r. Therefore, the potential energy of a pair of unlike charges is negative.

SOLUTIONS TO SELECTED PROBLEMS

3. (a) Calculate the speed of a proton that is accelerated from rest through a potential difference of 120 V. (b) Calculate the speed of an electron that is accelerated through the same potential difference.

Solution

Conceptualize: Since 120 V is only a modest potential difference, we might expect that the final speed of the particles will be substantially less than the speed of light. We should also expect the speed of the electron to be significantly greater than the proton because, with $m_e \ll m_p$, an equal force on both particles will result in a much greater acceleration for the electron.

Categorize: Conservation of energy of the proton-field system can be applied to this problem to find the final speed from the kinetic energy of the particles. (Review this work-energy theory of motion from Chapter 8 if necessary.)

Analyze:

(a) Energy is conserved as the proton moves from high to low potential, which can be defined for this problem as moving from 120 V down to 0 V:

$$K_i + U_i + \Delta E_{nc} = K_f + U_f: \quad 0 + qV_i + 0 = \frac{1}{2}mv_p^2 + 0$$

$$(1.60 \times 10^{-19} \text{ C})(120 \text{ V})\left(\frac{1 \text{ J}}{1 \text{ V} \cdot \text{C}}\right) = \frac{1}{2}(1.67 \times 10^{-27} \text{ kg})v_p^2$$

$$v_p = 1.52 \times 10^5 \text{ m/s} \qquad \diamond$$

(b) The electron will gain speed in moving the other way, from $V_i = 0$ to $V_f = 120$ V :

$$K_i + U_i + \Delta E_{nc} = K_f + U_f: \quad 0 + 0 + 0 = \frac{1}{2}mv_e^2 + qV$$

$$0 = \frac{1}{2}(9.11 \times 10^{-31} \text{ kg})v_e^2 + (-1.60 \times 10^{-19} \text{ C})(120 \text{ J/C})$$

$$v_e = 6.49 \times 10^6 \text{ m/s} \qquad \diamond$$

Finalize: Both of these speeds are significantly less than the speed of light as expected, which also means that we were justified in not using the relativistic kinetic energy formula. (For precision to three significant digits, the relativistic formula is only needed if v is greater than about $0.1c$.)

7. An electron moving parallel to the x axis has an initial speed of 3.70×10^6 m/s at the origin. Its speed is reduced to 1.40×10^5 m/s at the point $x = 2.00$ cm. Calculate the potential difference between the origin and that point. Which point is at the higher potential?

Solution Use the energy version of the isolated system model to equate the energy of the electron-field system when the electron is at $x = 0$ and $x = 2$ cm.

The unknown will be the difference in potential $\qquad V_f - V_i$

Thus, $\quad K_i + U_i + \Delta E_{mech} = K_f + U_f \qquad$ becomes $\qquad \frac{1}{2}mv_i^2 + qV_i + 0 = \frac{1}{2}mv_f^2 + qV_f$

or $\quad \frac{1}{2}m(v_i^2 - v_f^2) = q(V_f - V_i) \qquad$ so $\qquad V_f - V_i = \frac{m(v_i^2 - v_f^2)}{2q}$

Noting that the electron's charge is negative, and evaluating the potential,

$$V_f - V_i = \frac{(9.11 \times 10^{-31}\ \text{kg})\left[(3.70 \times 10^6\ \text{m/s})^2 - (1.40 \times 10^5\ \text{m/s})^2\right]}{2(-1.60 \times 10^{-19}\ \text{C})} = -38.9\ \text{V} \qquad \lozenge$$

The negative sign means that the 2.00-cm location is lower in potential than the origin. A positive charge would slow in free flight toward higher voltage, but the negative electron slows as it moves into lower potential. The 2.00-cm distance was unnecessary information for this problem. If the field were uniform, we could find the x component from $\Delta V = -E_x d$.

---

17. At a certain distance from a point charge, the magnitude of the electric field is 500 V/m and the electric potential is –3.00 kV. (a) What is the distance to the charge? (b) What is the magnitude of the charge?

Solution At a distance r from a point charge,

$$V = \frac{k_e q}{r} \quad \text{and} \quad E = \frac{k_e |q|}{r^2}; \quad \text{thus,} \quad E = \frac{r|V|}{r^2} = \frac{|V|}{r}$$

(a) $\qquad r = \frac{|V|}{E} = \frac{3\,000\ \text{V}}{500\ \text{N/C}} = 6.00\ \frac{\text{N·m/C}}{\text{N/C}} = 6.00\ \text{m} \qquad \lozenge$

(b) $\qquad q = \frac{rV}{k_e} = \frac{(6.00\ \text{m})(-3\,000\ \text{V})}{8.99 \times 10^9\ \text{N·m}^2/\text{C}^2} = -2.00\ \mu\text{C} \qquad \lozenge$

---

19. The three charges in Figure P25.19 are at the vertices of an isosceles triangle. Calculate the electric potential at the midpoint of the base, taking $q = 7.00 \ \mu C$.

Solution Let $q_1 = q$ and $q_2 = q_3 = -q$.

The charges are at distances of

$$r_1 = \sqrt{(0.040\ 0 \text{ m})^2 - (0.010\ 0 \text{ m})^2} = 3.87 \times 10^{-2} \text{ m}$$

and $r_2 = r_3 = 0.010\ 0 \text{ m}$

The potential at point P is

$$V_P = \frac{k_e q_1}{r_1} + \frac{k_e q_2}{r_2} + \frac{k_e q_3}{r_3} = k_e(q) \left(\frac{1}{r_1} + \frac{-1}{r_2} + \frac{-1}{r_3} \right)$$

so $V_P = (8.99 \times 10^9 \text{ N·m}^2 / \text{C}^2)(7.00 \times 10^{-6} \text{ C}) \left(\dfrac{1}{0.038\ 7 \text{ m}} - \dfrac{1}{0.010\ 0 \text{ m}} - \dfrac{1}{0.010\ 0 \text{ m}} \right)$

and $V_P = -11.0 \times 10^6 \text{ V}$ ◊

Figure P25.19 (modified)

Related Calculation: Calculate the electric field vector at the same point due to the three charges. The separate fields of the two negative charges are in opposite directions and add to zero:

$$\mathbf{E}_P = \frac{k_e q_1}{r_1^2} \hat{\mathbf{r}}_1 = \frac{(8.99 \times 10^9 \text{ N·m}^2 / \text{C}^2)(7.00 \times 10^{-6} \text{ C})}{(0.040\ 0 \text{ m})^2 - (0.010\ 0 \text{ m})^2} \text{ down}$$

$$\mathbf{E}_P = (42.0 \times 10^6 \text{ N} / \text{C})(-\hat{\mathbf{j}})$$

23. Show that the amount of work required to assemble four identical point charges of magnitude Q at the corners of a square of side s is $5.41 k_e Q^2 / s$.

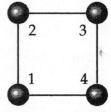

Solution The work required equals the sum of the potential energies for all pairs of charges. No energy is involved in placing q_4 at a given position in empty space.

When q_3 is brought from far away and placed close to q_4, the system potential energy can be expressed as $q_3 V_4$, where V_4 is the potential at the position of q_3 established by charge q_4. When q_2 is brought into the system, it interacts with two other charges, so we have two additional terms $q_2 V_3$ and $q_2 V_4$ in the total potential energy.

Finally, when we bring the fourth charge q_1 into the system, it interacts with three other charges, giving us three more energy terms. Thus, the complete expression for the energy is:

$$U = q_1 V_2 + q_1 V_3 + q_1 V_4 + q_2 V_3 + q_2 V_4 + q_3 V_4$$

$$U = \frac{q_1 k_e q_2}{r_{12}} + \frac{q_1 k_e q_3}{r_{13}} + \frac{q_1 k_e q_4}{r_{14}} + \frac{q_2 k_e q_3}{r_{23}} + \frac{q_2 k_e q_4}{r_{24}} + \frac{q_3 k_e q_4}{r_{34}}$$

$$U = \frac{Q k_e Q}{s} + \frac{Q k_e Q}{s\sqrt{2}} + \frac{Q k_e Q}{s} + \frac{Q k_e Q}{s} + \frac{Q k_e Q}{s\sqrt{2}} + \frac{Q k_e Q}{s}$$

Evaluating, $\quad U = \frac{k_e Q^2}{s}\left(4 + \frac{2}{\sqrt{2}}\right) = 5.41 k_e \frac{Q^2}{s}$ ◊

---

39. Over a certain region of space, the electric potential is $V = 5x - 3x^2 y + 2yz^2$. Find the expressions for the x, y, and z components of the electric field over this region. What is the magnitude of the field at the point P that has coordinates $(1, 0, -2)$ m?

Solution First, we find the x, y, and z components of the field; then, we evaluate them at point P. (We assume that V is given in volts, as a function of distances in meters.)

$E_x = -\dfrac{\partial V}{\partial x} = -5 + 6xy \qquad E_y = -\dfrac{\partial V}{\partial y} = 3x^2 - 2z^2 \qquad E_z = -\dfrac{\partial V}{\partial z} = -4yz$ ◊

At point P, $\qquad E_x = -5 + 6(1.00 \text{ m})(0 \text{ m}) = -5.00 \text{ N}/\text{C}$

At point P, $\qquad E_y = 3(1.00 \text{ m})^2 - 2(-2.00 \text{ m})^2 = -5.00 \text{ N}/\text{C}$

At point P, $\qquad E_z = -4(0 \text{ m})(-2.00 \text{ m}) = 0 \text{ N}/\text{C}$

At P, the field's magnitude is $\quad E = \sqrt{(-5.00 \text{ N}/\text{C})^2 + (-5.00 \text{ N}/\text{C})^2 + 0^2} = 7.07 \text{ N}/\text{C}$ ◊

---

41. It is shown in Example 25.7 that the potential at a point P a distance a above one end of a uniformly charged rod of length ℓ lying along the x axis is

$$V = \frac{k_e Q}{\ell} \ln\left(\frac{\ell + \sqrt{\ell^2 + a^2}}{a}\right)$$

Use this result to derive an expression for the y component of the electric field at P. (*Suggestion*: Replace a with y.)

Solution Replacing a with y as directed, we can differentiate the potential as shown:

$$E_y = -\frac{\partial V}{\partial y} = -\frac{k_e Q}{\ell}\frac{d}{dy}\left[\ln\left(\ell + \sqrt{\ell^2 + y^2}\right) - \ln y\right] = -\frac{k_e Q}{\ell}\left[\frac{2y/2\sqrt{\ell^2 + y^2}}{\ell + \sqrt{\ell^2 + y^2}} - \frac{1}{y}\right] = \frac{k_e Q}{y\sqrt{\ell^2 + y^2}} \quad \lozenge$$

---

43. A rod of length L (Fig. P25.43) lies along the x axis with its left end at the origin. It has a nonuniform charge density $\lambda = \alpha x$, where α is a positive constant. (a) What are the units of α? (b) Calculate the electric potential at A.

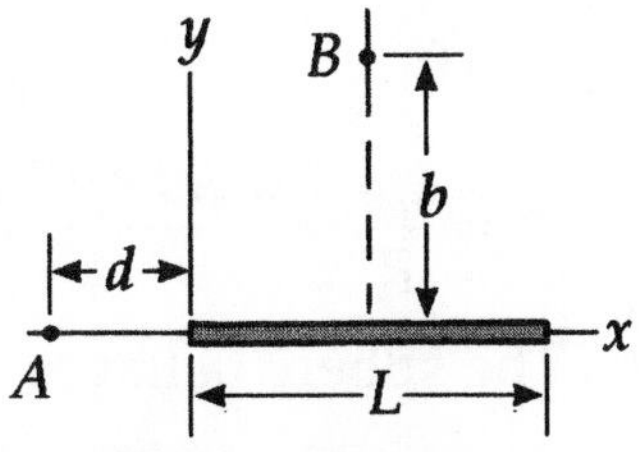

Figure P25.43

Solution

(a) As a linear charge density, λ has units of C/m.
   So $\alpha = \lambda / x$ must have units of C/m^2. $\quad \lozenge$

(b) Consider a small segment of the rod at location x and of length dx. The amount of charge on it is $\lambda\,dx = (\alpha x)dx$. Its distance from A is $d + x$, so its contribution to the electric potential at A is

$$dV = k_e\frac{dq}{r} = k_e\frac{\alpha x\,dx}{d + x}$$

We must integrate all these contributions for the whole rod, from $x = 0$ to $x = L$:

$$V = \int_{\text{all }q} dV = \int_0^L \frac{k_e\alpha x}{d + x}\,dx$$

To perform the integral, make a change of variables to

$$u = d + x, \quad du = dx, \quad u\,(\text{at }x=0) = d, \quad \text{and} \quad u\,(\text{at }x=L) = d + L:$$

$$V = \int_d^{d+L}\frac{k_e\alpha(u - d)}{u}\,du = k_e\alpha\int_d^{d+L} du - k_e\alpha d\int_d^{d+L}\left(\frac{1}{u}\right)du$$

[Keep track of symbols: the unknown is V. The values k_e, α, d, and L are known and constant. Also note that x and u are variables, and will not appear in the answer.]

$$V = k_e\alpha u\big|_d^{d+L} - k_e\alpha d\ln u\big|_d^{d+L} = k_e\alpha(d + L - d) - k_e\alpha d\left(\ln(d + L) - \ln d\right)$$

$$V = k_e\alpha L - k_e\alpha d\ln\left(\frac{d + L}{d}\right) \quad \lozenge$$

We have the answer when the unknown is expressed in terms of the d, L, and α mentioned in the problem and the universal constant k_e.

---

49. A spherical conductor has a radius of 14.0 cm and charge of 26.0 μC. Calculate the electric field and the electric potential (a) $r = 10.0$ cm, (b) $r = 20.0$ cm, and (c) $r = 14.0$ cm from the center.

Solution

(a) Inside a conductor when charges are not moving, the electric field is zero and the potential is uniform, the same as on the surface.

$$\mathbf{E} = 0 \qquad \qquad \Diamond$$

$$V = \frac{k_e q}{R} = \frac{(8.99 \times 10^9 \text{ N} \cdot \text{m}^2 / \text{C}^2)(26.0 \times 10^{-6} \text{ C})}{0.140 \text{ m}} = 1.67 \times 10^6 \text{ V} \qquad \Diamond$$

(b) The sphere behaves like a point charge at its center when you stand outside.

$$\mathbf{E} = \frac{k_e q}{r^2} \hat{\mathbf{r}} = \frac{(8.99 \times 10^9 \text{ N} \cdot \text{m}^2 / \text{C}^2)(26.0 \times 10^{-6} \text{ C})}{(0.200 \text{ m})^2} \hat{\mathbf{r}} = (5.84 \times 10^6 \text{ N} / \text{C}) \hat{\mathbf{r}} \qquad \Diamond$$

$$V = \frac{k_e q}{r} = 1.17 \times 10^6 \text{ V} \qquad \Diamond$$

(c) $$\mathbf{E} = \frac{k_e q}{r^2} \hat{\mathbf{r}} = (11.9 \times 10^6 \text{ N} / \text{C}) \hat{\mathbf{r}} \qquad \Diamond$$

$$V = 1.67 \times 10^6 \text{ V} \quad \text{as in part (a)} \qquad \Diamond$$

---

51. Lightning can be studied with a Van de Graaff generator, essentially consisting of a spherical dome on which charge is continuously deposited by a moving belt. Charge can be added until the electric field at the surface of the dome becomes equal to the dielectric strength of air. Any more charge leaks off in sparks, as shown in Figure P25.51 of the main text. Assume the dome has a diameter of 30.0 cm and is surrounded by dry air with dielectric strength 3.00×10^6 V/m. (a) What is the maximum potential of the dome? (b) What is the maximum charge on the dome?

Solution

Conceptualize: Van de Graaff generators produce voltages that can make your hair stand on end, on the order of 100 kV. With these high voltages, the maximum charge on the dome is probably more than typical point charge values of about 1 μC.

The maximum potential and charge will be limited by the electric field strength at which the air surrounding the dome will ionize. This critical value is determined by the **dielectric strength** of air which, from Table 26.1, is $E_{critical} = 3 \times 10^6$ V/m. An electric field stronger than this will cause the air to act like a conductor instead of an insulator. This process is called dielectric breakdown and may be seen as a spark.

Categorize: From the maximum allowed electric field, we can find the charge and potential that would create this situation. Since we are only given the diameter of the dome, we will assume that the conductor is spherical, which allows us to use the electric field and potential equations for a spherical conductor. With these equations, it will be easier to do part (b) first and use the result for part (a).

Analyze:

At the surface of a spherical conductor with total charge Q,

(b) $\quad |E| = \dfrac{k_e Q}{r^2}: \quad Q = \dfrac{E r^2}{k_e} = \dfrac{(3.00 \times 10^6 \text{ V/m})(0.150 \text{ m})^2}{8.99 \times 10^9 \text{ N} \cdot \text{m}^2/\text{C}^2}(1 \text{ N} \cdot \text{m/V} \cdot \text{C}) = 7.51 \; \mu\text{C}$ ◊

(a) $\qquad\qquad V = \dfrac{k_e Q}{r} = \dfrac{(8.99 \times 10^9 \text{ N} \cdot \text{m}^2/\text{C}^2)(7.51 \times 10^{-6} \text{ C})}{0.150 \text{ m}} = 450 \text{ kV}$ ◊

Finalize: These calculated results seem reasonable based on our predictions. The voltage is about 4000 times larger than the 120 V found from common electrical outlets, but the charge is similar in magnitude to many of the static charge problems we have solved earlier. This implies that most of these charge configurations would have to be in a vacuum because the electric field near these point charges would be strong enough to cause sparking in air. (Example: A charged ball with $Q = 1 \; \mu\text{C}$ and $r = 1$ mm would have an electric field near its surface of

$$E = \dfrac{k_e Q}{r^2} = \dfrac{(9 \times 10^9 \text{ N} \cdot \text{m}^2/\text{C}^2)(1 \times 10^{-6} \text{ C})}{(0.001 \text{ m})^2} = 9 \times 10^9 \text{ V/m}$$

which is well beyond the dielectric breakdown of air!)

---

53. The liquid-drop model of the atomic nucleus suggests that high-energy oscillations of certain nuclei can split the nucleus into two unequal fragments plus a few neutrons. The fission products acquire kinetic energy from their mutual Coulomb repulsion. Calculate the electric potential energy (in electron volts) of two spherical fragments from a uranium nucleus having the following charges and radii: $38e$ and 5.50×10^{-15} m; $54e$ and 6.20×10^{-15} m. Assume that the charge is distributed uniformly throughout the volume of each spherical fragment and that just before separating they are at rest with their surfaces in contact. The electrons surrounding the nucleus can be ignored.

Solution The problem is equivalent to finding the potential energy of a point charge $38e$ at a distance 11.7×10^{-15} m (the distance between their centers when in contact) from a point charge $54e$.

$$U = qV = k_e \frac{q_1 q_2}{r_{12}} = (8.99 \times 10^9 \text{ J} \cdot \text{m}/\text{C}^2) \frac{(38)(1.60 \times 10^{-19} \text{ C})(54)(1.60 \times 10^{-19} \text{ C})}{(5.50 \times 10^{-15} \text{ m} + 6.20 \times 10^{-15} \text{ m})}$$

$$U = 4.04 \times 10^{-11} \text{ J} = 253 \text{ MeV}$$ ◊

---

55. The Bohr model of the hydrogen atom states that the single electron can exist only in certain allowed orbits around the proton. The radius of each Bohr orbit is $r = n^2 (0.052\,9 \text{ nm})$ where $n = 1, 2, 3, \ldots$. Calculate the electric potential energy of a hydrogen atom when the electron is in the (a) first allowed orbit, with $n = 1$; (b) second allowed orbit, $n = 2$; and (c) when the electron has escaped from the atom, with $r = \infty$. Express your answers in electron volts.

Solution

Conceptualize: We may remember from chemistry that the lowest energy level for hydrogen is $E_1 = -13.6$ eV, and higher energy levels can be found from $E_n = E_1/n^2$, so that $E_2 = -3.40$ eV and $E_\infty = 0$ eV (see Section 42.2). Since these are the total energies (potential plus kinetic), the electric potential energy alone should be lower (more negative) because the kinetic energy of the electron must be positive.

Categorize: The electric potential energy is given by $U = k_e \frac{q_1 q_2}{r}$

Analyze:

(a) For the first allowed Bohr orbit,

$$U = (8.99 \times 10^9 \text{ N} \cdot \text{m}^2/\text{C}^2) \frac{(-1.60 \times 10^{-19} \text{ C})(1.60 \times 10^{-19} \text{ C})}{(0.052\,9 \times 10^{-9} \text{ m})}$$

$$U = -4.35 \times 10^{-18} \text{ J} = \frac{-4.35 \times 10^{-18} \text{ J}}{1.60 \times 10^{-19} \text{ J}/\text{eV}} = -27.2 \text{ eV}$$ ◊

(b) For the second allowed orbit,

$$U = (8.99 \times 10^9 \text{ N} \cdot \text{m}^2/\text{C}^2) \frac{(-1.60 \times 10^{-19} \text{ C})(1.60 \times 10^{-19} \text{ C})}{2^2 (0.052\,9 \times 10^{-9} \text{ m})}$$

$$U = -1.088 \times 10^{-18} \text{ J} = -6.80 \text{ eV}$$ ◊

(c) When the electron is at $r = \infty$,

$$U = (8.99 \times 10^9 \text{ N} \cdot \text{m}^2/\text{C}^2) \frac{(-1.60 \times 10^{-19} \text{ C})(1.60 \times 10^{-19} \text{ C})}{\infty \text{ m}} = 0 \text{ J}$$ ◊

Finalize: The potential energies appear to be twice the magnitude of the total energy values, so apparently the kinetic energy of the electron has the same absolute magnitude as the total energy.

---

59. Calculate the work that must be done to charge a spherical shell of radius R to a total charge Q.

Solution When the potential of the shell is V due to a charge q, the work required to add an additional increment of charge dq is

$$dW = V \, dq \qquad \text{where} \quad V = \frac{k_e q}{R}$$

$$dW = \left(\frac{k_e q}{R}\right) dq \quad \text{and} \quad W = \frac{k_e}{R}\int_0^Q q \, dq \, ; \qquad \text{therefore,} \qquad W = \left(\frac{k_e}{R}\right)\left(\frac{Q^2}{2}\right) \quad \Diamond$$

---

63. From Gauss's law, the electric field set up by a uniform line of charge is

$$\mathbf{E} = \left(\frac{\lambda}{2\pi\epsilon_0 r}\right)\hat{\mathbf{r}}$$

where $\hat{\mathbf{r}}$ is a unit vector pointing radially away from the line and λ is the charge density along the line. Derive an expression for the potential difference between $r = r_1$ and $r = r_2$.

Solution In Equation 25.3, $V_2 - V_1 = \Delta V = -\int_1^2 \mathbf{E} \cdot d\mathbf{s}$, think about stepping from distance r_1 out to the larger distance r_2 away from the charged line. Then $d\mathbf{s} = dr \, \hat{\mathbf{r}}$, and we can make r the variable of integration:

$$V_2 - V_1 = -\int_{r_1}^{r_2} \frac{\lambda}{2\pi\epsilon_0 r} \hat{\mathbf{r}} \cdot dr \, \hat{\mathbf{r}} \qquad \text{with} \qquad \hat{\mathbf{r}} \cdot \hat{\mathbf{r}} = 1 \cdot 1 \cos 0° = 1$$

The potential difference is

$$V_2 - V_1 = -\frac{\lambda}{2\pi\epsilon_0}\int_{r_1}^{r_2}\frac{dr}{r} = -\frac{\lambda}{2\pi\epsilon_0}\ln r \Big]_{r_1}^{r_2}$$

and

$$V_2 - V_1 = -\frac{\lambda}{2\pi\epsilon_0}(\ln r_2 - \ln r_1) = -\frac{\lambda}{2\pi\epsilon_0}\ln\frac{r_2}{r_1} \quad \Diamond$$

If $r_2 > r_1$, then $V_2 - V_1$ is negative. This means the potential decreases as we move away from a positively-charged filament.

---

69. An electric dipole is located along the y axis as shown in Figure P25.69. The magnitude of its electric dipole moment is defined as $p=2qa$. (a) At a point P, which is far from the dipole $(r\gg a)$, show that the electric potential is

$$V = \frac{k_e p \cos\theta}{r^2}$$

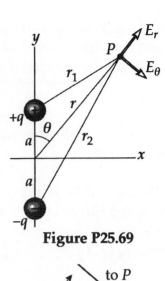

(b) Calculate the radial component E_r and the perpendicular component E_θ of the associated electric field. Note that $E_\theta = -(1/r)(\partial V/\partial\theta)$. Do these results seem reasonable for $\theta=90°$ and $0°$? for $r=0$? (c) For the dipole arrangement shown, express V in terms of Cartesian coordinates using $r=(x^2+y^2)^{1/2}$ and

$$\cos\theta = \frac{y}{(x^2+y^2)^{1/2}}$$

Using these results and again taking $r\gg a$, calculate the field components E_x and E_y.

Figure P25.69

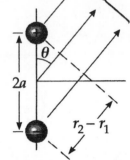

Solution

(a) The total potential is
$$V = \frac{k_e q}{r_1} - \frac{k_e q}{r_2} = \frac{k_e q}{r_1 r_2}(r_2 - r_1)$$

From the diagram for $r\gg a$, $r_2-r_1 \approx 2a\cos\theta$, so $V \approx \frac{k_e q}{(r)(r)}(2a\cos\theta) = \frac{k_e p\cos\theta}{r^2}$ ◊

(b)
$$E_r = -\frac{\partial V}{\partial r} = -\frac{\partial}{\partial r}\left(\frac{k_e p\cos\theta}{r^2}\right) = \frac{2k_e p\cos\theta}{r^3} = E_r \qquad ◊$$

In spherical coordinates, $\quad E_\theta = -\frac{1}{r}\left(\frac{\partial V}{\partial\theta}\right) = -\frac{1}{r}\frac{\partial}{\partial\theta}\left(\frac{k_e p\cos\theta}{r^2}\right) = \frac{k_e p\sin\theta}{r^3} = E_\theta \quad ◊$

The limits of these values at $\theta=0$ and $\theta=90°$ correspond to an electric field that points outward from the positive charge on top, and loops to the negative charge below. This is reasonable, but it has no meaning at $r=0$. ◊

(c) For $r = \sqrt{x^2+y^2}$, $\qquad \cos\theta = \frac{y}{\sqrt{x^2+y^2}} \qquad$ and $\qquad V = \frac{k_e py}{(x^2+y^2)^{3/2}}$

$$E_x = -\frac{\partial V}{\partial x} = -\frac{\partial}{\partial x}\left(\frac{k_e py}{(x^2+y^2)^{3/2}}\right) = \frac{3k_e pxy}{(x^2+y^2)^{5/2}} = E_x \qquad ◊$$

$$E_y = -\frac{\partial V}{\partial y} = -\frac{\partial}{\partial y}\left(\frac{k_e py}{(x^2+y^2)^{3/2}}\right) = \frac{k_e p(2y^2-x^2)}{(x^2+y^2)^{5/2}} \qquad ◊$$

Chapter 26
CAPACITANCE AND DIELECTRICS

EQUATIONS AND CONCEPTS

The **capacitance of a capacitor** is defined as the ratio of the magnitude of the charge on either conductor (or plate) to the magnitude of the potential difference between the conductors. *Capacitance is always a positive quantity and has SI units of farads (F).*

$$C \equiv \frac{Q}{\Delta V} \tag{26.1}$$

$$1\,F = 1\,C\,/\,V$$

An **air-filled parallel-plate capacitor** has a capacitance that is proportional to the area of the plates and inversely proportional to the separation of the plates. Capacitance expressions for other geometric arrangements of the conducting plates are given in Table 26.2 of the textbook.

$$C = \frac{\epsilon_0 A}{d} \tag{26.3}$$

When a **dielectric material** (insulator) completely fills the region between the plates of a capacitor, the capacitance increases by a factor κ. *Kappa, called the dielectric constant, is dimensionless and is characteristic of a particular material.*

$$C = \kappa C_0 \tag{26.14}$$

For capacitors connected in parallel:

Each capacitor has two circuit points in common with each of the other capacitors in the group.

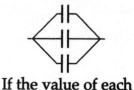

If the value of each capacitor is C, $C_{eq} = 3C$.

The total capacitance equals the algebraic sum of the individual capacitances.

$$C_{eq} = C_1 + C_2 + C_3 + \ldots \tag{26.8}$$

The total charge is the sum of the charges on the individual capacitors.

$$Q_{total} = Q_1 + Q_2 + Q_3 + \ldots$$

The potential difference across each capacitor is the same.

$$\Delta V_1 = \Delta V_2 = \Delta V_3 = \ldots$$

For capacitors connected in series:

Adjacent capacitors have one circuit point in common.

If each capacitor has value C, $C_{eq} = C/3$

The inverse of the equivalent capacitance equals the algebraic sum of the inverses of the capacitances of the individual capacitors.

$$\frac{1}{C_{eq}} = \frac{1}{C_1} + \frac{1}{C_2} + \frac{1}{C_3} + \ldots \qquad (26.10)$$

The potential difference across a series group of capacitors equals the sum of the potential differences across the individual capacitors.

$$\Delta V_{total} = \Delta V_1 + \Delta V_2 + \Delta V_3 \ldots$$

The total charge on the group is equal to the charge on each capacitor.

$$Q_{total} = Q_1 = Q_2 = Q_3 \ldots$$

For **two capacitors in series,** the equivalent capacitance of the pair is equal to the ratio of the product to the sum of their individual capacitance values.

$$C_{eq} = \frac{C_1 C_2}{C_1 + C_2}$$

A **series-parallel combination** of capacitors can be arranged using three or more capacitors. Using Equations 26.8 and 26.10 in sequence, each combination can be reduced to a single equivalent capacitor.

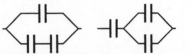

If each capacitor has value C, then in the first circuit $C_{eq} = 3/2C$ and in the second circuit $C_{eq} = 2/3C$.

The **electrostatic energy** stored in the electrostatic field of a charged capacitor equals the work done (by a battery or other source) in charging the capacitor from $q = 0$ to $q = Q$.

$$U = \frac{Q^2}{2C} = \frac{1}{2}Q\Delta V = \frac{1}{2}C(\Delta V)^2 \qquad (26.11)$$

The **energy density** at any point in the electrostatic field of a charged capacitor is proportional to the square of the electric field intensity at that point. *This is true for any electric field, not just that of a parallel plate capacitor.*

$$u_E = \frac{1}{2}\epsilon_0 E^2 \qquad (26.13)$$

An **electric dipole** consists of two equal magnitude and opposite sign charges separated by a distance $2a$.

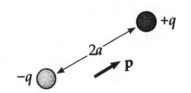

An **electric dipole moment** is characterized by a vector **p** that depends on the value of q and the distance of separation of the two charges. *The direction of the vector **p** is from the negative charge to the positive charge.*

$$p \equiv 2aq \qquad (26.16)$$

The **net torque** on an electric dipole in a uniform electric field can be expressed as a vector cross product. *The torque will be maximum when the direction of the dipole moment is perpendicular to the direction of the electric field.*

$$\boldsymbol{\tau} = \mathbf{p} \times \mathbf{E} \qquad (26.18)$$

The **potential energy** associated with a dipole-electric field system can be expressed as a dot product.

$$U = -\mathbf{p} \cdot \mathbf{E} \qquad (26.20)$$

SUGGESTIONS, SKILLS, AND STRATEGIES

PROBLEM-SOLVING HINTS FOR CAPACITANCE

- When analyzing a series-parallel combination of capacitors to determine the equivalent capacitance, you should make a sequence of circuit diagrams which show the successive steps in the simplification of the circuit; combine at each step those capacitors which are in simple-parallel or simple-series relationship to each other and use appropriate equations for series or parallel capacitors at each step of the simplification. At each step, you know two of the three quantities: Q, ΔV, and C. You will be able to determine the remaining quantity using the relation $Q = C\Delta V$.

- When calculating capacitance, be careful with your choice of units. To calculate capacitance in farads, make sure that distances are in meters and use the SI value of ϵ_0. When checking consistency of units, remember that the electric field has units of newtons per coulomb (N/C) or volts per meter (V/m).

- When two or more unequal capacitors are connected in series, they carry the same charge, but their potential differences are not the same. The capacitances add as reciprocals, and the equivalent capacitance of the combination is always less than the smallest individual capacitor.

- When two or more capacitors are connected in parallel, the potential differences across them are the same. The charge on each capacitor is proportional to its capacitance; hence, the capacitances add directly to give the equivalent capacitance of the parallel combination.

- A dielectric increases capacitance by the factor κ (the dielectric constant) because induced surface charges on the dielectric reduce the electric field inside the material from E to E/κ.

- Be careful about problems in which you may be connecting or disconnecting a battery to a capacitor. It is important to note whether modifications to the capacitor are being made while the capacitor is connected to the battery or after it is disconnected. If the capacitor remains connected to the battery, the voltage across the capacitor necessarily remains the same (equal to the battery voltage), and the charge is proportional to the capacitance. The capacitor may, however, be modified (for example, by the insertion of a dielectric). On the other hand, if you disconnect the capacitor from the battery before making any modifications to the capacitor, then its charge remains the same. In this case, as you vary the capacitance, the voltage across the plates changes in inverse proportion to capacitance, according to $\Delta V = Q/C$.

REVIEW CHECKLIST

You should be able to:

▷ Use the basic definition of capacitance and the equation for finding the potential difference between two points in an electric field in order to calculate the capacitance of a capacitor for cases of relatively simple geometry – parallel plates, cylindrical, spherical. (Section 26.2)

▷ Determine the equivalent capacitance of a network of capacitors in series-parallel combination and calculate the final charge on each capacitor and the potential difference across each when a known potential is applied across the combination. (Section 26.3)

▷ Make calculations involving the relationships among potential, charge, capacitance, stored energy, and energy density for capacitors, and apply these results to the particular case of a parallel plate capacitor. (Section 26.4)

▷ Calculate the capacitance, potential difference, and stored energy of a capacitor which is partially or completely filled with a dielectric. (Section 26.5)

▷ Calculate the dipole moment, torque, and potential energy associated with an electric dipole in an external electric field. (Section 26.6)

ANSWERS TO SELECTED QUESTIONS

11. If the potential difference across a capacitor is doubled, by what factor does the energy stored change?

Answer   Since $U = C(\Delta V^2)/2$, doubling ΔV will quadruple the stored energy.

□   □   □   □

20. If you were asked to design a capacitor where small size and large capacitance were required, what factors would be important in your design?

Answer   You should use a dielectric filled capacitor whose dielectric constant is very large. Furthermore, you should make the dielectric as thin as possible, keeping in mind that dielectric breakdown must also be considered.

□   □   □   □

SOLUTIONS TO SELECTED PROBLEMS

3. An isolated charged conducting sphere of radius 12.0 cm creates an electric field of 4.90×10^4 N/C at a distance 21.0 cm from its center. (a) What is its surface charge density? (b) What is its capacitance?

Solution

(a) The electric field outside a spherical charge distribution of radius R is $E = k_e q / r^2$. Therefore, $q = E r^2 / k_e$. Since the surface charge density is $\sigma = q / A$,

$$\sigma = \frac{E r^2}{k_e 4\pi R^2} = \frac{(4.90 \times 10^4 \text{ N/C})(0.210 \text{ m})^2}{(8.99 \times 10^9 \text{ N} \cdot \text{m}^2/\text{C}^2)(4\pi)(0.120 \text{ m})^2} = 1.33 \ \mu\text{C/m}^2 \qquad \diamond$$

(b) For an isolated charged sphere of radius R,

$$C = 4\pi\epsilon_0 R = (4\pi)(8.85 \times 10^{-12} \text{ C}^2/\text{N} \cdot \text{m}^2)(0.120 \text{ m}) = 13.3 \text{ pF} \qquad \diamond$$

---

7. An air-filled capacitor consists of two parallel plates, each with an area of 7.60 cm², separated by a distance of 1.80 mm. A 20.0-V potential difference is applied to these plates. Calculate (a) the electric field between the plates, (b) the surface charge density, (c) the capacitance, and (d) the charge on each plate.

Solution

(a) The potential difference between two points in a uniform electric field is $\Delta V = Ed$, so:

$$E = \frac{\Delta V}{d} = \frac{20.0 \text{ V}}{1.80 \times 10^{-3} \text{ m}} = 1.11 \times 10^4 \text{ V/m} \qquad \diamond$$

(b) The electric field between capacitor plates is $E = \dfrac{\sigma}{\epsilon_0}$, so $\sigma = \epsilon_0 E$:

$$\sigma = (8.85 \times 10^{-12} \text{ C}^2/\text{N} \cdot \text{m}^2)(1.11 \times 10^4 \text{ V/m}) = 9.83 \times 10^{-8} \text{ C/m}^2 = 98.3 \text{ nC/m}^2 \qquad \diamond$$

(c) For a parallel-plate capacitor, $C = \dfrac{\epsilon_0 A}{d}$:

$$C = \frac{(8.85 \times 10^{-12} \text{ C}^2/\text{N} \cdot \text{m}^2)(7.60 \times 10^{-4} \text{ m}^2)}{1.80 \times 10^{-3} \text{ m}} = 3.74 \times 10^{-12} \text{ F} = 3.74 \text{ pF} \qquad \diamond$$

(d) The charge on each plate is $Q = C\Delta V$:

$$Q = (3.74 \times 10^{-12} \text{ F})(20.0 \text{ V}) = 7.47 \times 10^{-11} \text{ C} = 74.7 \text{ pC} \qquad \diamond$$

---

9.   When a potential difference of 150 V is applied to the plates of a parallel-plate capacitor, the plates carry a surface charge density of 30.0 nC/cm^2. What is the spacing between the plates?

Solution   We have $Q = C\Delta V$ with $C = \epsilon_0 A / d$.

Thus, $Q = \epsilon_0 A \Delta V / d$

The surface charge density on each plate is the same in magnitude,

so $\sigma = \dfrac{Q}{A} = \dfrac{\epsilon_0 \Delta V}{d}$

Thus, $d = \dfrac{\epsilon_0 \Delta V}{Q/A} = \dfrac{(8.85 \times 10^{-12} \ C^2/N \cdot m^2)(150 \ V)}{(30.0 \times 10^{-9} \ C/cm^2)}$

$d = \left(4.42 \times 10^{-2} \ \dfrac{V \cdot C \cdot cm^2}{N \cdot m^2} \right) \left(\dfrac{1 \ m^2}{10^4 \ cm^2} \right) \left(\dfrac{J}{V \cdot C} \right) \left(\dfrac{N \cdot m}{J} \right) = 4.42 \ \mu m$ ◊

11.   A 50.0-m length of coaxial cable has an inner conductor that has a diameter of 2.58 mm and carries a charge of 8.10 μC. The surrounding conductor has an inner diameter of 7.27 mm and a charge of –8.10 μC. (a) What is the capacitance of this cable? (b) What is the potential difference between the two conductors? Assume the region between the conductors is air.

Solution

(a) $C = \dfrac{\ell}{2 k_e \ln(b/a)} = \dfrac{50.0 \ m}{2(8.99 \times 10^9 \ N \cdot m^2/C^2) \ln(7.27/2.58)} = 2.68 \times 10^{-9} \ F$ ◊

(b) $\Delta V = \dfrac{Q}{C} = \dfrac{8.10 \times 10^{-6} \ C}{2.68 \times 10^{-9} \ F} = 3.02 \ kV$ ◊

13.   An air-filled spherical capacitor is constructed with inner and outer shell radii of 7.00 and 14.0 cm, respectively. (a) Calculate the capacitance of the device. (b) What potential difference between the spheres results in a charge of 4.00 μC on the capacitor?

Solution

Conceptualize: Since the separation between the inner and outer shells is much larger than a typical electronic capacitor with $d \sim 0.1$ mm and capacitance in the microfarad range, we might expect the capacitance of this spherical configuration to be on the order of picofarads (based on a factor of about 700 times larger spacing between the conductors). The potential difference should be sufficiently low to prevent sparking through the air that separates the shells.

Categorize: The capacitance can be found from the equation for spherical shells, and the voltage can be found from $Q = C\Delta V$.

Analyze:

(a) For a spherical capacitor with inner radius a and outer radius b,

$$C = \frac{ab}{k_e(b-a)} = \frac{(0.0700 \text{ m})(0.140 \text{ m})}{(8.99 \times 10^9 \text{ N} \cdot \text{m}^2/\text{C}^2)(0.140 \text{ m} - 0.0700 \text{ m})} = 1.56 \times 10^{-11} \text{ F} = 15.6 \text{ pF} \quad \lozenge$$

(b) $\quad \Delta V = \frac{Q}{C} = \frac{4.00 \times 10^{-6} \text{ C}}{1.56 \times 10^{-11} \text{ F}} = 2.56 \times 10^5 \text{ V} = 256 \text{ kV} \quad \lozenge$

Finalize: The capacitance agrees with our prediction, but the voltage seems rather high. We can check this voltage by approximating the configuration as the electric field between two charged parallel plates separated by $d = 7.00$ cm, so

$$E \sim \frac{\Delta V}{d} = \frac{2.56 \times 10^5 \text{ V}}{0.0700 \text{ m}} = 3.66 \times 10^6 \text{ V/m}$$

This electric field barely exceeds the dielectric breakdown strength of air $(3 \times 10^6 \text{ V/m})$, so it may not even be possible to place 4.00 μC of charge on this capacitor!

---

21. Four capacitors are connected as shown in Figure P26.21. (a) Find the equivalent capacitance between points a and b. (b) Calculate the charge on each capacitor if $\Delta V_{ab} = 15.0$ V.

Solution

(a) We simplify the circuit of Figure P26.21 in three steps as shown in Figures (a), (b), and (c).

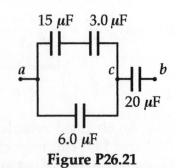

Figure P26.21

First, the 15.0 μF and 3.00 μF in series are equivalent to

$$\frac{1}{(1/15.0\ \mu F)+(1/3.00\ \mu F)} = 2.50\ \mu F$$

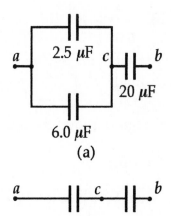

(a)

Next, 2.50 μF combines in parallel with 6.00 μF, creating an equivalent capacitance of 8.50 μF.

At last, 8.50 μF and 20.0 μF are in series, equivalent to

$$\frac{1}{(1/8.50\ \mu F)+(1/20.0\ \mu F)} = 5.96\ \mu F \quad \lozenge$$

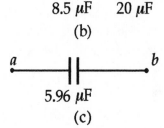

(b)

(b) We find the charge on and the voltage across each capacitor by working backwards through solution figures (c)-(a), alternately applying $Q=C\Delta V$ and $\Delta V=Q/C$ to every capacitor, real or equivalent. For the 5.96-μF capacitor, we have

(c)

$$Q = C\Delta V = (5.96\ \mu F)(15.0\ V) = 89.5\ \mu C$$

Thus, if a is higher in potential than b, just 89.5 μC flows between the wires and the plates to charge the capacitors in each picture. In (b) we have, for the 8.5-μF capacitor,

$$\Delta V_{ac} = \frac{Q}{C} = \frac{89.5\ \mu C}{8.50\ \mu F} = 10.5\ V$$

and for the 20.0 μF in (b), (a), and the original circuit, we have $Q_{20} = 89.5\ \mu C$ $\lozenge$

$$\Delta V_{cb} = \frac{Q}{C} = \frac{89.5\ \mu C}{20.0\ \mu F} = 4.47\ V$$

Next, (a) is equivalent to (b), so $\qquad \Delta V_{cb} = 4.47\ V \qquad$ and $\qquad \Delta V_{ac} = 10.5\ V$

For the 2.50 μF, $\Delta V = 10.5\ V \qquad$ and $\qquad Q = C\Delta V = (2.50\ \mu F)(10.5\ V) = 26.3\ \mu C$

For the 6.00 μF, $\Delta V = 10.5\ V \qquad$ and $\qquad Q_6 = C\Delta V = (6.00\ \mu F)(10.5\ V) = 63.2\ \mu C$ $\lozenge$

Now, 26.3 μC having flowed in the upper parallel branch in (a),
back in the original circuit we have $\qquad Q_{15} = 26.3\ \mu C \quad$ and $\quad Q_3 = 26.3\ \mu C$ $\lozenge$

Related Calculation: An exam problem might also ask for the voltage across each:

$$\Delta V_{15} = \frac{Q}{C} = \frac{26.3\ \mu C}{15.0\ \mu F} = 1.75\ V \quad \text{and} \quad \Delta V_3 = \frac{Q}{C} = \frac{26.3\ \mu C}{3.00\ \mu F} = 8.77\ V$$

23. Consider the circuit shown in Figure P26.23, where $C_1 = 6.00\ \mu F$, $C_2 = 3.00\ \mu F$, and $\Delta V = 20.0$ V. Capacitor C_1 is first charged by the closing of switch S_1. Switch S_1 is then opened, and the charged capacitor is connected to the uncharged capacitor by the closing of S_2. Calculate the initial charge acquired by C_1 and the final charge on each capacitor.

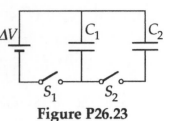

Figure P26.23

Solution

When S_1 is closed, the charge on C_1 will be $Q_1 = C_1 \Delta V_1 = (6.00\ \mu F)(20.0\ V) = 120\ \mu C$ ◊

When S_1 is opened and S_2 is closed, the total charge will remain constant and be shared by the two capacitors: $Q_1' = 120\ \mu C - Q_2'$.   The potential differences across the two capacitors will be equal.

$$\Delta V' = \frac{Q_1'}{C_1} = \frac{Q_2'}{C_2} \qquad \text{or} \qquad \frac{120\ \mu C - Q_2'}{6.00\ \mu F} = \frac{Q_2'}{3.00\ \mu F}$$

and $Q_2' = 40.0\ \mu C$ $Q_1' = 120\ \mu C - 40.0\ \mu C = 80.0\ \mu C$ ◊

---

25. A group of identical capacitors is connected first in series and then in parallel. The combined capacitance in parallel is 100 times larger than for the series connection. How many capacitors are in the group?

Solution

Conceptualize: Since capacitors in parallel add and ones in series add as inverses, 2 capacitors in parallel would have a capacitance 4 times greater than if they were in series, and 3 capacitors would give a ratio $C_p/C_s = 9$, so maybe $n = \sqrt{C_p/C_s} = \sqrt{100} = 10$.

Categorize: The ratio reasoning above seems like an efficient way to solve this problem, but we should check the answer with a more careful analysis based on the general relationships for series and parallel combinations of capacitors.

Analyze: Call C the capacitance of one capacitor and n the number of capacitors. The equivalent capacitance for n capacitors in parallel is

$$C_p = C_1 + C_2 + \ldots + C_n = nC$$

The relationship for n capacitors in series is $\dfrac{1}{C_s} = \dfrac{1}{C_1} + \dfrac{1}{C_2} + \ldots + \dfrac{1}{C_n} = \dfrac{n}{C}$

Therefore $\dfrac{C_p}{C_s} = \dfrac{nC}{C/n} = n^2$ or $n = \sqrt{C_p/C_s} = \sqrt{100} = 10$ ◊

Finalize: Our prediction appears to be correct. A qualitative reason that $C_p/C_s = n^2$ is because the amount of charge that can be stored on the capacitors increases according to the area of the plates for a parallel combination, but the total charge remains the same for a series combination.

---

37. A parallel-plate capacitor has a charge Q and plates of area A. What force acts on one plate to attract it toward the other plate? Because the electric field between the plates is $E = Q/A\epsilon_0$, you might think that the force is $F = QE = Q^2/A\epsilon_0$. This is wrong, because the field E includes contributions from both plates, and the field created by the positive plate cannot exert any force on the positive plate. Show that the force exerted on each plate is actually $F = Q^2/2\epsilon_0 A$. (*Suggestion:* Let $C = \epsilon_0 A/x$ for an arbitrary plate separation x; then require that the work done in separating the two charged plates be $W = \int F\, dx$.) The force exerted on one charged plate by another is sometimes used in a machine shop to hold a workpiece stationary.

Solution   The electric field in the space between the plates is
$$E = \frac{\sigma}{\epsilon_0} = \frac{Q}{A\epsilon_0}$$

You might think that the force on one plate is
$$F = QE = \frac{Q^2}{A\epsilon_0}$$

but this is two times too large, because neither plate exerts a force on itself. The force on one plate is exerted by the other, through its electric field:
$$E = \frac{\sigma}{2\epsilon_0} = \frac{Q}{2A\epsilon_0}$$

The force on each plate is:
$$F = (Q_{\text{self}})(E_{\text{other}}) = \frac{Q^2}{2A\epsilon_0}$$

To prove this, we follow the hint, and calculate that the work done in separating the plates, which equals the potential energy stored in the charged capacitor:
$$U = \frac{1}{2}\frac{Q^2}{C} = \int F\, dx$$

From the fundamental theorem of calculus, $dU = F\, dx$

and
$$F = \frac{d}{dx}U = \frac{d}{dx}\left(\frac{Q^2}{2C}\right) = \frac{1}{2}\frac{d}{dx}\left(\frac{Q^2}{A\epsilon_0/x}\right)$$

Solving,
$$F = \frac{1}{2}\frac{d}{dx}\left(\frac{Q^2 x}{A\epsilon_0}\right) = \frac{1}{2}\left(\frac{Q^2}{A\epsilon_0}\right) \qquad \lozenge$$

---

45. A commercial capacitor is to be constructed as shown in Figure 26.17a. This particular capacitor is made from two strips of aluminum separated by a strip of paraffin-coated paper. Each strip of foil and paper is 7.00 cm wide. The foil is 0.004 00 mm thick, and the paper is 0.025 0 mm thick and has a dielectric constant of 3.70. What length should the strips have, if a capacitance of 9.50×10^{-8} F is desired before the capacitor is rolled up? (Adding a second strip of paper and rolling the capacitor effectively doubles its capacitance, by allowing charge storage on both sides of each strip of foil.)

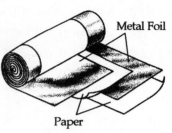

Metal Foil

Paper

Figure 26.17a

Solution $\quad C = \dfrac{\kappa \epsilon_0 A}{d} = \dfrac{(3.70)(8.85 \times 10^{-12} \ \text{C}^2/\text{N} \cdot \text{m}^2)(0.0700 \ \text{m})L}{2.50 \times 10^{-5} \ \text{m}} = 9.50 \times 10^{-8} \ \text{F}$

Solving, $\quad L = 1.04$ m $\qquad\qquad\qquad\qquad\qquad\qquad\qquad\qquad\qquad\qquad\qquad\qquad\qquad\quad \Diamond$

The distance between positive and negative charges is the thickness of one sheet of paper because these charges sit on adjacent surfaces of the metal foils. The electric field between the surfaces of the foil is zero and the thickness of the foil does not affect the capacitance.

---

59. A parallel-plate capacitor is constructed using a dielectric material whose dielectric constant is 3.00 and whose dielectric strength is 2.00×10^8 V/m. The desired capacitance is 0.250 μF, and the capacitor must withstand a maximum potential difference of 4 000 V. Find the minimum area of the capacitor plates.

Solution $\quad E_{max} = 2.00 \times 10^8 \ \text{V/m} = \dfrac{\Delta V_{max}}{d} \qquad$ so $\qquad d = \dfrac{\Delta V_{max}}{E_{max}}$

For $\qquad C = \dfrac{\kappa \epsilon_0 A}{d} = 0.250 \times 10^{-6} \ \text{F} \qquad$ and $\qquad \kappa = 3.00,$

$$A = \frac{Cd}{\kappa \epsilon_0} = \frac{C \Delta V_{max}}{\kappa \epsilon_0 E_{max}} = \frac{(0.250 \times 10^{-6} \ \text{F})(4000 \ \text{V})}{(3.00)(8.85 \times 10^{-12} \ \text{F/m})(2.00 \times 10^8 \ \text{V/m})} = 0.188 \ \text{m}^2 \quad \Diamond$$

---

67. An isolated capacitor of unknown capacitance has been charged to a potential difference of 100 V. When the charged capacitor is then connected in parallel to an uncharged 10.0-μF capacitor, the potential difference across the combination is 30.0 V. Calculate the unknown capacitance.

Solution

Conceptualize: The voltage of the combination will be reduced according to the size of the added capacitance. (Example: If the unknown capacitance were $C = 10.0 \ \mu F$, then $\Delta V_1 = 50.0$ V because the charge is now distributed evenly between the two capacitors.) Since the final voltage is less than half the original, we might guess that the unknown capacitor is about $5.00 \ \mu F$.

Categorize: We can use the relationships for capacitors in parallel to find the unknown capacitance, along with the requirement that the initial charge on the unknown capacitor must be the same as the final total charge on the two capacitors in parallel.

Analyze: We name our ignorance and call the unknown capacitance C_u. The charge originally deposited on **each** plate, + on one, − on the other, is

$$Q = C_u \Delta V = C_u (100 \ \text{V})$$

Now in the new connection this same conserved charge redistributes itself between the two capacitors according to $Q = Q_1 + Q_2$.

$$Q_1 = C_u (30.0 \ \text{V}) \qquad \text{and} \qquad Q_2 = (10.0 \ \mu F)(30.0 \ \text{V}) = 300 \ \mu C$$

We can eliminate Q and Q_1 by substitution:

$$C_u (100 \ \text{V}) = C_u (30.0 \ \text{V}) + 300 \ \mu C \quad \text{so} \quad C_u = \frac{300 \ \mu C}{70.0 \ \text{V}} = 4.29 \ \mu F \qquad \Diamond$$

Finalize: The calculated capacitance is close to what we expected, so our result seems reasonable. In this and other capacitance combination problems, it is important not to confuse the charge and voltage of the system with those of the individual components, especially if they have different values. Careful attention must be given to the subscripts to avoid this confusion. It is also important to not confuse the variable "C" for capacitance with the unit of charge, "C" for coulombs.

69. A parallel-plate capacitor of plate separation d is charged to a potential difference ΔV_0. A dielectric slab of thickness d and dielectric constant κ is introduced between the plates while the battery remains connected to the plates. (a) Show that the ratio of energy stored after the dielectric is introduced to the energy stored in the empty capacitor is $U/U_0 = \kappa$. Give a physical explanation for this increase in stored energy. (b) What happens to the charge on the capacitor? (Note that this situation is not the same as Example 26.7, in which the battery was removed from the circuit before the dielectric was introduced.)

Solution

(a) The capacitance changes from, say, C_0 to κC_0. The battery will maintain constant voltage across it by pumping out extra charge.

The original energy is
$$U_0 = \tfrac{1}{2}C_0(\Delta V)^2$$

and the final energy is
$$U = \tfrac{1}{2}\kappa C_0(\Delta V)^2 \qquad \text{so} \qquad U/U_0 = \kappa$$

The extra energy comes from (part of the) electrical work done by the battery in separating extra charge. ◊

(b) The original charge is
$$Q_0 = C_0 \Delta V$$

and the final value is
$$Q = \kappa C_0 \Delta V$$

so the charge increases by the factor κ. ◊

---

73. The inner conductor of a coaxial cable has a radius of 0.800 mm, and the outer conductor's inside radius is 3.00 mm. The space between the conductors is filled with polyethylene, which has a dielectric constant of 2.30 and a dielectric strength of 18.0×10^6 V/m. What is the maximum potential difference that this cable can withstand?

Solution We can increase the potential difference between the core and the sheath until the electric field in the polyethylene starts to punch a hole through it, passing a spark. This will first happen at the surface of the inner conductor, where the electric field is strongest. From Example 26.2, when there is a vacuum between the conductors, the voltage between them is

$$\Delta V = |V_b - V_a| = 2k_e \lambda \ln\!\left(\frac{b}{a}\right) = \frac{\lambda}{2\pi\epsilon_0}\ln\!\left(\frac{b}{a}\right)$$

With a dielectric, a factor $1/\kappa$ must be included, and the equation becomes

$$\Delta V = \frac{\lambda}{2\pi\kappa\epsilon_0}\ln\!\left(\frac{b}{a}\right)$$

So when $E = E_{\max}$ at $r = a$,

$$\frac{\lambda_{\max}}{2\pi\kappa\epsilon_0} = E_{\max}a \qquad \text{and} \qquad \Delta V_{\max} = \frac{\lambda_{\max}}{2\pi\kappa\epsilon_0}\ln\!\left(\frac{b}{a}\right) = E_{\max}a\ln\!\left(\frac{b}{a}\right)$$

Thus, $\quad \Delta V_{\max} = (18.0\times10^6 \text{ V}/\text{m})(0.800\times10^{-3} \text{ m})\ln\!\left(\dfrac{3.00 \text{ mm}}{0.800 \text{ mm}}\right) = 19.0 \text{ kV}$ ◊

Chapter 27
CURRENT AND RESISTANCE

EQUATIONS AND CONCEPTS

Electric current, I, is defined as the rate at which charge flows through the cross-sectional area of a conductor. *Under the influence of an electric field, electric charges will move through gases, liquids, and solid conductors.*

$$I \equiv \frac{dQ}{dt} \qquad (27.2)$$

The **direction of the current** is assigned to be the direction of the flow of positive charge. The SI unit of current is the **ampere** (A); 1 A of current is equivalent to 1 C of charge passing through a cross section of surface in 1 s.

$$1\,A = 1\,C/1\,s \qquad (27.3)$$

The **average current** in a conductor can be related to microscopic quantities in the conductor: the number of mobile charge carriers per unit volume n, the quantity of charge q associated with each carrier, and the drift velocity v_d of the carriers.

$$I_{av} = \frac{\Delta Q}{\Delta t} = nqv_d A \qquad (27.4)$$

Current density in a conductor is defined as current per unit area. The magnitude of the current density as expressed in Equation 27.5 is valid when the surface of cross-sectional area A is perpendicular to the direction of the current and the current density is uniform.

$$J \equiv \frac{I}{A} = nqv_d \qquad (27.5)$$

Current density is a vector quantity which is in the direction of motion of positive charge carriers.

$$\mathbf{J} = nq\mathbf{v}_d \qquad (27.6)$$

In **ohmic materials** (e.g. most metals), current density is proportional to the electric field in the conductor. *The conductivity of a conductor, σ, is characteristic of the conducting material and, in most cases, independent of the electric field.*

$$\mathbf{J} = \sigma\mathbf{E} \qquad (27.7)$$

Resistance is defined as the ratio of the potential difference across a conductor to the current in the conductor.

$$R \equiv \frac{\Delta V}{I} \qquad (27.8)$$

The **SI unit of resistance** is the ohm (Ω). *If a potential difference of 1 V across a conductor results in a current of 1 A, the resistance of the conductor is 1 Ω.*

$$1\,\Omega = 1\,\text{V/A} \qquad (27.9)$$

Resistivity, ρ, is the inverse of conductivity and is an intrinsic property of the material of which a conductor is made. The SI units of resistivity are ohm-meters. *Resistivity is not a property of a specific conducting element.*

$$\rho = \frac{1}{\sigma} \qquad (27.10)$$

The **resistance of a given conductor** of uniform cross section depends on the length, cross-sectional area, and resistivity of the conducting material.

$$R = \rho\frac{\ell}{A} \qquad (27.11)$$

The **drift velocity** of an electron in a conductor is due to the electric field produced by the potential difference across the conductor.

$$\mathbf{v}_d = \frac{q\mathbf{E}}{m_e}\tau \qquad (27.14)$$

In an **ohmic conductor** (one that obeys Ohm's law), conductivity and resistivity do not depend on the magnitude of the electric field. *Equations 27.16 and 27.17 express σ and τ in terms of microscopic properties of the conductor.*

$$\sigma = \frac{nq^2\tau}{m_e} \qquad (27.16)$$

$$\rho = \frac{1}{\sigma} = \frac{m_e}{nq^2\tau} \qquad (27.17)$$

The **average time interval between collisions** is related to the average distance between collisions (the mean free path). This characteristic time, denoted by τ, can be related to drift velocity (Equation 27.14) or resistivity (Equation 27.11).

$$\tau = \frac{\ell}{\overline{v}} \qquad (27.18)$$

The **temperature coefficient of resistivity** determines the rate at which resistivity (and resistance) increases or decreases with a change in the temperature of a conductor. T_0 is a stated reference temperature (usually 20 °C).

$$\rho = \rho_0\left[1 + \alpha(T - T_0)\right] \qquad (27.19)$$

$$R = R_0\left[1 + \alpha(T - T_0)\right] \qquad (27.21)$$

Power is the rate at which energy is delivered to a resistor or other circuit element when a potential difference is maintained between the terminals of the circuit element. The SI unit of power is the watt (W). For a device that obeys Ohm's law, the power dissipated can be expressed in alternative forms.

$$\mathcal{P} = I\Delta V \qquad (27.22)$$

$$\mathcal{P} = I^2R = \frac{(\Delta V)^2}{R} \qquad (27.23)$$

SUGGESTIONS, SKILLS, AND STRATEGIES

Equation 27.11, $R = \rho \ell / A$, can be used directly to calculate the resistance of a conductor of uniform cross-sectional area and constant resistivity. For those cases in which the area, resistivity, or both vary along the length of the conductor, the resistance must be determined as an integral of dR.

The conductor is subdivided into elements of length dx over which ρ and A may be considered constant in value and the total resistance is

$$R = \int \frac{\rho}{A}\, dx$$

Consider, for example, the case of a truncated cone of constant resistivity, radii a and b and height h. The conductor should be subdivided into disks of thickness dx, radius r, area $= \pi r^2$ and oriented parallel to the faces of the cone as shown in the figure below. Note from the geometry that

$$x = \left(\frac{r-a}{b-a}\right)h$$

so that $\qquad r = \dfrac{x}{h}(b-a) + a$

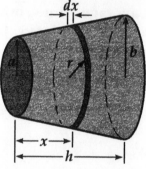

and $\qquad R = \displaystyle\int_0^h \frac{\rho}{\pi r^2}\, dx$

The remainder of this calculation is left as a problem for you to work out (see Problem 27.72 of the text).

REVIEW CHECKLIST

You should be able to:

▷ Calculate electron drift speed, and quantity of charge passing a point in a given time interval in a specified current-carrying conductor. (Section 27.1)

▷ Determine the resistance of a conductor using Ohm's law. Also, calculate the resistance based on the physical characteristics of a conductor. (Section 27.2)

▷ Make calculations of the variation of resistance with temperature. (Section 27.4)

▷ Calculate the power, current, resistance, or potential difference when energy is delivered to a circuit element. (Section 27.6)

ANSWERS TO SELECTED QUESTIONS

5. Two wires A and B of circular cross section are made of the same metal and have equal lengths, but the resistance of wire A is three times greater than that of wire B. What is the ratio of their cross-sectional areas? How do their radii compare?

Answer Since $R = \dfrac{\rho \ell}{\text{Area}}$, the ratio of resistances is given by $\dfrac{R_A}{R_B} = \dfrac{\text{Area}_B}{\text{Area}_A}$.

Hence, the ratio of their areas is three to one; that is, the area of wire B is three times that of wire A. From the ratio of the areas, we can calculate that the radius of wire B is $\sqrt{3}$ times the radius of wire A.

☐   ☐   ☐   ☐

11. Use the atomic theory of matter to explain why the resistance of a material should increase as its temperature increases.

Answer As the temperature increases, the amplitude of atomic vibrations increases. This makes it more likely that the drifting electrons will be scattered by atomic vibrations, and makes it more difficult for charges to participate in organized motion inside the conductor.

☐   ☐   ☐   ☐

17. If charges flow very slowly through a metal, why does it not require several hours for a light to come on when you throw a switch?

Answer Individual electrons move with a small average velocity through the conductor, but as soon as the voltage is applied, electrons all along the conductor start to move. Actually the current does not flow "immediately", but is limited by the speed of light.

☐   ☐   ☐   ☐

SOLUTIONS TO SELECTED PROBLEMS

3. Suppose that the current through a conductor decreases exponentially with time according to the equation $I(t) = I_0 e^{-t/\tau}$ where I_0 is the initial current (at $t=0$), and τ is a constant having dimensions of time. Consider a fixed observation point within the conductor. (a) How much charge passes this point between $t=0$ and $t=\tau$? (b) How much charge passes this point between $t=0$ and $t=10\tau$? (c) **What if?** How much charge passes this point between $t=0$ and $t=\infty$?

Solution

From $I = \dfrac{dQ}{dt}$, we have

$$dQ = I\, dt$$

From this, we derive a general integral

$$Q = \int dQ = \int I\, dt$$

In all three cases, define an end-time, T.

$$Q = \int_0^T I_0 e^{-t/\tau}\, dt$$

Integrating from time $t=0$ to time $t=T$,

$$Q = \int_0^T (-I_0\tau) e^{-t/\tau}\left(-\frac{dt}{\tau}\right)$$

Setting $Q=0$ at $t=0$,

$$Q = -I_0\tau\left(e^{-T/\tau} - e^0\right) = I_0\tau\left(1 - e^{-T/\tau}\right)$$

(a) If $T = \tau$,

$$Q = I_0\tau(1 - e^{-1}) = 0.6321 I_0\tau \qquad \Diamond$$

(b) If $T = 10\tau$,

$$Q = I_0\tau(1 - e^{-10}) = 0.99995 I_0\tau \qquad \Diamond$$

(c) If $T = \infty$,

$$Q = I_0\tau(1 - e^{-\infty}) = I_0\tau \qquad \Diamond$$

9. The electron beam emerging from a certain high-energy electron accelerator has a circular cross-section of radius 1.00 mm. (a) The beam current is 8.00 μA. Find the current density in the beam, assuming that it is uniform throughout. (b) The speed of the electrons is so close to the speed of light that their speed can be taken as $c = 3.00 \times 10^8$ m/s with negligible error. Find the electron density in the beam. (c) How long does it take for Avogadro's number of electrons to emerge from the accelerator?

Solution

(a) The current density is $\qquad J = \dfrac{I}{A} = \dfrac{8.00 \times 10^{-6} \text{ A}}{\pi (1.00 \times 10^{-3} \text{ m})^2} = 2.55 \text{ A} / \text{m}^2 \qquad \lozenge$

(b) From $J = nev_d$, we have $\qquad n = \dfrac{J}{ev_d} = \dfrac{2.55 \text{ A} / \text{m}^2}{(1.60 \times 10^{-19} \text{ C})(3.00 \times 10^8 \text{ m} / \text{s})}$

$$n = 5.31 \times 10^{10} \text{ m}^{-3} \qquad \lozenge$$

(c) From $I = \Delta Q / \Delta t$, we have $\qquad \Delta t = \dfrac{\Delta Q}{I} = \dfrac{N_A e}{I} = \dfrac{(6.02 \times 10^{23})(1.60 \times 10^{-19} \text{ C})}{8.00 \times 10^{-6} \text{ A}}$

$$\Delta t = 1.20 \times 10^{10} \text{ s} \text{ (or about 382 years!)} \qquad \lozenge$$

---

15. A 0.900-V potential difference is maintained across a 1.50-m length of tungsten wire that has a cross-sectional area of 0.600 mm^2. What is the current in the wire?

Solution

From Ohm's law $\qquad\qquad I = \dfrac{\Delta V}{R} \qquad\qquad$ where $\qquad\qquad R = \dfrac{\rho \ell}{A}$

Therefore, $\qquad\qquad I = \dfrac{\Delta V A}{\rho \ell} = \dfrac{(0.900 \text{ V})(6.00 \times 10^{-7} \text{ m}^2)}{(5.6 \times 10^{-8} \ \Omega \cdot \text{m})(1.50 \text{ m})} = 6.43 \text{ A} \qquad \lozenge$

---

17. Suppose that you wish to fabricate a uniform wire out of 1.00 g of copper. If the wire is to have a resistance of $R = 0.500 \ \Omega$, and if all of the copper is to be used, what will be (a) the length and (b) the diameter of this wire?

Solution Don't mix up symbols! Call the density ρ_d and the resistivity ρ_r. From $\rho_d = m / V$, the volume is $V = A\ell = m / \rho_d$. The resistance is $R = \rho_r \ell / A$.

(a) We can solve for ℓ by eliminating A:

$$A = \dfrac{m}{\ell \rho_d} \qquad\qquad R = \dfrac{\rho_r \ell}{m / \ell \rho_d} \qquad\qquad R = \dfrac{\rho_r \rho_d \ell^2}{m}$$

$$\ell = \sqrt{\dfrac{mR}{\rho_r \rho_d}} = \sqrt{\dfrac{(1.00 \times 10^{-3} \text{ kg})(0.500 \ \Omega)}{(1.70 \times 10^{-8} \ \Omega \cdot \text{m})(8.92 \times 10^3 \text{ kg} / \text{m}^3)}} = 1.82 \text{ m} \qquad \lozenge$$

(b) To have a single diameter, the wire has a circular cross section:

$$A = \pi r^2 = \pi \left(\frac{d}{2}\right)^2 = \frac{m}{\ell \rho_d}$$

$$d = \sqrt{\frac{4m}{\pi \ell \rho_d}} = \sqrt{\frac{4(1.00 \times 10^{-3} \text{ kg})}{\pi (1.82 \text{ m})(8.92 \times 10^3 \text{ kg}/\text{m}^3)}} = 2.80 \times 10^{-4} \text{ m} = 0.280 \text{ mm} \qquad \lozenge$$

25. If the magnitude of the drift velocity of free electrons in a copper wire is 7.84×10^{-4} m/s, what is the electric field in the conductor?

Solution

Conceptualize: For electrostatic cases, we learned that the electric field inside a conductor is always zero. On the other hand, if there is a current, a non-zero electric field must be maintained by a battery or other source to make the charges flow. Therefore, we might expect the electric field to be small, but definitely **not** zero.

Categorize: The drift velocity of the electrons can be used to find the current density, which can be used with Ohm's law to find the electric field inside the conductor.

Analyze: We first need the electron density in copper, which from Example 27.1 is $n = 8.49 \times 10^{28} \text{ e}^-/\text{m}^3$. The current density in this wire is then

$$J = nqv_d = (8.49 \times 10^{28} \text{ e}^-/\text{m}^3)(1.60 \times 10^{-19} \text{ C}/\text{e}^-)(7.84 \times 10^{-4} \text{ m}/\text{s})$$

$$J = 1.06 \times 10^7 \text{ A}/\text{m}^2$$

Ohm's law can be stated as

$$J = \sigma E = E/\rho \text{ where } \rho = 1.7 \times 10^{-8} \ \Omega \cdot \text{m for copper,}$$

so then $\quad E = \rho J = (1.70 \times 10^{-8} \ \Omega \cdot \text{m})(1.06 \times 10^7 \text{ A}/\text{m}^2) = 0.181 \text{ V}/\text{m} \qquad \lozenge$

Finalize: This electric field is certainly smaller than typical static values outside charged objects. The direction of the electric field should be along the length of the conductor, otherwise the electrons would be forced to leave the wire! The reality is that excess charges arrange themselves on the surface of the wire to create an electric field that "steers" the free electrons to flow along the length of the wire from low to high potential (opposite the direction of a positive test charge). It is also interesting to note that when the electric field is being established it travels at the speed of light; but the drift velocity of the electrons is literally at a "snail's pace"!

31. An aluminum wire with a diameter of 0.100 mm has a uniform electric field of 0.200 V/m imposed along its entire length. The temperature of the wire is 50.0 °C. Assume one free electron per atom. (a) Use the information in Table 27.1 and determine the resistivity. (b) What is the current density in the wire? (c) What is the total current in the wire? (d) What is the drift speed of the conduction electrons? (e) What potential difference must exist between the ends of a 2.00-m length of the wire to produce the stated electric field?

Solution The resistivity is found from $\rho = \rho_0[1 + \alpha(T - T_0)]$

(a) $\rho = (2.82 \times 10^{-8}\ \Omega \cdot m)[1 + (3.90 \times 10^{-3}\ °C^{-1})(30.0\ °C)] = 3.15 \times 10^{-8}\ \Omega \cdot m$ ◊

(b) $J = \sigma E = \dfrac{E}{\rho} = \left(\dfrac{0.200\ V/m}{3.15 \times 10^{-8}\ \Omega \cdot m}\right)\left(\dfrac{1\ \Omega \cdot A}{V}\right) = 6.35 \times 10^6\ A/m^2$ ◊

(c) $J = \dfrac{I}{A} = \dfrac{I}{\pi r^2}$

$I = J\pi r^2 = (6.35 \times 10^6\ A/m^2)[\pi(5.00 \times 10^{-5}\ m)^2] = 49.9\ mA$ ◊

(d) The mass density gives the number-density of free electrons; we assume that each atom donates one free electron:

$$n = \left(\dfrac{2.70 \times 10^3\ kg}{m^3}\right)\left(\dfrac{1\ mol}{26.98\ g}\right)\left(\dfrac{10^3\ g}{kg}\right)\left(\dfrac{6.02 \times 10^{23}\ free\ e^-}{1\ mol}\right) = 6.02 \times 10^{28}\ e^-/m^3$$

Now $J = nqv_d$ gives:

$v_d = \dfrac{J}{nq} = \dfrac{(6.35 \times 10^6\ A/m^2)}{(6.02 \times 10^{28}\ e^-/m^3)(-1.60 \times 10^{-19}\ C/e^-)} = -6.59 \times 10^{-4}\ m/s$ ◊

The sign indicates that the electrons drift opposite to the field and current.

(e) $\Delta V = E\ell = (0.200\ V/m)(2.00\ m) = 0.400\ V$ ◊

33. What is the fractional change in the resistance of an iron filament when its temperature changes from 25.0 °C to 50.0 °C?

Solution $R = R_0[1 + \alpha \Delta T]$ or $R - R_0 = R_0 \alpha \Delta T$

The fractional change in resistance is $f = (R - R_0)/R_0$

Therefore, $f = \dfrac{R_0 \alpha \Delta T}{R_0} = \alpha \Delta T = (5.00 \times 10^{-3}\ °C^{-1})(50.0\ °C - 25.0\ °C) = 0.125$ ◊

39. What is the required resistance of an immersion heater that increases the temperature of 1.50 kg of water from 10.0 °C to 50.0 °C in 10.0 min while operating at 110 V?

Solution   Assume that $E_{(thermal)} = E_{(electrical)}$

Since $E_{(thermal)} = mc\Delta T$ and $E_{(electrical)} = \mathscr{P}\Delta t = (\Delta V)^2 \Delta t / R$

where $c = 4186 \text{ J}/\text{kg} \cdot °\text{C}$

the resistance is $R = \dfrac{(\Delta V)^2 \Delta t}{cm\Delta T} = \dfrac{(110 \text{ V})^2(600 \text{ s})}{(4186 \text{ J}/\text{kg}\cdot°\text{C})(1.50 \text{ kg})(40.0 °\text{C})} = 28.9 \ \Omega$ ◊

41. Suppose that a voltage surge produces 140 V for a moment. By what percentage does the power output of a 120-V, 100-W light bulb increase? Assume that its resistance does not change.

Solution

Conceptualize: The voltage increases by about 20%, but since $\mathscr{P} = (\Delta V)^2 / R$, the power will increase as the square of the voltage:

$$\frac{\mathscr{P}_f}{\mathscr{P}_i} = \frac{(\Delta V_f)^2 / R}{(\Delta V_i)^2 / R} = \frac{(140 \text{ V})^2}{(120 \text{ V})^2} = 1.361 \text{ or a 36.1\% increase.}$$

Categorize: We have already found an answer to this problem by reasoning in terms of ratios, but we can also calculate the power explicitly for the bulb and compare with the original power by using Ohm's law and the equation for electrical power. To find the power, we must first find the resistance of the bulb, which should remain relatively constant during the power surge (we can check the validity of this assumption later).

Analyze: From $\mathscr{P} = (\Delta V)^2 / R$, we find that $R = \dfrac{(\Delta V_i)^2}{\mathscr{P}} = \dfrac{(120 \text{ V})^2}{100 \text{ W}} = 144 \ \Omega$

The final current is $I_f = \dfrac{\Delta V_f}{R} = \dfrac{140 \text{ V}}{144 \ \Omega} = 0.972 \text{ A}$

The power during surge is $\mathscr{P} = \dfrac{(\Delta V_f)^2}{R} = \dfrac{(140 \text{ V})^2}{144 \ \Omega} = 136 \text{ W}$

So the percentage increase is $\dfrac{136 \text{ W} - 100 \text{ W}}{100 \text{ W}} = 0.361 = 36.1\%$ ◊

Finalize: Our result tells us that this 100-W light bulb momentarily acts like a 136-W light bulb, which explains why it would suddenly get brighter. Some electronic devices (like computers) are sensitive to voltage surges like this, which is the reason that **surge protectors** are recommended to protect these devices from being damaged.

In solving this problem, we assumed that the resistance of the bulb did not change during the voltage surge, but we should check this assumption. Let us assume that the filament is made of tungsten and that its resistance will change linearly with temperature according to Equation 27.21. Let us further assume that the increased voltage lasts for a time long enough so that the filament comes to a new equilibrium temperature. The temperature change can be estimated from the power surge according to Stefan's law (Equation 20.18), assuming that all the power loss is due to radiation. By this law, $T \propto \sqrt[4]{\mathcal{P}}$ so that a 36% change in power should correspond to only about a 8% increase in temperature. A typical operating temperature of a white light bulb is about 3000 °C, so $\Delta T \approx 0.08(3273 \text{ K}) = 260$ °C. Then the increased resistance would be roughly

$$R = R_0\left[1 + \alpha(T - T_0)\right] = (144 \ \Omega)\left[1 + (4.5\times10^{-3})(260)\right] \approx 310 \ \Omega$$

It appears that the resistance could double from 144 Ω. On the other hand, if the voltage surge lasts only a very short time, the 136 W we calculated originally accurately describes the conversion of electrical into internal energy in the filament.

---

49. Compute the cost per day of operating a lamp that draws 1.70 A from a 110-V line. Assume the cost of electrical energy is $0.060 0/kWh.

Solution

The power of the lamp is $\mathcal{P} = \Delta V I = U/\Delta t$, where U is the energy transformed. Then the energy you buy, in standard units, is

$$U = \Delta V I \Delta t = (110 \text{ V})(1.70 \text{ A})(1 \text{ day})\left(\frac{24 \text{ h}}{1 \text{ day}}\right)\left(\frac{3600 \text{ s}}{\text{h}}\right)\left(\frac{1 \text{ J}}{\text{V} \cdot \text{C}}\right)\left(\frac{1 \text{ C}}{\text{A} \cdot \text{s}}\right) = 16.2 \text{ MJ}$$

In kilowatt hours,

$$U = \Delta V I \Delta t = (110 \text{ V})(1.70 \text{ A})(1 \text{ day})\left(\frac{24 \text{ h}}{1 \text{ day}}\right)\left(\frac{\text{J}}{\text{V} \cdot \text{C}}\right)\left(\frac{\text{C}}{\text{A} \cdot \text{s}}\right)\left(\frac{\text{W} \cdot \text{s}}{\text{J}}\right) = 4.49 \text{ kW} \cdot \text{h}$$

So the lamp costs $(4.49 \text{ kWh})(\$0.0600/\text{kWh}) = 26.9 \text{ cents}$ ◊

---

51. A certain toaster has a heating element made of Nichrome resistance wire. When the toaster is first connected to a 120-V source of potential difference (and the wire is at a temperature of 20.0 °C) the initial current is 1.80 A. However, the current begins to decrease as the resistive element warms up. When the toaster has reached its final operating temperature, the current has dropped to 1.53 A. (a) Find the power delivered to the toaster when it is at its operating temperature. (b) What is the final temperature of the heating element?

Solution

Conceptualize: Most toasters are rated at about 1000 W (usually stamped on the bottom of the unit), so we might expect this one to have a similar power rating. The temperature of the heating element should be hot enough to toast bread but low enough that the nickel-chromium alloy element does not melt. (The melting point of nickel is 1455 °C, and chromium melts at 1907 °C.)

Categorize: The power can be calculated directly by multiplying the current and the voltage. The temperature can be found from the linear resistivity equation for Nichrome, with $\alpha = 0.4 \times 10^{-3}\ °\text{C}^{-1}$ from Table 27.1.

Analyze:

(a) $\mathcal{P} = \Delta VI = (120\ \text{V})(1.53\ \text{A}) = 184\ \text{W}$ ◊

(b) The resistance at 20.0 °C is $R_0 = \dfrac{\Delta V}{I} = \dfrac{120\ \text{V}}{1.80\ \text{A}} = 66.7\ \Omega$

At operating temperature, $R = \dfrac{120\ \text{V}}{1.53\ \text{A}} = 78.4\ \Omega$

Neglecting thermal expansion,

we have $R = \dfrac{\rho \ell}{A} = \dfrac{\rho_0[1 + \alpha(T - T_0)]\ell}{A} = R_0[1 + \alpha(T - T_0)]$

$T = T_0 + \dfrac{R/R_0 - 1}{\alpha} = 20.0\ °\text{C} + \dfrac{78.4\ \Omega/66.7\ \Omega - 1}{0.4 \times 10^{-3}\ °\text{C}^{-1}} = 461\ °\text{C}$ ◊

Finalize: Although this toaster appears to use significantly less power than most, the temperature seems high enough to toast a piece of bread in a reasonable amount of time. In fact, the temperature of a typical 1000-W toaster would not be vastly higher because Stefan's radiation law (Equation 20.18) tells us that (assuming all power is lost through radiation) $T \propto \sqrt[4]{\mathcal{P}}$, so that the temperature might be about 850 °C. In either case, the operating temperature is well below the melting point of the heating element.

57. A more general definition of the temperature coefficient of resistivity is

$$\alpha = \frac{1}{\rho}\frac{d\rho}{dT}$$

where ρ is the resistivity at temperature T. (a) Assuming that α is constant, show that

$$\rho = \rho_0 e^{\alpha(T-T_0)}$$

where ρ_0 is the resistivity at temperature T_0. (b) Using the series expansion $e^x \approx 1+x$ for $x \ll 1$, show that the resistivity is given approximately by the expression $\rho = \rho_0[1+\alpha(T-T_0)]$ for $\alpha(T-T_0) \ll 1$.

Solution

(a) We are given

$$\alpha = \frac{1}{\rho}\frac{d\rho}{dT}$$

Separating variables,

$$\int_{\rho_0}^{\rho}\frac{d\rho}{\rho} = \int_{T_0}^{T}\alpha\,dT$$

Integrating both sides,

$$\ln(\rho/\rho_0) = \alpha(T-T_0)$$

Thus

$$\rho = \rho_0 e^{\alpha(T-T_0)} \qquad\qquad\qquad \Diamond$$

(b) From the series expansion $e^x \approx 1+x$, with x much less than 1,

$$\rho = \rho_0\left[1+\alpha(T-T_0)\right] \qquad\qquad \Diamond$$

===

59. An experiment is conducted to measure the electrical resistivity of Nichrome in the form of wires with different lengths and cross-sectional areas. For one set of measurements, a student uses 30-gauge wire, which has a cross-sectional area of 7.30×10^{-8} m^2. The student measures the potential difference across the wire and the current in the wire with a voltmeter and an ammeter, respectively. For each of the measurements given in the table taken on wires of three different lengths, calculate the resistance of the wires and the corresponding values of the resistivity. What is the average value of the resistivity, and how does this value compare with the value given in Table 27.1?

ℓ (m)	ΔV (V)	I (A)	$R(\Omega)$	$\rho\,(\Omega\cdot\text{m})$
0.540	5.22	0.500		
1.028	5.82	0.276		
1.543	5.94	0.187		

Solution   For each row in the table, first find the resistance from $R = \Delta V / I$, and then find the resistivity from $R = \rho \ell / A$ or $\rho = RA / \ell$.

In the first row,
$$R = \frac{\Delta V}{I} = \frac{5.22 \text{ V}}{0.500 \text{ A}} = 10.4 \; \Omega$$

and
$$\rho = \frac{RA}{\ell} = \frac{(10.4 \; \Omega)(7.30 \times 10^{-8} \text{ m}^2)}{0.540 \text{ m}} = 1.41 \times 10^{-6} \; \Omega \cdot \text{m}$$

	ℓ (m)	$R(\Omega)$	$\rho \; (\Omega \cdot \text{m})$
Applying this	0.540	10.4	1.41×10^{-6}
to each entry,	1.028	21.1	1.50×10^{-6}
we obtain:	1.543	31.8	1.50×10^{-6}

Thus the average resistivity is $\rho = 1.47 \times 10^{-6} \; \Omega \cdot \text{m}$ ◊

This differs from the tabulated $1.50 \times 10^{-6} \; \Omega \cdot \text{m}$ by 2%. The difference is accounted for by the experimental uncertainty, which we may estimate as $(1.47 - 1.41)/1.47 = 4\%$.

---

65.   An electric car is designed to run off a bank of 12.0-V batteries with total energy storage of 2.00×10^7 J. (a) If the electric motor draws 8.00 kW, what is the current delivered to the motor? (b) If the electric motor draws 8.00 kW as the car moves at a steady speed of 20.0 m/s, how far will the car travel before it is "out of juice"?

Solution

(a)   Since $\mathcal{P} = \Delta V I$

$$I = \frac{\mathcal{P}}{\Delta V} = \frac{8.00 \times 10^3 \text{ W}}{12.0 \text{ V}} = 667 \text{ A}$$ ◊

(b)   The time the car runs is

$$\Delta t = \frac{U}{\mathcal{P}} = \left(\frac{2.00 \times 10^7 \text{ J}}{8.00 \times 10^3 \text{ W}} \right) \left(\frac{1 \text{ W} \cdot \text{s}}{\text{J}} \right) = 2.50 \times 10^3 \text{ s}$$

So it moves a distance of

$$\Delta x = v \Delta t = (20.0 \text{ m/s})(2.50 \times 10^3 \text{ s}) = 50.0 \text{ km}$$ ◊

---

71. Material with uniform resistivity ρ is formed into a wedge as shown in Figure P27.71. Show that the resistance between face A and face B of this wedge is

$$R = \rho \frac{L}{w(y_2 - y_1)} \ln\left(\frac{y_2}{y_1}\right)$$

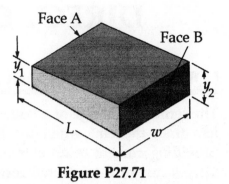

Face A

Face B

y_1

y_2

L

w

Figure P27.71

Solution

The current flows generally parallel to L. Consider a slice of the material perpendicular to this current, of thickness dx, and at distance x from face A. Then the other dimensions of the slice are w and y,

where $\quad \dfrac{y - y_1}{x} = \dfrac{y_2 - y_1}{L} \quad$ by proportion,

so $\quad y = y_1 + (y_2 - y_1)\dfrac{x}{L}$

The bit of resistance which this slice contributes is

$$dR = \frac{\rho\, dx}{A} = \frac{\rho\, dx}{wy} = \frac{\rho\, dx}{w\left(y_1 + (y_2 - y_1)(x/L)\right)}$$

and the whole resistance is that of all the slices:

$$R = \int_{x=0}^{L} dR = \int_{0}^{L} \frac{\rho\, dx}{w\left(y_1 + (y_2 - y_1)(x/L)\right)} = \frac{\rho}{w} \frac{L}{y_2 - y_1} \int_{x=0}^{L} \frac{((y_2 - y_1)/L)\, dx}{y_1 + (y_2 - y_1)(x/L)}$$

With $\quad u = y_1 + (y_2 - y_1)\dfrac{x}{L}, \quad$ this is of the form $\displaystyle\int \frac{du}{u}$,

so $\quad R = \dfrac{\rho L}{w(y_2 - y_1)} \left. \ln\left(y_1 + (y_2 - y_1)(x/L)\right) \right]_{x=0}^{L}$

$$R = \frac{\rho L}{w(y_2 - y_1)}(\ln y_2 - \ln y_1) = \frac{\rho L}{w(y_2 - y_1)} \ln \frac{y_2}{y_1} \qquad \Diamond$$

Chapter 28
DIRECT CURRENT CIRCUITS

EQUATIONS AND CONCEPTS

The **terminal voltage of a battery** will be less than the emf when the battery is providing current to an external circuit. This is due to the **internal resistance** of the battery. *The terminal voltage is equal to the emf when the current is zero (this is the open circuit voltage).*

$$\Delta V = \mathcal{E} - Ir \qquad (28.1)$$

The **current, I, delivered by a battery** in a simple **DC** circuit, depends on the value of the emf of the source, $\mathcal{E}$; the total load resistance in the circuit, R; and the internal resistance of the source, r.

$$I = \frac{\mathcal{E}}{R + r} \qquad (28.3)$$

The **total or equivalent resistance of a series combination of resistors** is equal to the numerical sum of the individual resistances. *The resistors are connected so that there is one common circuit point per adjacent pair of resistors. The current has the same value in each resistor in a series group.*

$$R_{eq} = R_1 + R_2 + R_3 + \ldots \qquad (28.6)$$

(series combination)

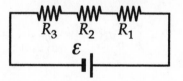

The **inverse of the equivalent resistance of several resistors connected in parallel** is equal to the sum of the inverses of the individual resistors. *The equivalent resistance is less than the smallest individual value of resistance in the group. Resistors in parallel are connected so that each resistor has two circuit points in common with all other resistors in the group. The potential difference is the same across each resistor in a parallel group.*

$$\frac{1}{R_{eq}} = \frac{1}{R_1} + \frac{1}{R_2} + \frac{1}{R_3} + \ldots \qquad (28.8)$$

(parallel combination)

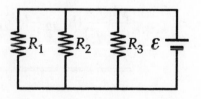

A **series-parallel combination of resistors** can usually be reduced to an equivalent single loop by making successive step-by-step combinations of subsets of resistors in the original circuit and using Equations 28.6 and 28.8. *See Examples 28.4, 28.5 and 28.6 in the textbook.*

Kirchhoff's rules can be used to analyze more complex multiloop circuits. *Review the procedure for applying Kirchhoff's rules as outlined in Suggestions, Skills, and Strategies.*

Junction rule. The sum of the currents entering any junction in a circuit must equal the sum of the currents leaving that junction.

$$\sum I_{in} = \sum I_{out} \qquad (28.9)$$

Loop rule. The algebraic sum of the potential differences across all elements (e.g. resistors and sources of emf) around a closed circuit loop must be zero.

$$\sum_{\substack{closed \\ loop}} \Delta V = 0 \qquad (28.10)$$

When a **capacitor is charged in series with a resistor** (see circuit inset in the figure) a quantity τ, called the time constant of the circuit, is used to describe the manner in which the charge on the capacitor varies with time. The charge on the capacitor increases from zero to 63% of its maximum value in a time interval equal to one time constant. *Also, during one time constant, the charging current decreases from its initial maximum value of $I_0 = \mathcal{E}/R$ to 37% of I_0.*

$$\tau = RC$$

$$q(t) = C\mathcal{E}\left[1 - e^{-t/RC}\right] \qquad (28.14)$$

$$I(t) = \frac{\mathcal{E}}{R}e^{-t/RC} \qquad (28.15)$$

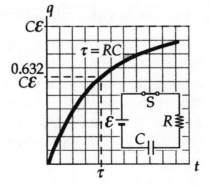

When a **capacitor is discharged through a resistor** (see circuit inset in the figure), the charge and current decrease exponentially in time. In these equations, the initial current is $I_0 = Q/RC$.

$$q(t) = Qe^{-t/RC} \tag{28.17}$$

$$I(t) = -\frac{Q}{RC}e^{-t/RC} \tag{28.18}$$

when $Q/RC = I_0$

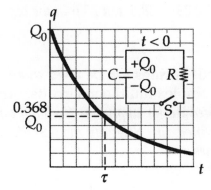

SUGGESTIONS, SKILLS, AND STRATEGIES

PROBLEM-SOLVING STRATEGY FOR RESISTORS

- When two or more resistors are connected in series, the current is the same in each resistor; but the potential difference is not the same across each resistor (unless they have equal values of resistance). The resistors add directly to give the equivalent resistance of the series combination.

- When two or more resistors are connected in parallel, the potential difference is the same across each resistor. Since the current is inversely proportional to the resistance, the currents through them are not the same (unless they have equal values of resistance). The equivalent resistance of a parallel combination of resistors is found through reciprocal addition, and the equivalent resistance is always less than the smallest individual resistor.

- A multiloop circuit consisting of a series-parallel combination of resistors can often be reduced to a simple loop circuit containing only one equivalent resistor. To do so, repeatedly examine the circuit and replace any resistors that are in series or in parallel with a single resistor of equivalent value using Equations 28.6 and 28.8. Sketch the new circuit after each set of changes has been made. Continue this process until a single equivalent resistance is found.

- If the current through, or the potential difference across, a resistor in the multiloop circuit is to be identified, start with the final equivalent circuit found in the last step above, and gradually work your way back through the circuits, using $V = IR$ to find the voltage drop across each equivalent resistor.

STRATEGY FOR USING KIRCHHOFF'S RULES

- First, draw the circuit diagram and assign labels and symbols to all the known and unknown quantities. You must assign directions to the currents in each part of the circuit. Do not be alarmed if you assume the incorrect direction of a current. The resulting value will be negative, but its magnitude will be correct. Although the assignment of current directions is arbitrary, you must stick with your assumption throughout a specific circuit as you apply Kirchhoff's rules.

- Apply the junction rule to any junction in the circuit. The junction rule may be applied as many times as a new current (one not used in a previous application) appears in the resulting equation. In general, the number of times the junction rule can be used is one fewer than the number of junction points in the circuit.

- Now apply Kirchhoff's loop rule to as many loops in the circuit as are needed to solve for the unknowns. Remember you must have as many equations as there are unknowns (I's, R's, and $\mathcal{E}$'s). In order to apply this rule, you must correctly identify the increase or decrease in potential as you cross each element in traversing the closed loop. Review the conditions stated below and watch out for signs!

Convenient "rules of thumb" which you may use to determine the increase or decrease in potential as you cross a resistor or source of emf in traversing a circuit loop are illustrated in the following figure. Notice that the potential decreases (changes by $-IR$) when the resistor is traversed in the direction of the current.

There is an increase in potential of $+IR$ if the direction of travel is opposite the direction of current. If a source of emf is traversed in the direction of the emf (from $-$ to $+$ on the battery), the potential increases by $\mathcal{E}$. If the direction of travel is from $+$ to $-$, the potential decreases by $\mathcal{E}$

$$\Delta V = V_b - V_a = -IR$$

$$\Delta V = V_b - V_a = IR$$

$$\Delta V = V_b - V_a = \mathcal{E}$$

$$\Delta V = V_b - V_a = -\mathcal{E}$$

- Finally, you must solve the equations simultaneously for the unknown quantities. Be careful in your algebraic steps, and check your numerical answers for consistency.

As an illustration of the use of Kirchhoff's rules, consider a three-loop circuit which has the general form shown in figure at the right. In this illustration, the actual circuit elements, R's and $\mathcal{E}$'s are not shown but assumed known. There are six possible different values of I in the circuit; therefore you will need six independent equations to solve for the six values of I. There are four junction points in the circuit (at points a, d, f, and h). The first rule applied at any three of these points will yield three equations. The circuit can be thought of as a group of three "blocks" as shown in the figure. Kirchhoff's second law, when applied to each of these loops (*abcda*, *ahfga*, and *defhd*), will yield three additional equations.

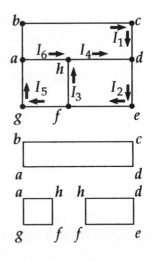

You can then solve the total of six equations simultaneously for the six values of I_1, I_2, I_3, I_4, I_5, and I_6. You can, of course, expect that the sum of the changes in potential difference around **any other closed loop** in the circuit will be zero (for example, *abcdefga* or *ahfedcba*); however the equations found by applying Kirchhoff's second rule to these additional loops **will not be independent** of the six equations found previously.

REVIEW CHECKLIST

You should be able to:

▷ Calculate the terminal potential difference and internal resistance of a battery. (Section 28.1)

▷ Calculate the equivalent resistance of a group of resistors in parallel, series, or series-parallel combination. (Section 28.2)

▷ Use Ohm's law to calculate the current in a circuit and the potential difference between any two points in a single loop circuit or a circuit that can be reduced to an equivalent single-loop circuit. (Sections 28.1 and 28.2)

▷ Apply Kirchhoff's rules to solve multiloop circuits; that is, find the current and the potential difference between any two points. (Section 28.3)

▷ Calculate the charging (discharge) current $I(t)$ and the accumulated (residual) charge $q(t)$ during charging (and discharge) of a capacitor in an RC circuit. (Section 28.4)

ANSWERS TO SELECTED QUESTIONS

3. Is the direction of current through a battery always from the negative terminal to the positive terminal? Explain.

Answer No. If there is one battery in a circuit, the current inside it will be from its negative to its positive terminal. Whenever a battery is delivering energy to a circuit, it will carry current in this direction. On the other hand, when another source of emf is charging the battery in question, it will have current pushed through it from its positive terminal to its negative terminal.

□   □   □   □

11. Why is it possible for a bird to sit on a high-voltage wire without being electrocuted?

Answer The resistance of the short segment of wire between the bird's feet is so small that the potential difference $\Delta V = IR$ between the feet is negligible. In order for the bird to be electrocuted, a potential difference is required. There is negligible potential difference between the bird's feet.

It could also be said like this: A small amount of current does flow through the bird's feet, according to the rule of parallel resistors. However, since the resistance of the bird's feet is much higher than that of the wire, the amount of current that flows through the bird is still not enough to harm it.

□   □   □   □

13. Referring to Figure Q28.13, describe what happens to the lightbulb after the switch is closed. Assume that the capacitor has a large capacitance and is initially uncharged, and assume that the light illuminates when connected directly across the battery terminals.

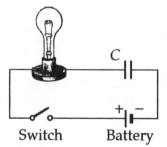

Figure Q28.13

Answer The bulb will light up for an instant as the capacitor is being charged and there is a current in the circuit. As soon as the capacitor is fully charged, the current in the circuit will drop to zero, and the bulb will cease to glow.

□   □   □   □

24. What advantage does 120-V operation offer over 240 V? What disadvantages?

Answer   Both 120-V and 240-V lines can deliver injurious or lethal shocks, but there is a somewhat better factor of safety with the lower voltage. To say it a different way, the insulation on a 120-V wire can be thinner. On the other hand, a 240-V line carries less current to operate a device with the same power, so the conductor itself can be thinner. Finally, as we will see in Chapter 33, the last step-down transformer can also be somewhat smaller if it has to go down only to 240 volts from the high voltage of the main power line.

□   □   □   □

SOLUTIONS TO SELECTED PROBLEMS

1.   A battery has an emf of 15.0 V. The terminal voltage of the battery is 11.6 V when it is delivering 20.0 W of power to an external load resistor R. (a) What is the value of R? (b) What is the internal resistance of the battery?

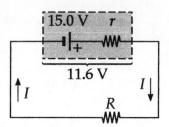

Solution

Conceptualize: The internal resistance of a battery usually is less than $1\,\Omega$, with physically larger batteries having less resistance due to the larger anode and cathode areas. The voltage of this battery drops significantly (23%), when the load resistance is added, so a sizable amount of current must be drawn from the battery. If we assume that the internal resistance is about $1\,\Omega$, then the current must be about 3 A to give the 3.4-V drop across the battery's internal resistance. If this is true, then the load resistance must be about $R \approx 12\text{ V}/3\text{ A} = 4\,\Omega$.

Categorize: We can find R exactly by using Joule's law for the power delivered to the load resistor when the voltage is 11.6 V. Then we can find the internal resistance of the battery by summing the electric potential differences around the circuit.

Analyze:

(a)   Combining Joule's law, $\mathcal{P} = \Delta V I$, and the definition of resistance, $\Delta V = IR$, gives

$$R = \frac{(\Delta V)^2}{\mathcal{P}} = \frac{(11.6\text{ V})^2}{20.0\text{ W}} = 6.73\,\Omega \qquad \Diamond$$

(b)   The electromotive force of the battery must equal the voltage drops across the resistances: $\mathcal{E} = IR + Ir$, where $I = \Delta V/R$.

$$r = \frac{\mathcal{E} - IR}{I} = \frac{(\mathcal{E} - \Delta V)R}{\Delta V} = \frac{(15.0\text{ V} - 11.6\text{ V})(6.73\,\Omega)}{11.6\text{ V}} = 1.97\,\Omega \qquad \Diamond$$

Finalize: The resistance of the battery is larger than 1 Ω, but it is reasonable for an old battery or for a battery consisting of several small electric cells in series. The load resistance agrees reasonably well with our prediction, despite the fact that the battery's internal resistance was about twice as large as we assumed. Note that in our initial guess we did not consider the power of the load resistance; however, there is not sufficient information to accurately solve this problem without this data.

---

9. Consider the circuit shown in Figure P28.9. Find (a) the current in the 20.0-Ω resistor and (b) the potential difference between points *a* and *b*.

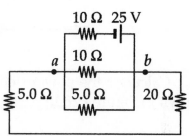

Solution

If we turn the given diagram on its side, we find that it is the same as figure (a). The 20.0-Ω and 5.00-Ω resistors are in series, so the first reduction is as shown in (b).

Figure P28.9

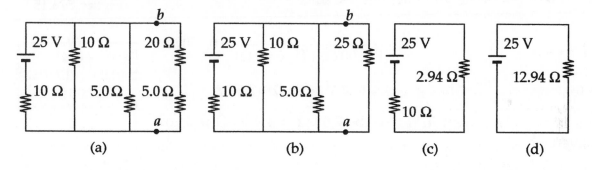

(a)   (b)   (c)   (d)

In addition, since the 10.0-Ω, 5.00-Ω, and 25.0-Ω resistors are then in parallel, we can solve for their equivalent resistance as:

$$R_{eq} = \frac{1}{(1/10.0\ \Omega)+(1/5.00\ \Omega)+(1/25.0\ \Omega)} = 2.94\ \Omega$$

This is shown in figure (c), which in turn reduces to the circuit shown in (d).

Next, we work backwards through the diagrams, applying $I = \Delta V/R$ and $\Delta V = IR$ alternately to each resistor, real or equivalent. The 12.94-Ω resistor is connected across 25.0-V, so the current through the voltage source in every diagram is

$$I = \frac{\Delta V}{R} = \frac{25.0\ V}{12.94\ \Omega} = 1.93\ A$$

In figure (c), this 1.93 A goes through the 2.94-Ω equivalent resistor to give a voltage drop of:

$$\Delta V = IR = (1.93\ A)(2.94\ \Omega) = 5.68\ V$$

From figure (b), we see that this voltage drop is the same across ΔV_{ab}, the 10-Ω resistor, and the 5.00-Ω resistor.

(b)   Therefore, $V_{ab} = 5.68$ V   ◊

(a)   Since the current through the 20-Ω resistor is also the current through the 25.0-Ω line ab,

$$I = \frac{\Delta V_{ab}}{R_{ab}} = \frac{5.68 \text{ V}}{25.0 \text{ }\Omega} = 0.227 \text{ A} \qquad ◊$$

---

15.   Calculate the power delivered to each resistor in the circuit shown in Figure P28.15.

Solution   To find the power we must find the current in each resistor, so we find the resistance seen by the battery. The given circuit reduces as shown in the figures, since

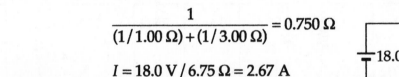

Figure P28.15

$$\frac{1}{(1/1.00 \text{ }\Omega) + (1/3.00 \text{ }\Omega)} = 0.750 \text{ }\Omega$$

In (b), $I = 18.0 \text{ V} / 6.75 \text{ }\Omega = 2.67 \text{ A}$

This is also the current in (a), so the 2.00-Ω and 4.00-Ω resistors convert

(a)

$$\mathcal{P}_2 = I\Delta V = I^2R = (2.67 \text{ A})^2(2.00 \text{ }\Omega) = 14.2 \text{ W} \qquad ◊$$

$$\mathcal{P}_4 = I^2R = (2.67 \text{ A})^2(4.00 \text{ }\Omega) = 28.4 \text{ W} \qquad ◊$$

The voltage across the 0.750-Ω resistor in (a), and across both the 3.00-Ω and the 1.00-Ω resistors in Figure P28.15, is

$$\Delta V = IR = (2.67 \text{ A})(0.750 \text{ }\Omega) = 2.00 \text{ V}$$

(b)

Then for the 3.00-Ω resistor, $I = \dfrac{\Delta V}{R} = \dfrac{2.00 \text{ V}}{3.00 \text{ }\Omega}$

and $\mathcal{P} = I\Delta V = \left(\dfrac{2.00 \text{ V}}{3.00 \text{ }\Omega}\right)(2.00 \text{ V}) = 1.33 \text{ W} \qquad ◊$

For the 1.00-Ω resistor, $I = \dfrac{2.00 \text{ V}}{1.00 \text{ }\Omega}$ and $\mathcal{P} = \left(\dfrac{2.00 \text{ V}}{1.00 \text{ }\Omega}\right)(2.00 \text{ V}) = 4.00 \text{ W} \qquad ◊$

---

21. Determine the current in each branch of the circuit shown in Figure P28.21.

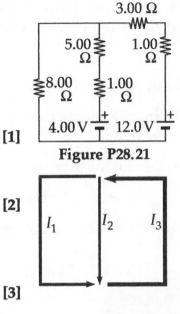

Figure P28.21

Solution   First, we arbitrarily define the initial current directions and names, as shown in the second figure below. The current rule then says that

$$I_3 = I_1 + I_2 \qquad [1]$$

By the voltage rule, clockwise around the left-hand loop,

$$+I_1(8.00\ \Omega) - I_2(5.00\ \Omega) - I_2(1.00\ \Omega) - 4.00\ \text{V} = 0 \qquad [2]$$

Clockwise around the right-hand loop, combining the 1 and 5 Ω resistors and the 1 and 3 Ω resistors,

$$4.00\ \text{V} + I_2(6.00\ \Omega) + I_3(4.00\ \Omega) - 12.0\ \text{V} = 0 \qquad [3]$$

We substitute $(I_1 + I_2)$ for I_3, and reduce our three equations to:

$$(8.00\ \Omega)I_1 - (6.00\ \Omega)I_2 - 4.00\ \text{V} = 0$$

and

$$4.00\ \text{V} + (6.00\ \Omega)I_2 + (4.00\ \Omega)(I_1 + I_2) - 12.0\ \text{V} = 0$$

Solving the top equation for I_2,

$$I_2 = \frac{(8.00\ \Omega)I_1 - 4.00\ \text{V}}{6.00\ \Omega}$$

and rearranging the bottom equation,

$$I_1 + \frac{10.0\ \Omega}{4.00\ \Omega}I_2 = \frac{8.00\ \text{V}}{4.00\ \Omega}$$

We can substitute for I_2,

$$I_1 + 3.33 I_1 - 1.67\ \text{V}/\Omega = 2.00\ \text{V}/\Omega$$

and solve for I_1, down the 8-Ω resistor:

$$I_1 = 0.846\ \text{V}/\Omega = 0.846\ \text{A} \qquad \Diamond$$

If we were solving the three simultaneous equations by determinants or by calculator matrix inversion, we would have to do much more work to find I_2 and I_3. Our method of solving by substitution is more generally useful, and makes it easy to work out the remaining answers after the first, just as when the cat has kittens. We substitute again, putting the value for I_1 into equations we already solved for the other unknowns.

Thus, $I_2 = \dfrac{(8.00\ \Omega)(0.846\ \text{A}) - 4.00\ \text{V}}{6.00\ \Omega} = 0.462\ \text{A}$ down in the middle branch $\Diamond$

$$I_3 = 0.846\ \text{A} + 0.462\ \text{A} = 1.31\ \text{A} \qquad\qquad \text{up in the right-hand branch} \qquad \Diamond$$

29. For the circuit shown in Figure P28.29, calculate (a) the current in the 2.00-Ω resistor and (b) the potential difference between points a and b.

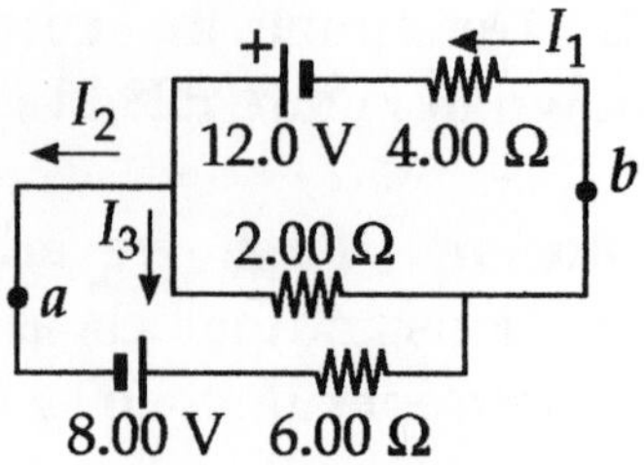

Figure P28.29

Solution

Arbitrarily choose current directions as labeled in the figure to the right.

(a) From the junction point rule, $I_1 = I_2 + I_3$ [1]

Traversing the top loop CCW, $12.0\text{ V} - (2.00\ \Omega)I_3 - (4.00\ \Omega)I_1 = 0$ [2]

Traversing the bottom loop CCW, $8.00\text{ V} - (6.00\ \Omega)I_2 + (2.00\ \Omega)I_3 = 0$ [3]

From Equation [2], $I_1 = \dfrac{12.0\text{ V} - (2.00\ \Omega)I_3}{4.00\ \Omega}$

From Equation [3], $I_2 = \dfrac{8.00\text{ V} + (2.00\ \Omega)I_3}{6.00\ \Omega}$

Substituting these values into Equation [1], we find that the current in the 2.00-Ω resistor is $I_3 = 0.909$ A.   ◊

(b) Through the center wire, $V_a - (0.909\text{ A})(2.00\ \Omega) = V_b$

Therefore, $V_a - V_b = 1.82\text{ V}$, with $V_a > V_b$ ◊

---

31. Consider a series RC circuit (see Figure 28.19) for which $R = 1.00\text{ M}\Omega$, $C = 5.00\ \mu\text{F}$, and $\mathcal{E} = 30.0$ V. Find (a) the time constant of the circuit and (b) the maximum charge on the capacitor after the switch is closed. (c) Find the current in the resistor 10.0 s after the switch is closed.

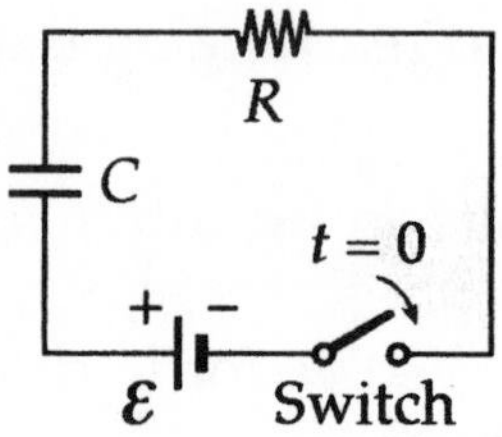

Figure P28.19

Solution

(a) $\tau = RC = (1.00 \times 10^6\ \Omega)(5.00 \times 10^{-6}\text{ F}) = 5.00\ \Omega \cdot \text{F} = 5.00\text{ s}$ ◊

(b)   After a long time, the capacitor is "charged to thirty volts", separating charges of

$Q = C\Delta V = (5.00 \times 10^{-6}\text{ F})(30.0\text{ V}) = 150\ \mu\text{C}$ ◊

(c) $I = I_0 e^{-t/\tau}$ where $I_0 = \dfrac{\mathcal{E}}{R}$ and $\tau = RC = 5.00\text{ s}$

$I = \dfrac{\mathcal{E}}{R}e^{-t/\tau} = \left(\dfrac{30.0\text{ V}}{1.00 \times 10^6\ \Omega}\right)e^{-10.0\,\text{s}/5.00\,\text{s}}$ $I = 4.06 \times 10^{-6}\text{ A} = 4.06\ \mu\text{A}$ ◊

37. The circuit in Figure P28.37 has been connected for a long time. (a) What is the voltage across the capacitor? (b) If the battery is disconnected, how long does it take the capacitor to discharge to one tenth of its initial voltage?

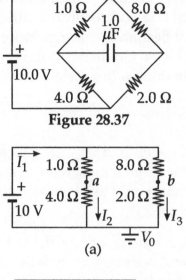

Figure 28.37

Solution

(a) After a long time the capacitor branch will carry negligible current. The current flow is as shown in figure (a). To find the voltage at point a, we first find the current, using the voltage rule:

$$10.0\ \text{V} - (1.00\ \Omega)I_2 - (4.00\ \Omega)I_2 = 0$$

$$I_2 = 2.00\ \text{A}$$

$$V_a - V_0 = (4.00\ \Omega)I_2 = 8.00\ \text{V}$$

Similarly, $10.0\ \text{V} - (8.00\ \Omega)I_3 - (2.00\ \Omega)I_3 = 0$

$$I_3 = 1.00\ \text{A}$$

At point b, $V_b - V_0 = (2.00\ \Omega)I_3 = 2.00\ \text{V}$

Thus, the voltage across the capacitor is $V_a - V_b = 8.00\ \text{V} - 2.00\ \text{V} = 6.00\ \text{V}$ ◊

(b) We suppose the battery is pulled out leaving an open circuit. We are left with figure (b), which can be reduced to equivalent circuits (c) and (d).

From (d), we can see that the capacitor discharges through a 3.60-Ω equivalent resistance.

According to $q = Qe^{-t/RC}$

we calculate that $qC = QCe^{-t/RC}$

and $\Delta V = \Delta V_i e^{-t/RC}$

Solving, $\frac{1}{10}\Delta V_i = \Delta V_i e^{-t/(3.60\ \Omega)(1.00\ \mu F)}$

$$e^{-t/3.60\ \mu s} = 0.100$$

$$(-t/3.60\ \mu s) = \ln(0.100) = -2.30$$

$$\frac{t}{3.60\ \mu s} = 2.30$$

$$t = (2.30)(3.60\ \mu s) = 8.29\ \mu s$$ ◊

(a) at right figure:

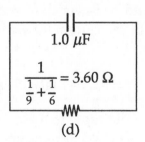

(b)

(c)

(d)

41. Assume that a galvanometer has an internal resistance of 60.0 Ω and requires a current of 0.500 mA to produce full-scale deflection. What resistance must be connected in parallel with the galvanometer if the combination is to serve as an ammeter that has a full-scale deflection for a current of 0.100 A?

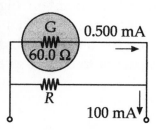

Solution

Conceptualize: An ammeter reads the flow of current in a portion of a circuit; therefore it must have a low resistance so that it does not significantly alter the current that would exist without the meter. Therefore, the resistance required is probably less than 1 Ω.

Categorize: From the values given for a full-scale reading, we can find the voltage across and the current through the shunt (parallel) resistor, and the resistance value can then be found from the definition of resistance.

Analyze: The voltage across the galvanometer must be the same as the voltage across the shunt resistor in parallel, so when the ammeter reads full scale,

$$\Delta V = (0.500 \text{ mA})(60.0 \text{ Ω}) = 30.0 \text{ mV}$$

Through the shunt resistor, $\qquad I = 100 \text{ mA} - 0.500 \text{ mA} = 99.5 \text{ mA}$

Therefore, $\qquad R = \dfrac{\Delta V}{I} = \dfrac{30.0 \text{ mV}}{99.5 \text{ mA}} = 0.302 \text{ Ω}$ ◊

Finalize: The shunt resistance is less than 1 Ω as expected. It is important to note that some caution would be necessary if this meter were used in a circuit that had a low resistance. For example, if the circuit resistance was 3 Ω, adding the ammeter to the circuit would reduce the current by about 10%, so the current displayed by the meter would be lower than without the meter. Problems 44 and 63 address a similar concern about measurement error when using electrical meters.

---

43. The same galvanometer described in the previous problem may be used to measure voltages. In this case a large resistor is wired in series with the galvanometer, as suggested in Figure 28.29. The effect is to limit the current in the galvanometer when large voltages are applied. Most of the potential drop occurs across the resistor placed in series. Calculate the value of the resistor that allows the galvanometer to measure an applied voltage of 25.0 V at full-scale deflection.

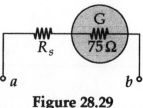

Figure 28.29 (modified)

Solution

Conceptualize: The problem states that the value of the resistor must be "large" in order to limit the current through the galvanometer, so we should expect a resistance of $k\Omega$ to $M\Omega$.

Categorize: The unknown resistance can be found by applying the definition of resistance to the portion of the circuit shown in Figure 28.29.

Analyze: $\Delta V_{ab} = 25.0$ V; from Problem 42, $I = 1.50$ mA and $R_g = 75.0\ \Omega$. For the two resistors in series, $R_{eq} = R_s + R_g$ so the definition of resistance gives us $\Delta V_{ab} = I(R_s + R_g)$.

Therefore $R_s = \dfrac{\Delta V_{ab}}{I} - R_g = \dfrac{25.0\ \text{V}}{1.50 \times 10^{-3}\ \text{A}} - 75.0\ \Omega = 16.6\ \text{k}\Omega$ ◊

Finalize: The resistance is relatively large, as expected. It is important to note that some caution would be necessary if this arrangement were used to measure the voltage across a circuit with a comparable resistance. For example, if the circuit resistance was 17 $k\Omega$, the voltmeter in this problem would cause a measurement inaccuracy of about 50%, because the meter would divert about half the current that normally would go through the resistor being measured. Problems 44 and 63 address a similar concern about measurement error when using electrical meters.

---

49. An electric heater is rated at 1 500 W, a toaster at 750 W, and an electric grill at 1 000 W. The three appliances are connected to a common 120-V household circuit. (a) How much current does each draw? (b) Is a circuit with a 25.0-A circuit breaker sufficient in this situation? Explain your answer.

Solution

(a) Heater: $I = \dfrac{\mathscr{P}}{\Delta V} = \left(\dfrac{1500\ \text{W}}{120\ \text{V}}\right)\left(\dfrac{1\,\text{J}/\text{s}}{1\,\text{W}}\right)\left(\dfrac{1\,\text{V}}{1\,\text{J}/\text{C}}\right)\left(\dfrac{1\,\text{A}}{1\,\text{C}/\text{s}}\right) = 12.5$ A ◊

   Toaster: $I = \dfrac{750\ \text{W}}{120\ \text{V}} = 6.25$ A ◊

   Grill: $I = \dfrac{1000\ \text{W}}{120\ \text{V}} = 8.33$ A ◊

(b) Together in parallel they pass current 12.5 A + 6.25 A + 8.33 A = 27.1 A, so 25-A wiring cannot provide power to the entire group of appliances at the same time. ◊

---

57. A battery has an emf $\mathcal{E}$ and internal resistance r. A variable load resistor R is connected across the terminals of the battery. (a) Determine the value of R such that the potential difference across the terminals is a maximum. (b) Determine the value of R so that the current in the circuit is a maximum. (c) Determine the value of R so that the power delivered to the load resistor is a maximum. Choosing the load resistance for maximum power transfer is a case of what is called *impedance matching* in general. Impedance matching is important in shifting gears on a bicycle, in connecting a loudspeaker to an audio amplifier, in connecting a battery charger to a bank of solar photoelectric cells, and in many other applications.

Solution

Conceptualize: If we consider the limiting cases, we can imagine that the **potential** across the battery will be a maximum when $R = \infty$ (open circuit), the **current** will be a maximum when $R = 0$ (short circuit), and the **power** will be a maximum when R is somewhere between these two extremes, perhaps when $R = r$.

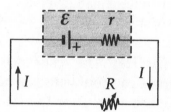

Categorize: We can use the definition of resistance to find the voltage and current as functions of R, and the power equation can be differentiated with respect to R.

Analyze:

(a) The battery has a voltage $\qquad \Delta V_{\text{terminal}} = \mathcal{E} - Ir = \dfrac{\mathcal{E}R}{R+r}$

   Taking the limit as $R \to \infty$, $\qquad \Delta V_{\text{terminal}} \to \mathcal{E}$ $\qquad\qquad \Diamond$

(b) The circuit's current is $\qquad I = \dfrac{\mathcal{E}}{R+r}$

   Therefore, as $R \to 0$, $\qquad I \to \mathcal{E}/r$ $\qquad\qquad\qquad\qquad \Diamond$

(c) The power delivered is $\qquad \mathcal{P} = I^2 R = \dfrac{\mathcal{E}^2 R}{(R+r)^2}$

   To maximize the power $\mathcal{P}$ as a function of R, we differentiate, with respect to R,

$$\frac{d\mathcal{P}}{dR} = \mathcal{E}^2 R(-2)(R+r)^{-3} + \mathcal{E}^2(R+r)^{-2}$$

   and require that $\qquad \dfrac{d\mathcal{P}}{dR} = \dfrac{-2\mathcal{E}^2 R}{(R+r)^3} + \dfrac{\mathcal{E}^2}{(R+r)^2} = 0$

   Then $2R = R + r$ $\qquad$ and $\qquad R = r$ $\qquad\qquad\qquad \Diamond$

Finalize: The results agree with our predictions. Making load resistance equal to the source resistance to maximize power transfer is called impedance matching.

63. The value of a resistor R is to be determined using the ammeter-voltmeter setup shown in Figure P28.63. The ammeter has a resistance of 0.500 Ω, and the voltmeter has a resistance of 20 000 Ω. Within what range of actual values of R will the measured values be correct to within 5.00% if the measurement is made using the circuit shown in (a) Figure P28.63a and (b) Figure P28.63b?

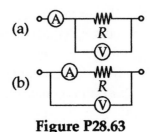

Figure P28.63

Solution

Conceptualize: An ideal ammeter has zero resistance, and an ideal voltmeter has infinite resistance, so that adding the meter does not alter the current or voltage of the existing circuit. For the non-ideal meters in this problem, a low value of R will give a large voltage measurement error in circuit (b), while a large value of R will give significant current measurement error in circuit (a). We could hope that these meters yield accurate measurements in either circuit for typical resistance values of 1 Ω to 1 MΩ.

Categorize: The definition of resistance can be applied to each circuit to find the minimum and maximum current and voltage allowed within the 5.00% tolerance range.

Analyze:

(a) In Figure P28.63a, at least a little current goes through the voltmeter, so less current flows through the resistor than the ammeter reports, and the resistance computed by dividing the voltage by the inflated ammeter reading will be too small. Thus, we require that $\Delta V/I \geq 0.950R$ where I is the current through the ammeter.

Call I_R the current through the resistor; then $I - I_R$ is the current in the voltmeter. Since the resistor and the voltmeter are in parallel, the voltage across the meter equals the voltage across the resistor. Applying the definition of resistance:

$$\Delta V = I_R R = (I - I_R)(20\,000\,\Omega) \quad \text{so} \quad I = \frac{I_R(R + 20\,000\,\Omega)}{20\,000\,\Omega}$$

Our requirement is

$$\frac{I_R R}{\left(\dfrac{I_R(R + 20\,000\,\Omega)}{20\,000\,\Omega} \right)} \geq 0.95R$$

Solving,

$$20\,000\,\Omega \geq 0.95(R + 20\,000\,\Omega) = 0.95R + 19\,000\,\Omega$$

and

$$R \leq \frac{1000\,\Omega}{0.95} \quad \text{or} \quad R \leq 1.05\ \text{k}\Omega \qquad \Diamond$$

(b) If R is too small, the resistance of an ammeter in series will significantly reduce the current that would otherwise flow through R. In Figure 28.63b, the voltmeter reading is $I(0.500\ \Omega) + IR$, at least a little larger than the voltage across the resistor. So the resistance computed by dividing the inflated voltmeter reading by the ammeter reading will be too large.

We require

$$\frac{\Delta V}{I} \leq 1.05R$$

$$\frac{I(0.500\ \Omega) + IR}{I} \leq 1.05R$$

Thus,

$$0.500\ \Omega \leq 0.0500R$$

and

$$R \geq 10.0\ \Omega \qquad \diamond$$

Finalize: The range of R values seems correct since the ammeter's resistance should be less than 5% of the smallest R value ($0.500\ \Omega \leq 0.05R$ means that R should be greater than $10\ \Omega$), and R should be less than 5% of the voltmeter's internal resistance ($R \leq 0.05 \times 20\ \text{k}\Omega = 1\ \text{k}\Omega$). Only for the restricted range between 10 ohms and 1000 ohms can we indifferently use either of the connections (a) and (b) for a reasonably accurate resistance measurement. For low values of the resistance R, circuit (a) must be used. Only circuit (b) can accurately measure a large value of R.

---

65. The values of the components in a simple series RC circuit containing a switch (Figure 28.19) are $C = 1.00\ \mu\text{F}$, $R = 2.00 \times 10^6\ \Omega$, and $\mathcal{E} = 10.0\ \text{V}$. At the instant 10.0 s after the switch is closed, calculate (a) the charge on the capacitor, (b) the current in the resistor, (c) the rate at which energy is being stored in the capacitor, and (d) the rate at which energy is being delivered by the battery.

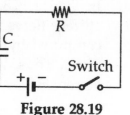

Figure 28.19

Solution

(a) $q = C\mathcal{E}\left(1 - e^{-t/RC}\right) = (1.00 \times 10^{-6}\ \text{F})(10.0\ \text{V})\left[1 - e^{-10.0\ \text{s}/\left((2.00 \times 10^6\ \Omega)(1.00 \times 10^{-6}\ \text{F})\right)}\right]$

$q = 9.93 \times 10^{-6}\ \text{C} = 9.93\ \mu\text{C} \qquad \diamond$

(b) $I = \dfrac{dq}{dt} = \dfrac{d}{dt}\left[C\mathcal{E}\left(1 - e^{-(t/RC)}\right)\right] = (\mathcal{E}/R)e^{-(t/RC)} = \left(\dfrac{10.0\ \text{V}}{2.00 \times 10^6\ \Omega}\right)e^{-(10.0/2.00)}$

$I = 3.37 \times 10^{-8}\ \text{A} \qquad \diamond$

(c) Since the energy stored in the capacitor is $U = q^2/2C$, the rate of storing energy is

$$\frac{dU}{dt} = \frac{q}{C}\frac{dq}{dt} = (q/C)I = \left(\frac{9.93\times10^{-6}\text{ C}}{1.00\times10^{-6}\text{ F}}\right)(3.37\times10^{-8}\text{ A}) = 3.34\times10^{-7}\text{ W} \qquad \lozenge$$

(d) $\mathcal{P}_{batt} = I\mathcal{E} = (3.37\times10^{-8}\text{ A})(10.0\text{ V}) = 3.37\times10^{-7}\text{ W}$ $\qquad\qquad\qquad\qquad \lozenge$

67. Three 60.0-W, 120-V lightbulbs are connected across a 120-V power source, as shown in Figure P28.67. Find (a) the total power delivered to the three bulbs and (b) the voltage across each. Assume that the resistance of each bulb is constant (even though in reality the resistance might increase markedly with current).

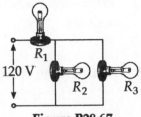

Figure P28.67

Solution

If the bulbs were all in parallel, the current in each would be

$$I = \frac{P}{\Delta V} = \frac{60.0\text{ W}}{120\text{ V}} = 0.500\text{ A}$$

Then in any connection each bulb has

$$R = \frac{\Delta V}{I} = \frac{120\text{ V}}{0.500\text{ A}} = 240\ \Omega$$

so R_2 and R_3 have equivalent resistance

$$\frac{1}{(1/240\ \Omega)+(1/240\ \Omega)} = 120\ \Omega$$

The three together have net resistance

$$240\ \Omega + 120\ \Omega = 360\ \Omega$$

The total current can be calculated as

$$I = \frac{\Delta V}{R} = \frac{120\text{ V}}{360\ \Omega} = 0.333\text{ A}$$

(a) Thus the power dissipated is $\mathcal{P} = \Delta VI$:

$$\mathcal{P} = (120\text{ V})(0.333\text{ A}) = 40.0\text{ W} \qquad \lozenge$$

(b) For bulb R_1, $\Delta V = IR$:

$$\Delta V = (0.333\text{ A})(240\ \Omega) = 80.0\text{ V} \qquad \lozenge$$

For bulb R_2 or R_3, the potential difference $\quad \Delta V = 120\text{ V} - 80.0\text{ V} = 40.0\text{ V} \qquad \lozenge$

71. In Figure P28.71, suppose the switch has been closed for a time sufficiently long for the capacitor to become fully charged. Find (a) the steady-state current in each resistor and (b) the charge Q on the capacitor. (c) The switch is now opened at $t=0$. Write an equation for the current I_{R2} through R_2 as a function of time and (d) find the time interval required for the charge on the capacitor to fall to one-fifth its initial value.

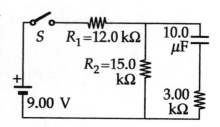

Figure P28.71

Solution

(a) When the capacitor is fully charged, no current flows in it or in the 3.00-kΩ resistor:

$$I_3 = 0 \qquad \lozenge$$

So the same current flows in resistors 1 and 2: $I_1 = I_2$, as given by the voltage rule,

$$+9.00 \text{ V} - (12.0 \text{ k}\Omega)I_1 - (15.0 \text{ k}\Omega)I_1 = 0$$

$$I_1 = 9.00 \text{ V} / 27.0 \text{ k}\Omega = 0.333 \text{ mA} = I_2 \qquad \lozenge$$

(b) For the right-hand loop, the voltage rule gives for the capacitor voltage:

$$(+15.0 \text{ k}\Omega)(0.333 \text{ mA}) - \Delta V_c - (3.00 \text{ k}\Omega)(0) = 0$$

Then $\Delta V_c = 5.00 \text{ V}$ and $Q = C\Delta V_c = (10.0 \ \mu\text{F})(5.00 \text{ V}) = 50.0 \ \mu\text{C} \qquad \lozenge$

(c) At $t=0$, the current in R_1 drops to zero. The capacitor, charged to 5.00 V with top plate positive, will drive a uniform current $I_2 = I_3$ counter-clockwise around the right-hand loop. At $t=0$, its value is given by the equation:

$$+5.00 \text{ V} - (15.0 \text{ k}\Omega)I_2 - (3.00 \text{ k}\Omega)I_2 = 0$$

$$I_2 = 5.00 \text{ V} / 18.0 \text{ k}\Omega = 0.278 \text{ mA}$$

Thereafter, it decays as it drains the capacitor's charge, with time constant

$$R_{eq}C = (18.0 \text{ k}\Omega)(10.0 \ \mu\text{F}) = 180 \text{ ms}$$

So its equation is $\qquad I_2 = (0.278 \text{ mA})e^{-t/(180 \text{ ms})} \qquad \lozenge$

(d) The charge decays according to $\qquad q = Q_0 e^{-t/RC}$

Substituting known values, $\qquad \frac{1}{5}(50.0 \ \mu\text{C}) = (50.0 \ \mu\text{C})e^{-t/(180 \text{ ms})}$

Solving for t, $\qquad 0.200 = e^{-t/(180 \text{ ms})}$

Thus, $-t/(180 \text{ ms}) = -1.61$ and $t = 1.61(180 \text{ ms}) = 290 \text{ ms} \qquad \lozenge$

Chapter 29
MAGNETIC FIELDS

EQUATIONS AND CONCEPTS

The **magnetic force on a charge** moving in a magnetic field can be stated as a vector cross product. *The experimental observations of charged particles moving in a magnetic field are stated in Section 29.1 of the textbook and are summarized in Eq. 29.1.*

$$\mathbf{F}_B = q\mathbf{v} \times \mathbf{B} \tag{29.1}$$

The **magnitude of the magnetic force** will be maximum when the charge moves perpendicular to the direction of the magnetic field.

$$F_B = |q|vB\sin\theta \tag{29.2}$$

The **magnetic field intensity B** at some point in space is defined in terms of the magnetic force exerted on a moving positive electric charge at that point. The SI unit of the magnetic field is the tesla (T) or weber per square meter (Wb/m^2). The cgs unit of magnetic field is the gauss (G).

$$B \equiv \frac{F_B}{|q|v\sin\theta}$$

$$1\,\text{T} = 1\frac{\text{N}}{\text{C}\cdot\text{m/s}}$$

$$1\,\text{T} = 10^4\,\text{G}$$

The **direction of the magnetic force** on a moving charge can be determined by applying the right-hand-rule (refer to figure): Hold your right hand open and point your fingers along the direction of **B** while pointing your thumb in the direction of **v**. The force on a positive charge will be directed out of the palm of your hand. *If the charge is negative, the direction of the force will be reversed. A second version of the right-hand rule is illustrated in Suggestions, Skills, and Strategies.*

The **force on a straight segment of current-carrying conductor** placed in a uniform magnetic field is given by Equation 29.3.

$$\mathbf{F}_B = I\mathbf{L}\times\mathbf{B} \qquad (29.3)$$

The **direction of the force is determined** by the right-hand rule. In this case, your thumb must point in the direction of the current. *Use the right-hand rule to confirm that the direction of* **B** *in the figure at right is into the plane of the page.*

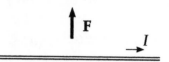

The **magnitude of the magnetic force** on the conductor depends on the angle between the direction of the conductor and the direction of the field. The **magnetic force will be maximum** when the conductor is directed perpendicular to the magnetic field.

$$F_B = BIL\sin\theta$$

$$F_{B,\,max} = BIL$$

The **magnetic force on a wire of arbitrary shape** is found by integrating over the length of the wire. In this equation the direction of *d***s** is that of the current. The magnetic force on a curved current-carrying wire in a magnetic field is equal to that of a straight wire connecting the end points and carrying the same current. Also, the net magnetic force on a closed current loop in a uniform field is zero.

$$\mathbf{F}_B = I\int_a^b d\mathbf{s}\times\mathbf{B} \qquad (29.5)$$

A **net torque** will be exerted on a current loop placed in an external magnetic field. In Equation 29.9, the vector **A** is normal to the plane of the loop and has a magnitude equal to the area of the loop. *When the fingers of the right hand are curled around the loop in the direction of the current, the thumb points in the direction of* **A.**

$$\boldsymbol{\tau} = I\mathbf{A}\times\mathbf{B} \qquad (29.9)$$

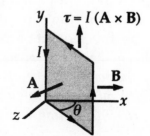

The **magnitude of the torque** depends on the angle between the direction of **B** and the direction of **A** and will be maximum when the magnetic field is parallel to the plane of the loop, that is, when **A** is perpendicular to **B**.

$$\tau = IAB\sin\theta$$

$$\tau_{max} = IAB \qquad (29.8)$$

The **direction of rotation of the loop** is in the direction of decreasing values of θ (i.e. vector **A** rotates toward the direction of the magnetic field). When **A** becomes parallel to **B** the torque on the loop will be zero. The loop shown in the figure will rotate counterclockwise as seen from above. *The direction of the vector torque τ is indicated by the thumb of the right hand when the fingers curl **A** in to the direction of* **B**.

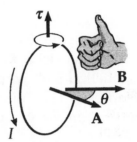

The **magnetic dipole moment** of a current loop is a vector quantity along the direction of **A**. The torque on a current loop can also be expressed in terms of the magnetic moment of the loop.

$$\mu = IA \qquad (29.10)$$

$$\tau = \mu \times B \qquad (29.11)$$

Motion of a charged particle entering a uniform magnetic field with the velocity vector initially perpendicular to the field has the following characteristics:

• The **path** of the particle will be circular and in a plane perpendicular to the direction of the field. The direction of rotation of the particle will be as determined by the right-hand rule.

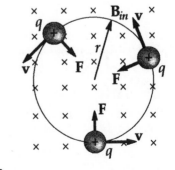

• The **radius** of the circular path will be proportional to the linear momentum of the charged particle.

$$r = \frac{mv}{qB} \qquad (29.13)$$

- The **angular frequency** (or cyclotron frequency) of the particle will be proportional to the ratio of charge to mass. *Note that the frequency, and hence the period of rotation do not depend on the radius of the path.*

$$\omega = \frac{qB}{m} \qquad (29.14)$$

$$T = \frac{2\pi m}{qB} \qquad (29.15)$$

If the **initial velocity is not perpendicular** to the field, the particle will move in a helical path whose axis is parallel to the field. The "pitch" of the helix will depend on the component of velocity parallel to the field.

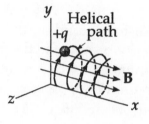

Applications of the motion of charged particles in a magnetic field include:

- **Velocity Selector** — When a beam of charged particles is directed into a region where uniform electric and magnetic fields are perpendicular to each other and to the initial direction of the particle beam, only those particles with a velocity $v = E/B$ will emerge undeflected.

$$v = \frac{E}{B} \qquad (29.17)$$

- **Mass Spectrometer** — If an ion beam, after passing through a velocity selector, is directed perpendicularly into a second uniform magnetic field, the ratio of charge to mass for the isotopic species can be determined by measuring the radius of curvature of the beam in the second field.

$$\frac{m}{q} = \frac{rB_0B}{E} \qquad (29.18)$$

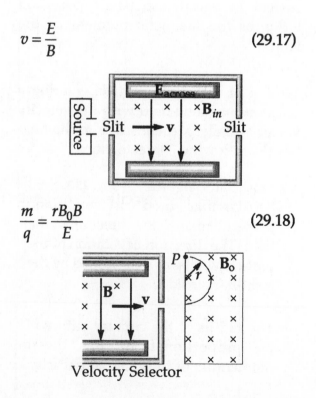

Velocity Selector

- **Cyclotron** — The maximum kinetic energy acquired by an ion in a cyclotron depends on the radius of the "dees" and the intensity of the magnetic field. This relationship holds until the ion reaches relativistic energies (≈ 20 MeV). *This energy limit is true for ions of proton mass or greater, but is not true for electrons.*

$$K = \tfrac{1}{2}mv^2 = \frac{q^2 B^2 R^2}{2m} \qquad (29.19)$$

The **Hall voltage** arises from the deflection of charge carriers in a current-carrying conductor when placed in a magnetic field. In Equation 29.22, d is the width of the conductor (the dimension that is perpendicular to the directions of both I and $\mathbf{B}$) and n is the charge-carrier density. *When the Hall Coefficient ($1/nq$) is known for a calibrated sample, magnetic field strengths can be determined by accurate measurements of the Hall voltage, ΔV_H. By measuring the voltage as + or −, the sign of the charge carriers can be determined.*

$$\Delta V_H = \frac{IBd}{nqA} \qquad (29.22)$$

SUGGESTIONS, SKILLS, AND STRATEGIES

ILLUSTRATING THE DIRECTION OF MAGNETIC FIELDS

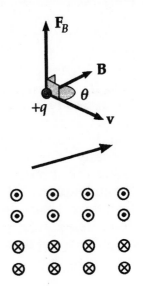

B can be **shown in perspective** along with other vector quantities (e.g velocity and magnetic force).

B lying in the plane of the page is shown by an arrow, or by multiple arrows emphasizing the field nature of **B**.

B directed out of the page is shown by dots representing tips of arrows coming outward.

B directed into the page is shown by crosses representing the feathers of arrows going inward.

MAGNETIC FIELDS AND THE CROSS-PRODUCT

Equation 29.1, $\mathbf{F} = q\mathbf{v} \times \mathbf{B}$, serves as the definition of the magnetic field vector $\mathbf{B}$. If the vectors $\mathbf{v}$ and $\mathbf{B}$ are given in unit vector notation then $\mathbf{v} \times \mathbf{B}$ can be written as

$$\mathbf{v} \times \mathbf{B} = \hat{\mathbf{i}}(v_y B_z - v_z B_y) + \hat{\mathbf{j}}(v_z B_x - v_x B_z) + \hat{\mathbf{k}}(v_x B_y - v_y B_x)$$

The components of the magnetic force are:

$$F_x = q(v_y B_z - v_z B_y) \qquad F_y = q(v_z B_x - v_x B_z) \qquad F_z = q(v_x B_y - v_y B_x)$$

USING THE RIGHT-HAND RULE

The right-hand rule is used to find the direction of the cross-product $\mathbf{F} = \mathbf{v} \times \mathbf{B}$. There are two versions of the right-hand rule; in order to avoid confusion, you should pick the version that suits you. In either version, note that the stated order of $\mathbf{v}$ and $\mathbf{B}$ is very important ($\mathbf{v} \times \mathbf{B} = -\mathbf{B} \times \mathbf{v}$).

Version 1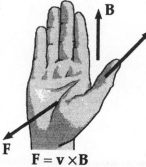

$\mathbf{F} = \mathbf{v} \times \mathbf{B}$

Hold your open right hand with your thumb pointing in the direction of $\mathbf{v}$ (*the first named vector quantity*) and your fingers pointing in the direction of $\mathbf{B}$ (*the second named vector quantity*). By necessity, your fingers and thumb cannot stretch farther than 180°, so if you do this, your hand will be oriented properly. The force vector $\mathbf{F}$ now is directed out of the palm of your hand. If your thumb and fingers are aligned, then the orientation of your hand is not determined; this is the case where the angle between $\mathbf{B}$ and $\mathbf{v}$ is 0° or 180° and there is no force.

Version 2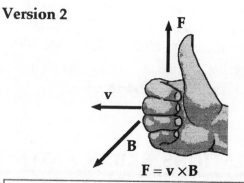

$\mathbf{F} = \mathbf{v} \times \mathbf{B}$

Orient your hand so that your fingers point in the direction of $\mathbf{v}$ (*the first named vector quantity*) and then curl your fingers to point in the direction of $\mathbf{B}$ (*the second named vector quantity*). Note that since your fingers cannot bend farther than 180°, you may have to flip your hand upside down to do this. Your thumb now points in the direction of $\mathbf{F}$, where $\mathbf{F}$ is perpendicular to both $\mathbf{v}$ and $\mathbf{B}$. If you do not need to curl your fingers because they already point along $\mathbf{B}$, then the angle between $\mathbf{B}$ and $\mathbf{v}$ is 0° and there is no force.

You may sometimes hear of a third version of the right-hand rule, involving the thumb, index finger and middle finger. However we recommend that you practice using one of the versions described above.

Å different right-hand rule applies in some specific situations, relating two quantities rather than the three quantities in a cross product. *One of the quantities involves circulation ands the other is a vector.* The rule looks similar to version 2 of the cross product rule: curl your fingers around in the direction of the circulation, and your thumb will point in the direction of the vector. This rule applies, for example, when finding the direction of the area vector for a current loop.

REVIEW CHECKLIST

You should be able to:

▷ Use the defining equation for a magnetic field **B** to determine the magnitude and direction of the magnetic force exerted on an electric charge moving in a region where there is a magnetic field. You should understand clearly the important differences between the forces exerted on electric charges by electric fields and those forces exerted on moving electric charges by magnetic fields. (Section 29.1)

▷ Calculate the magnitude and direction of the magnetic force on a current-carrying conductor when placed in an external magnetic field. (Section 29.2)

▷ Determine the magnitude and direction of the torque exerted on a closed current loop in an external magnetic field. You should understand how to correctly designate the direction of the area vector corresponding to a given current loop, and to incorporate the magnetic moment of the loop into the calculation of the torque on the loop. (Section 29.3)

▷ Calculate the period and radius of the circular orbit of a charged particle moving in a uniform magnetic field. (Section 29.4)

▷ Understand the essential features of the velocity selector and mass spectrometer and make appropriate quantitative calculations regarding the operation of these instruments. (Section 29.5)

ANSWERS TO SELECTED QUESTIONS

2. Two charged particles are projected into a magnetic field perpendicular to their velocities. If the charges are deflected in opposite directions, what can you say about them?

Answer We know the magnetic field is constant, and the velocity vectors are the same, but one force is the negative of the other. From $\mathbf{F}_B = q(\mathbf{v} \times \mathbf{B})$, we can conclude that the only thing that could cause the force to be of opposite sign is if the charges were of opposite sign.

☐   ☐   ☐   ☐

11. Is it possible to orient a current loop in a uniform magnetic field such that the loop does not tend to rotate? Explain.

Answer Yes. If the magnetic field is perpendicular to the plane of the loop, the forces on opposite sides will be equal and opposite, but will produce no net torque.

☐   ☐   ☐   ☐

13. How can a current loop be used to determine the presence of a magnetic field in a given region of space?

Answer The loop can be mounted free to rotate around an axis. The loop will rotate about this axis when placed in an external magnetic field for some arbitrary orientation. As the current through the loop is increased, the torque on it will increase.

☐   ☐   ☐   ☐

SOLUTIONS TO SELECTED PROBLEMS

1. Determine the initial direction of the deflection of charged particles as they enter the magnetic fields as shown in Figure P29.1.

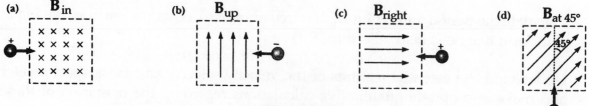

Figure P29.1

Solution   Both versions of the right-hand rule, open-handed and curled, are shown here. Use the set of figures that you have chosen, and ignore the other set.

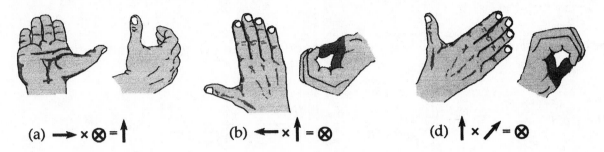

(a) $\rightarrow \times \otimes = \uparrow$ (b) $\leftarrow \times \uparrow = \otimes$ (d) $\uparrow \times \nearrow = \otimes$

(a)   By solution figure (a), $\mathbf{v} \times \mathbf{B}$ is (right) × (away) = up.   ◊

(b)   By solution figure (b), $\mathbf{v} \times \mathbf{B}$ is (left) × (up) = away.
      Since the charge is negative, $q\mathbf{v} \times \mathbf{B}$ is toward you.   ◊

(c) $\mathbf{v} \times \mathbf{B}$ is zero since the angle between $\mathbf{v}$ and $\mathbf{B}$ is 180° and $\sin 180° = 0$.
      There is no deflection.   ◊

(d) $\mathbf{v} \times \mathbf{B}$ is (up) × (up and right), or away from you.   ◊

---

5.   A proton moves perpendicular to a uniform magnetic field $\mathbf{B}$ at 1.00×10^7 m/s and experiences an acceleration of 2.00×10^{13} m/s^2 in the +x direction when its velocity is in the +z direction. Determine the magnitude and direction of the field.

Solution

By Newton's 2nd law, $\sum F = ma = (1.67 \times 10^{-27} \text{ kg})(2.00 \times 10^{13} \text{ m/s}^2)$

The magnetic force $F = 3.34 \times 10^{-14}$ N $= qvB \sin 90°$

Rearranging, $$B = \frac{F}{qv} = \frac{3.34 \times 10^{-14} \text{ N}}{(1.60 \times 10^{-19} \text{ C})(1.00 \times 10^7 \text{ m/s})} = 2.09 \times 10^{-2} \text{ T}$$ ◊

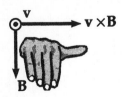

By the right-hand rule, $\mathbf{B}$ must be in the $-y$ direction.   ◊

This yields a force on the proton in the +x direction when v points in the +z direction.

105

7.   A proton moving at 4.00×10^6 m/s through a magnetic field of 1.70 T experiences a magnetic force of magnitude 8.20×10^{-13} N. What is the angle between the proton's velocity and the field?

Solution   Since the magnitude of the force on a moving charge in a magnetic field is $F_B = qvB\sin\theta$,

$$\theta = \sin^{-1}\left[\frac{F_B}{qvB}\right]: \quad \theta = \sin^{-1}\left[\frac{8.20 \times 10^{-13} \text{ N}}{(1.60 \times 10^{-19} \text{ C})(4.00 \times 10^6 \text{ m/s})(1.70 \text{ T})}\right] = 48.9° \text{ or } 131° \qquad \Diamond$$

9.   A proton moves with a velocity of $\mathbf{v} = (2\hat{\mathbf{i}} - 4\hat{\mathbf{j}} + \hat{\mathbf{k}})$ m/s in a region in which the magnetic field is $\mathbf{B} = (\hat{\mathbf{i}} + 2\hat{\mathbf{j}} - 3\hat{\mathbf{k}})$ T. What is the magnitude of the magnetic force this charge experiences?

Solution

The force on a charge is proportional to the vector product of the velocity and the magnetic field:

$$\mathbf{F}_B = q\mathbf{v} \times \mathbf{B} = (1.60 \times 10^{-19} \text{ C})\left((2\hat{\mathbf{i}} - 4\hat{\mathbf{j}} + \hat{\mathbf{k}}) \text{ m/s}\right) \times \left((\hat{\mathbf{i}} + 2\hat{\mathbf{j}} - 3\hat{\mathbf{k}}) \text{ T}\right)$$

Since $1 \text{ C} \cdot \text{m} \cdot \text{T/s} = 1$ N, we can write this in determinant form as:

$$\mathbf{F}_B = (1.60 \times 10^{-19} \text{ N})\begin{vmatrix} \hat{\mathbf{i}} & \hat{\mathbf{j}} & \hat{\mathbf{k}} \\ 2 & -4 & 1 \\ 1 & 2 & -3 \end{vmatrix}$$

Expanding the determinant as described in Equation 11.8, we have

$$\mathbf{F}_{B,x} = (1.60 \times 10^{-19} \text{ N})[(-4)(-3) - (1)(2)]\hat{\mathbf{i}}$$

$$\mathbf{F}_{B,y} = (1.60 \times 10^{-19} \text{ N})[(1)(1) - (2)(-3)]\hat{\mathbf{j}}$$

$$\mathbf{F}_{B,z} = (1.60 \times 10^{-19} \text{ N})[(2)(2) - (1)(-4)]\hat{\mathbf{k}}$$

Again in vector notation,

$$\mathbf{F}_B = (1.60 \times 10^{-19} \text{ N})(10\hat{\mathbf{i}} + 7\hat{\mathbf{j}} + 8\hat{\mathbf{k}}) = (16.0\hat{\mathbf{i}} + 11.2\hat{\mathbf{j}} + 12.8\hat{\mathbf{k}}) \times 10^{-19} \text{ N}$$

$$|\mathbf{F}_B| = \left(\sqrt{16.0^2 + 11.2^2 + 12.8^2}\right) \times 10^{-19} \text{ N} = 23.4 \times 10^{-19} \text{ N} \qquad \Diamond$$

11. A wire having a mass per unit length of 0.500 g/cm carries a 2.00-A current horizontally to the south. What are the direction and magnitude of the minimum magnetic field needed to lift this wire vertically upward?

Solution

Conceptualize: Since $I = 2.00$ A south, **B** must be to the east to make **F** upward according to the right-hand rule for currents in a magnetic field.

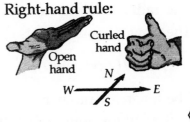

Right-hand rule:

As before, in viewing the diagrams to the right, use the version of the right-hand rule you have chosen, and ignore the other version.

The magnitude of **B** should be significantly greater than the earth's magnetic field (~ 50 μT), since we do not typically see wires levitating when current flows through them.

Categorize: The force on a current-carrying wire in a magnetic field is $\mathbf{F}_B = I\boldsymbol{\ell} \times \mathbf{B}$, from which we can find **B**.

Analyze: With I to the south and **B** to the east, the force on the wire is simply $F_B = I\ell B \sin 90°$, which must oppose the weight of the wire, mg.

So, $$B = \frac{F_B}{I\ell} = \frac{mg}{I\ell} = \frac{g}{I}\left(\frac{m}{\ell}\right) = \left(\frac{9.80 \text{ m/s}^2}{2.00 \text{ A}}\right)\left(0.500 \frac{\text{g}}{\text{cm}}\right)\left(\frac{10^2 \text{ cm/m}}{10^3 \text{ g/kg}}\right) = 0.245 \text{ T} \qquad \diamond$$

Finalize: The required magnetic field is about 5000 times stronger than the earth's magnetic field. Thus it was reasonable to ignore the earth's magnetic field in this problem. In other situations the earth's field can have a significant effect.

---

17. A nonuniform magnetic field exerts a net force on a magnetic dipole. A strong magnet is placed under a horizontal conducting ring of radius r that carries current I as shown in Figure P29.17. If the magnetic field **B** makes an angle θ with the vertical at the ring's location, what are the magnitude and direction of the resultant force on the ring?

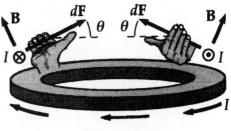

Figure P29.17

Solution

The magnetic force on each bit of ring is inward and upward, at an angle θ above the radial line, according to:

$$|d\mathbf{F}| = I|d\mathbf{s}\times\mathbf{B}| = I\,ds\,B$$

The radially inward components tend to squeeze the ring, but cancel out as forces. The upward components $I\,ds\,B\sin\theta$ all add to

$$\mathbf{F} = I(2\pi r)B\sin\theta \quad \text{up} \qquad \diamond$$

The magnetic moment of the ring is down. This problem is a model for the force on a dipole in a nonuniform magnetic field, or for the force that one magnet exerts on another magnet.

---

23. A rectangular coil consists of $N=100$ closely wrapped turns and has dimensions $a=0.400$ m and $b=0.300$ m. The coil is hinged along the y axis, and its plane makes an angle $\theta=30.0°$ with the x axis (Fig. P29.23). What is the magnitude of the torque exerted on the coil by a uniform magnetic field $B=0.800$ T directed along the x axis when the current is $I=1.20$ A in the direction shown? What is the expected direction of rotation of the coil?

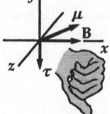

Figure P29.23

Solution

The magnetic moment of the coil is $\mu=NIA$, perpendicular to its plane and making a $60°$ angle with the x axis as shown to the right. The torque on the dipole is then

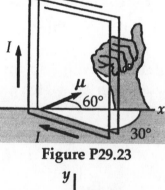

$$\boldsymbol{\tau}=\boldsymbol{\mu}\times\mathbf{B}=NBAI\sin\phi \quad \text{down}$$

having a magnitude $\tau=NBAI\sin\phi$

$$\tau=(100)(0.800\text{ T})(0.400\times0.300\text{ m}^2)(1.20\text{ A})(\sin 60.0°)=9.98\text{ N·m} \quad \diamond$$

Note that ϕ is the angle between the magnetic moment and the **B** field. We model the coil as a rigid body under a net torque; the coil will rotate so as to align the magnetic moment with the **B** field. Looking down along the y axis, we will see the coil rotate in the clockwise direction. $\qquad \diamond$

33. A proton (charge $+e$, mass m_p), a deuteron (charge $+e$, mass $2m_p$), and an alpha particle (charge $+2e$, mass $4m_p$) are accelerated through a common potential difference ΔV. Each of the particles enters a uniform magnetic field **B**, with its velocity in a direction perpendicular to **B**. The proton moves in a circular path of radius r_p. Determine the radii of the circular orbits for the deuteron, r_d, and the alpha particle, r_α, in terms of r_p.

Solution

Conceptualize: In general, particles with greater speed, more mass, and less charge will have larger radii as they move in a circular path due to a constant magnetic force. Since the effects of mass and charge have opposite influences on the path radius, it is somewhat difficult to predict which particle will have the larger radius. However, since the mass and charge ratios of the three particles are all similar in magnitude within a factor of four, we should expect that the radii also fall within a similar range.

Categorize: The radius of each particle's path can be found by applying Newton's second law, where the force causing the centripetal acceleration is the magnetic force: $\mathbf{F} = q\mathbf{v} \times \mathbf{B}$. The speed of the particles can be found from the kinetic energy resulting from the change in electric potential given.

Analyze:

An electric field changes the speed of each particle according to $(K + U)_i = (K + U)_f$. Therefore, assuming that the particles start from rest,

we can write
$$q\Delta V = \tfrac{1}{2}mv^2$$

The magnetic field changes their direction as described by $\Sigma \mathbf{F} = m\mathbf{a}$:

$$qvB\sin 90° = \frac{mv^2}{r}; \qquad \text{thus} \qquad r = \frac{mv}{qB} = \frac{m}{qB}\sqrt{\frac{2q\Delta V}{m}} = \frac{1}{B}\sqrt{\frac{2m\Delta V}{q}}$$

For the protons,
$$r_p = \frac{1}{B}\sqrt{\frac{2m_p\Delta V}{e}}$$

For the deuterons,
$$r_d = \frac{1}{B}\sqrt{\frac{2(2m_p)\Delta V}{e}} = \sqrt{2}\,r_p \qquad\qquad \Diamond$$

For the alpha particles,
$$r_\alpha = \frac{1}{B}\sqrt{\frac{2(4m_p)\Delta V}{2e}} = \sqrt{2}\,r_p \qquad\qquad \Diamond$$

Finalize: Somewhat surprisingly, the radii of the deuterons and alpha particles are the same and are only 41% greater than for the protons.

37. A cosmic-ray proton in interstellar space has an energy of 10.0 MeV and executes a circular orbit having a radius equal to that of Mercury's orbit around the Sun (5.80×10^{10} m). What is the magnetic field in that region of space?

Solution Think of the proton as having accelerated through a potential difference $\Delta V = 10^7$ V. We use the energy version of the isolated system model, applied to the particle and the electric field that made it speed up from rest. The proton's kinetic energy is

$$E = \tfrac{1}{2}mv^2 = e\Delta V \qquad \text{so its speed is} \qquad v = \sqrt{\frac{2e\Delta V}{m}}$$

Now we use the particle in a magnetic field model and the particle in uniform circular motion model.

$\sum F = ma$ becomes $\dfrac{mv^2}{R} = evB\sin 90°$

So $\qquad B = \dfrac{mv}{eR} = \dfrac{m}{eR}\sqrt{\dfrac{2e\Delta V}{m}} = \dfrac{1}{R}\sqrt{\dfrac{2m\Delta V}{e}}$

and $\qquad B = \dfrac{1}{5.80 \times 10^{10} \text{ m}}\sqrt{\dfrac{2(1.67 \times 10^{-27} \text{ kg})(10^7 \text{ V})}{1.60 \times 10^{-19} \text{ C}}} = 7.88 \times 10^{-12} \text{ T}$ ◊

43. A cyclotron designed to accelerate protons has a magnetic field of magnitude 0.450 T over a region of radius 1.20 m. What are (a) the cyclotron frequency and (b) the maximum speed acquired by the protons?

Solution

(a) The cyclotron frequency is

$$\omega = \frac{qB}{m}$$

For protons, $\qquad \omega = \dfrac{(1.60 \times 10^{-19} \text{ C})(0.450 \text{ T})}{1.67 \times 10^{-27} \text{ kg}} = 4.31 \times 10^7 \text{ rad/s}$ ◊

(b) $R = \dfrac{mv}{Bq}$: $\qquad v = \dfrac{BqR}{m} = \dfrac{(0.450 \text{ T})(1.60 \times 10^{-19} \text{ C})(1.20 \text{ m})}{1.67 \times 10^{-27} \text{ kg}} = 5.17 \times 10^7 \text{ m/s}$ ◊

47. The picture tube in a television uses magnetic deflection coils rather than electric deflection plates. Suppose an electron beam is accelerated through a 50.0-kV potential difference and then through a region of uniform magnetic field 1.00 cm wide. The screen is located 10.0 cm from the center of the coils and is 50.0 cm wide. When the field is turned off, the electron beam hits the center of the screen. What field magnitude is necessary to deflect the beam to the side of the screen? Ignore relativistic corrections.

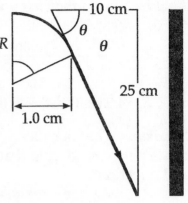

Solution   The beam is deflected by the angle $\theta = \tan^{-1}\left(\dfrac{25.0 \text{ cm}}{10.0 \text{ cm}}\right) = 68.2°$

The two angles θ shown are equal because their sides are perpendicular, right side to right side and left side to left side. The radius of curvature of the electrons in the field is

$$R = \frac{1.00 \text{ cm}}{\sin 68.2°} = 1.077 \text{ cm}$$

Now $\frac{1}{2}mv^2 = q\Delta V$ so $v = \sqrt{\dfrac{2q\Delta V}{m}} = 1.33 \times 10^8 \text{ m/s}$

$\sum \mathbf{F} = m\mathbf{a}$ becomes $\dfrac{mv^2}{R} = |q|vB\sin 90°$

$$B = \frac{mv}{|q|R} = \frac{(9.11 \times 10^{-31} \text{ kg})(1.33 \times 10^8 \text{ m/s})}{(1.60 \times 10^{-19} \text{ C})(1.077 \times 10^{-2} \text{ m})} = 70.1 \text{ mT} \qquad \diamond$$

---

51. In an experiment designed to measure the Earth's magnetic field using the Hall effect, a copper bar 0.500 cm thick is positioned along an east-west direction. If a current of 8.00 A in the conductor results in a Hall voltage of 5.10×10^{-12} V, what is the magnitude of the Earth's magnetic field? (Assume that $n = 8.49 \times 10^{28}$ electrons/m³ and that the plane of the bar is rotated to be perpendicular to the direction of **B**.)

Solution

Conceptualize: The Earth's magnetic field is about 50 μT (see Table 29.1), so we should expect a result of that order of magnitude.

Categorize: The magnetic field can be found from the Hall effect voltage:

$$\Delta V_H = \frac{IB}{nqt} \qquad \text{or} \qquad B = \frac{nqt\Delta V_H}{I}$$

Analyze: From the Hall voltage,

$$B = \frac{(8.49\times10^{28} \text{ e}^-/\text{m}^3)(1.60\times10^{-19} \text{ C}/\text{e}^-)(0.00500 \text{ m})(5.10\times10^{-12} \text{ V})}{8.00 \text{ A}}$$

$B = 4.33\times10^{-5} \text{ T} = 43.3 \ \mu\text{T}$ ◊

Finalize: The calculated magnetic field is slightly less than we expected but is reasonable considering that the Earth's local magnetic field varies in both magnitude and direction.

53. Sodium melts at 99 °C. Liquid sodium, an excellent thermal conductor, is used in some nuclear reactors to cool the reactor core. The liquid sodium is moved through pipes by pumps that exploit the force on a moving charge in a magnetic field. The principle is as follows. Assume the liquid metal to be in an electrically insulating pipe having a rectangular cross section of width w and height h. A uniform magnetic field perpendicular to the pipe affects a section of length L (Fig. P29.53). An electric current directed perpendicular to the pipe and to the magnetic field produces a current density J in the liquid sodium. (a) Explain why this arrangement produces on the liquid a force that is directed along the length of the pipe. (b) Show that the section of liquid in the magnetic field experiences a pressure increase JLB.

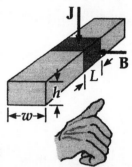

Figure P29.53

Solution

(a) By the right-hand rule, the electric current carried by the material experiences a force $I\mathbf{h}\times\mathbf{B}$ along the pipe, into the page. ◊

(b) The sodium, consisting of ions and electrons, flows along the pipe transporting no net charge. But inside the section of length L, electrons drift upward to constitute downward electric current $J(\text{area}) = JLw$. The current feels magnetic force $I\mathbf{h}\times\mathbf{B} = JLwhB\sin 90°$. This force along the pipe axis will make the fluid move, exerting pressure

$$\frac{F}{\text{area}} = \frac{JLwhB}{hw} = JLB$$ ◊

The hand in the figure shows that the fluid moves away from you, into the page.

57. A positive charge $q = 3.20 \times 10^{-19}$ C moves with a velocity $\mathbf{v} = (2\hat{\mathbf{i}} + 3\hat{\mathbf{j}} - \hat{\mathbf{k}})$ m/s through a region where both a uniform magnetic field and a uniform electric field exist. (a) Calculate the total force on the moving charge (in unit-vector notation), taking $\mathbf{B} = (2\hat{\mathbf{i}} + 4\hat{\mathbf{j}} + \hat{\mathbf{k}})$ T and $\mathbf{E} = (4\hat{\mathbf{i}} - \hat{\mathbf{j}} - 2\hat{\mathbf{k}})$ V/m. (b) What angle does the force vector make with the positive x axis?

Solution   The total force is the Lorentz force,

(a) $\mathbf{F} = q\mathbf{E} + q(\mathbf{v} \times \mathbf{B}) = q(\mathbf{E} + \mathbf{v} \times \mathbf{B})$

$\mathbf{F} = q\left[(4\hat{\mathbf{i}} - \hat{\mathbf{j}} - 2\hat{\mathbf{k}}) \text{ V/m} + (2\hat{\mathbf{i}} + 3\hat{\mathbf{j}} - \hat{\mathbf{k}}) \text{ m/s} \times (2\hat{\mathbf{i}} + 4\hat{\mathbf{j}} + \hat{\mathbf{k}}) \text{ T}\right]$

$\mathbf{F} = q\left[(4\hat{\mathbf{i}} - \hat{\mathbf{j}} - 2\hat{\mathbf{k}}) \text{ V/m} + (7\hat{\mathbf{i}} - 4\hat{\mathbf{j}} + 2\hat{\mathbf{k}}) \text{ T·m/s}\right]$

$\mathbf{F} = q\left[(11\hat{\mathbf{i}} - 5\hat{\mathbf{j}}) \text{ V/m}\right] = q\left[(11\hat{\mathbf{i}} - 5\hat{\mathbf{j}}) \text{ N/C}\right]$

$\mathbf{F} = (3.20 \times 10^{-19} \text{ C})\left[(11\hat{\mathbf{i}} - 5\hat{\mathbf{j}}) \text{ N/C}\right] = (3.52\hat{\mathbf{i}} - 1.60\hat{\mathbf{j}}) \times 10^{-18} \text{ N}$ ◊

(b) $\mathbf{F} \cdot \hat{\mathbf{i}} = F\cos\theta = F_x$: $\theta = \cos^{-1}(F_x / F) = \cos^{-1}(3.52 / 3.87) = 24.4°$ ◊

---

67. Consider an electron orbiting a proton and maintained in a fixed circular path of radius $R = 5.29 \times 10^{-11}$ m by the Coulomb force. Treating the orbiting charge as a current loop, calculate the resulting torque when the system is in a magnetic field of 0.400 T directed perpendicular to the magnetic moment of the electron.

Solution

Conceptualize: Since the mass of the electron is very small ($\sim 10^{-30}$ kg), we should expect that the torque on the orbiting charge will be very small as well, perhaps $\sim 10^{-30}$ N·m.

Categorize: The torque on a current loop that is perpendicular to a magnetic field can be found from $|\boldsymbol{\tau}| = IAB\sin\theta$. The magnetic field is given, $\theta = 90°$, the area of the loop can be found from the radius of the circular path, and the current can be found from the centripetal acceleration that results from the Coulomb force that attracts the electron to proton.

Analyze:

The area of the loop is $A = \pi r^2 = \pi (5.29 \times 10^{-11} \text{ m})^2 = 8.79 \times 10^{-21} \text{ m}^2$.

If v is the speed of the electron, then the period of its circular motion will be

$$T = \frac{2\pi r}{v}$$

and the effective current due to the orbiting electron is

$$I = \frac{\Delta Q}{\Delta t} = \frac{e}{T}$$

Applying Newton's second law with the Coulomb force acting as the central force

gives
$$\sum F = \frac{k_e q^2}{R^2} = \frac{mv^2}{R}$$

so that
$$v = q\sqrt{\frac{k_e}{mR}} \quad \text{and} \quad T = 2\pi\sqrt{\frac{mR^3}{q^2 k_e}}$$

$$T = 2\pi\sqrt{\frac{(9.11\times10^{-31} \text{ kg})(5.29\times10^{-11} \text{ m})^3}{(1.60\times10^{-19} \text{ C})^2(8.99\times10^9 \text{ N}\cdot\text{m}^2/\text{C}^2)}} = 1.52\times10^{-16} \text{ s}$$

The torque is
$$|\tau| = \left(\frac{q}{T}\right) AB$$

$$|\tau| = \left(\frac{1.60\times10^{-19} \text{ C}}{1.52\times10^{-16} \text{ s}}\right)\left[\pi(5.29\times10^{-11} \text{ m})^2\right](0.400 \text{ T}) = 3.70\times10^{-24} \text{ N}\cdot\text{m} \lozenge$$

Finalize: The torque is certainly small, but a million times larger than we guessed. This torque will cause the atom to precess with a frequency proportional to the applied magnetic field. A similar process on the nuclear, rather than the atomic, level leads to nuclear magnetic resonance (NMR), which is used for magnetic resonance imaging (MRI) scans employed for medical diagnostic testing (see Section 44.8).

Chapter 30

SOURCES OF THE MAGNETIC FIELD

EQUATIONS AND CONCEPTS

The **Biot–Savart law** gives the magnetic field at a point in space due to an element of conductor ds which carries a current I and is at a distance r away from the point. *The vector $d\mathbf{B}$ is perpendicular to both $d\mathbf{s}$ (which points in the direction of the current) and to the unit vector $\hat{\mathbf{r}}$ (which points from the current element toward the point where the field is to be determined).* Confirm that at point P' shown in the figure, the direction of $d\mathbf{B}$ from element $d\mathbf{s}$ is *out* of the page, but the direction of $d\mathbf{B}$ from the closest segment of wire is *into* the page.

$$d\mathbf{B} = \frac{\mu_0}{4\pi} \frac{I\,d\mathbf{s} \times \hat{\mathbf{r}}}{r^2} \tag{30.1}$$

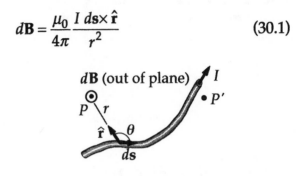

The **permeability of free space** is a constant.

$$\mu_0 = 4\pi \times 10^{-7}\ \text{T} \cdot \text{m} / \text{A} \tag{30.2}$$

The **total magnetic field** at a point in the vicinity of a current of finite length is found by integrating the Biot–Savart law expression over the entire current distribution. *Remember, the integrand is a vector quantity.*

$$\mathbf{B} = \frac{\mu_0 I}{4\pi} \int \frac{d\mathbf{s} \times \hat{\mathbf{r}}}{r^2} \tag{30.3}$$

The magnitude of the magnetic field due to several current distributions are shown below.

B at a **perpendicular distance a from a straight wire of finite length**, where θ_1 and θ_2 are as shown in the figure to the right.

$$B = \frac{\mu_0 I}{4\pi a}(\cos\theta_1 - \cos\theta_2) \qquad (30.4)$$

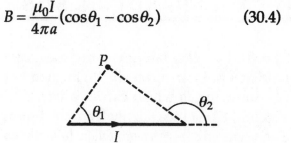

B at a **distance a from a long straight conductor**, carrying a current I.

$$B = \frac{\mu_0 I}{2\pi a} \qquad (30.5)$$

B at the **center of an arc of radius R** which **subtends an angle θ (in radians) at the center of the arc.**

$$B = \frac{\mu_0 I}{4\pi R}\theta \qquad (30.6)$$

B_x on the **axis of a circular loop** of radius R and at a **distance x from the plane** of the loop.

$$B_x = \frac{\mu_0 I R^2}{2(x^2 + R^2)^{3/2}} \qquad (30.7)$$

At the **center of a current loop** of radius R.

$$B = \frac{\mu_0 I}{2R} \qquad (30.8)$$

On the **axis of a current loop when** $x \gg R$.

$$B = \frac{\mu_0 I R^2}{2x^3} \qquad (30.9)$$

The **magnetic force per unit length** between very long parallel conductors depends on the distance a between the conductors and the magnitudes of the two currents. Parallel conductors carrying currents in the same direction attract each other, and parallel conductors carrying currents in opposite directions repel each other. *The forces on the two conductors will be equal in magnitude regardless of the relative magnitude of the two currents.*

$$\frac{F_B}{\ell} = \frac{\mu_0 I_1 I_2}{2\pi a}$$

(30.12)

Ampère's law describes the creation of magnetic fields by continuous current configurations. The integral of the tangential component of the magnetic field around any closed path is equal to the total current passing through a surface bounded by the closed path. This technique is most useful in calculating the magnetic field due to currents that have a high degree of symmetry. Examples are the magnetic field **B**:

$$\oint \mathbf{B} \cdot d\mathbf{s} = \mu_0 I$$

(30.13)

inside a toroid having N turns and at a distance r from the center of the toroid;

$$B = \frac{\mu_0 N I}{2\pi r}$$

(30.16)

within, and far from the ends of a solenoid of n turns per unit length.

$$B = \mu_0 \frac{N}{\ell} I = \mu_0 n I$$

(30.17)

The **magnetic flux** through a surface is the integral of the normal component of the field over the surface. *The magnetic flux through any closed surface is equal to zero.*

$$\Phi_B \equiv \int \mathbf{B} \cdot d\mathbf{A}$$

(30.18)

The **direction of the magnetic field due to current in a long wire** is determined by using the right-hand rule. Hold the conductor in the right hand with the thumb pointing in the direction of the conventional current. The fingers will then wrap around the wire in the direction of the magnetic field lines. *The magnetic field is tangent to the circular field lines at every point in the region around the conductor.*

The **direction of the magnetic field at the center of a current loop** is perpendicular to the plane of the loop and directed in the sense given by the right-hand rule.

Within a solenoid, the magnetic field is parallel to the axis of the solenoid and pointing in a sense determined by applying the right-hand rule to one of the coils.

The **orbital magnetic moment** of an electron is proportional to its orbital angular momentum L. The orbital angular momentum is quantized and equals integer multiples of $\hbar$ ($h/2\pi$).

$$\mu = \frac{e}{2m_e}L \qquad (30.25)$$

$$L = 0, \hbar, 2\hbar, 3\hbar, \ldots$$

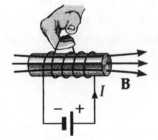

The **Bohr magneton** is numerically equal to the spin magnetic moment of an electron.

$$\mu_B = 9.27 \times 10^{-24} \text{ J/T} \qquad (30.28)$$

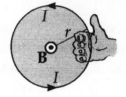

The **total magnetic field vector B** can be expressed in terms of magnetization vector, **M** (due to the presence of magnetic materials) and magnetic field strength vector, **H** (due to external currents). The SI units of **H** and **M** are amperes per meter.

$$\mathbf{B} = \mu_0(\mathbf{H} + \mathbf{M}) \qquad (30.30)$$

The **magnetization vector, M,** describes the magnetic state of a substance. The magnitude of **M** is defined as the magnetic moment per unit volume. In a vacuum, **M** = 0 because no magnetic material is present. *For paramagnetic and diamagnetic substances, the magnetization is proportional to the magnetic field strength. The proportionality constant c is called the magnetic susceptibility of the material.*

$$\mathbf{M} = \chi\mathbf{H} \qquad (30.32)$$

The **total field** can also be expressed in terms of the permeability μ_m of the substance.

$$\mathbf{B} = \mu_m\mathbf{H} \qquad (30.33)$$

The **magnetic permeability** μ_m of a magnetic substance is related to its magnetic susceptibility. Substances are:
- Paramagnetic if $\mu_m > \mu_0$
- Diamagnetic if $\mu_m < \mu_0$

$$\mu_m = \mu_0(1 + \chi) \qquad (30.34)$$

Curie's law states that the magnetization of a paramagnetic substance is proportional to the applied field and inversely proportional to the absolute temperature. The constant C is called Curie's constant.

$$M = C\frac{B_0}{T} \qquad (30.35)$$

REVIEW CHECKLIST

You should be able to:

▷ Use the Biot-Savart law to calculate the magnetic induction **B** at a specified point in the vicinity of a current element, and by integration find the total magnetic field due to a number of important geometric configurations. Use of the Biot-Savart law must include a clear understanding of the direction of the magnetic field contribution relative to the direction of the current element which produces it and the direction of the vector which locates the point at which the field is to be calculated. (Section 30.1)

▷ Calculate the force between two parallel current-carrying conductors. (Section 30.2)

▷ Use Ampère's law to calculate the magnetic field due to steady current configurations which have a sufficiently high degree of symmetry such as a long straight conductor, a long solenoid, and a toroidal coil. (Sections 30.3 and 30.4)

▷ Calculate the magnetic flux through a surface area placed in either a uniform or nonuniform magnetic field. (Section 30.5)

▷ Explain the generalized form of Ampère's law (Ampere-Maxwell law): magnetic fields are produced both by **conduction currents** and by **displacement currents produced by changing electric fields.** (Section 30.7)

▷ Make calculations of magnetic field strength, **H** (distinguished from magnetic flux density, **B**). (Section 30.8)

ANSWERS TO SELECTED QUESTIONS

3. Explain why two parallel wires carrying currents in opposite directions repel each other.

Answer The figure at the right will help you understand this result. The magnetic field due to wire 2 at the position of wire 1 is directed out of the paper. Hence, the magnetic force on wire 1, given by $I_1\mathbf{L}_1\times\mathbf{B}_2$, must be directed to the left since $\mathbf{L}_1\times\mathbf{B}_2$ is directed to the left. Likewise, you can show that the magnetic force on wire 2 due to the field of wire 1 is directed towards the right.

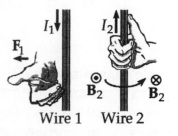

□   □   □   □

9. A hollow copper tube carries a current along its length. Why is **B**=0 inside the tube? Is **B** nonzero outside the tube?

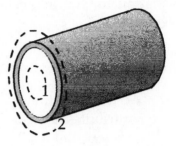

Answer   Let us apply Ampere's law to the closed path labeled 1 in this figure. Since there is no current through this path, and because of the symmetry of the configuration, we see that the magnetic field inside the tube must be zero. On the other hand, the net current through the path labeled 2 is *I*, the current carried by the conductor. Therefore, the field outside the tube is nonzero.

□   □   □   □

SOLUTIONS TO SELECTED PROBLEMS

3.   (a) A conductor in the shape of a square loop of edge length $\ell = 0.400$ m carries a current $I = 10.0$ A as in Fig. P30.3. Calculate the magnitude and direction of the magnetic field at the center of the square. (b) **What if?** If this conductor is formed into a single circular turn and carries the same current, what is the value of the magnetic field at the center?

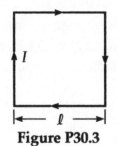

Figure P30.3

Solution

Conceptualize: As shown in the diagram to the right, the magnetic field at the center is directed into the page from the clockwise current. If we consider the sides of the square to be sections of four infinite wires, then we could expect the magnetic field at the center of the square to be a little less than four times the strength of the field at a point $\ell / 2$ away from an infinite wire with current I.

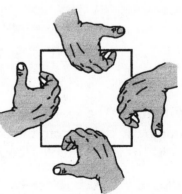

$$B < 4\frac{\mu_0 I}{2\pi a} = 4\left(\frac{(4\pi\times10^{-7}\text{ T·m}/\text{A})(10.0\text{ A})}{2\pi(0.200\text{ m})}\right) = 40.0\ \mu\text{T}$$

Forming the wire into a circle should not significantly change the magnetic field at the center since the average distance of the wire from the center will not be much different.

Categorize: Each side of the square is simply a section of a thin, straight conductor, so the solution derived from the Biot-Savart law in Example 30.1 can be applied to part (a) of this problem. For part (b), the Biot-Savart law can also be used to derive the equation for the magnetic field at the center of a circular current loop as shown in Example 30.3.

Analyze:

(a) We use Equation 30.4 for the field created by each side of the square. Each side contributes a field away from you at the center, so together they produce a magnetic field:

$$B = \frac{4\mu_0 I}{4\pi a}\left(\cos\frac{\pi}{4} - \cos\frac{3\pi}{4}\right) = \frac{4(4\pi\times10^{-7}\ \text{T·m/A})(10.0\ \text{A})}{4\pi(0.200\ \text{m})}\left(\frac{\sqrt{2}}{2}+\frac{\sqrt{2}}{2}\right)$$

so at the center of the square,

$$\mathbf{B} = 2.00\sqrt{2}\times10^{-5}\ \text{T} = 28.3\ \mu\text{T} \quad \text{perpendicularly into the page} \qquad \Diamond$$

(b) As in the first part of the problem, the direction of the magnetic field will be into the page. The new radius is found from the length of wire: $4\ell = 2\pi R$, so $R = 2\ell/\pi = 0.255$ m. Equation 30.8 gives the magnetic field at the center of a circular current loop:

$$B = \frac{\mu_0 I}{2R} = \frac{(4\pi\times10^{-7}\ \text{T·m/A})(10.0\ \text{A})}{2(0.255\ \text{m})} = 2.47\times10^{-5}\ \text{T} = 24.7\ \mu\text{T} \qquad \Diamond$$

Caution! If you use your calculator, it may not understand the keystrokes ⁤4⁤ ⁤×⁤ ⁤π⁤ ⁤EXP⁤ ⁤+/−⁤ ⁤7⁤. To get the right answer, you may need to use ⁤4⁤ ⁤EXP⁤ ⁤+/−⁤ ⁤7⁤ ⁤×⁤ ⁤π⁤.

Finalize: The magnetic field in part (a) is less than 40 μT as we predicted. Also, the magnetic fields from the square and circular loops are similar in magnitude, with the field from the circular loop being about 15% less than from the square loop.

Quick Tip: A simple way to use your right hand to find the magnetic field due to a current loop is to curl the fingers of your right hand in the direction of the current. Your extended thumb will then point in the direction of the magnetic field within the loop or solenoid.

5. Determine the magnetic field at a point P located a distance x from the corner of an infinitely long wire bent at a right angle, as shown in Figure P30.5. The wire carries a steady current I.

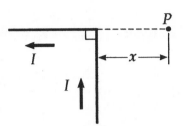

Figure P30.5

Solution The vertical section of wire constitutes one half of an infinitely long straight wire at distance x from P, so it creates a field equal to

$$B = \frac{1}{2}\left(\frac{\mu_0 I}{2\pi x}\right)$$

Hold your right hand with extended thumb in the direction of the current; the field is away from you, into the paper. For each bit of the horizontal section of wire $d\mathbf{s}$ is to the left and $\hat{\mathbf{r}}$ is to the right, so $d\mathbf{s}\times\hat{\mathbf{r}} = 0$. The horizontal current produces zero field at P.

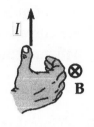

Thus, $\qquad\qquad \mathbf{B} = \dfrac{\mu_0 I}{4\pi x}$ into the paper $\qquad\qquad\qquad \lozenge$

---

17. In Figure P30.17, the current in the long, straight wire is $I_1 = 5.00$ A and the wire lies in the plane of the rectangular loop, which carries the current $I_2 = 10.0$ A. The dimensions are $c = 0.100$ m, $a = 0.150$ m, and $\ell = 0.450$ m. Find the magnitude and direction of the net force exerted on the loop by the magnetic field created by the wire.

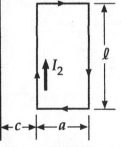

Figure P30.17

Solution

Conceptualize: Even though there are forces in opposite directions on the loop, we must remember that the magnetic field is stronger near the wire than it is farther away. By symmetry the forces exerted on sides 2 and 4 (the horizontal segments of length a) are equal and opposite, and therefore cancel. The magnetic field in the plane of the loop is directed into the page to the right of I_1. By the right-hand rule, $\mathbf{F} = I\boldsymbol{\ell}\times\mathbf{B}$ is directed toward the **left** for side 1 of the loop and a smaller force is directed toward the **right** for side 3. Therefore, we should expect the net force to be to the left, possibly in the μN range for the currents and distances given.

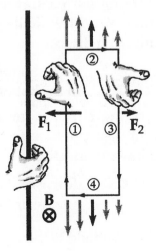

Categorize: The magnetic force between two parallel wires can be found from Equation 30.11, which can be applied to sides 1 and 3 of the loop to find the net force resulting from these opposing force vectors.

Analyze:
$$\mathbf{F} = \mathbf{F}_1 + \mathbf{F}_2 = \frac{\mu_0 I_1 I_2 \ell}{2\pi}\left(\frac{1}{c+a} - \frac{1}{c}\right)\hat{\mathbf{i}} = \frac{\mu_0 I_1 I_2 \ell}{2\pi}\left(\frac{-a}{c(c+a)}\right)\hat{\mathbf{i}}$$

$$\mathbf{F} = \frac{(4\pi\times10^{-7}\ \text{N}/\text{A}^2)(5.00\ \text{A})(10.0\ \text{A})(0.450\ \text{m})}{2\pi}\left(\frac{-0.150\ \text{m}}{(0.100\ \text{m})(0.250\ \text{m})}\right)\hat{\mathbf{i}}$$

$$\mathbf{F} = (-2.70\times10^{-5}\hat{\mathbf{i}})\ \text{N} \quad \text{or} \quad F = 2.70\times10^{-5}\ \text{N} \quad \text{toward the left} \qquad \lozenge$$

Finalize: The net force is to the left and in the μN range as we expected. The symbolic representation of the net force on the loop shows that the net force would be zero if either current disappeared, if either dimension of the loop became very small ($a \to 0$ or $\ell \to 0$), or if the magnetic field were uniform ($c \to \infty$).

---

21. Four long, parallel conductors carry equal currents of $I = 5.00$ A. Figure P30.21 is an end view of the conductors. The current direction is into the page at points A and B (indicated by the crosses) and out of the page at C and D (indicated by the dots). Calculate the magnitude and direction of the magnetic field at point P, located at the center of the square of edge length 0.200 m.

A ⊗- - - - - - - - -⊙ C
0.200 m
• 0.200 m
P
B ⊗- - - - - - - -⊙ D

Figure P30.21

Solution Each wire is distant from P by $(0.200\ \text{m})\cos 45.0° = 0.141$ m

Each wire produces a field at P of equal magnitude:

$$B = \frac{\mu_0 I}{2\pi a} = \frac{(2.00\times10^{-7}\ \text{T}\cdot\text{m}/\text{A})(5.00\ \text{A})}{0.141\ \text{m}} = 7.07\ \mu\text{T}$$

Carrying currents away from you, the left-hand wires produce fields at P of 7.07 μT, in the following directions:

 A: to the bottom and left, at 225°
 B: to the bottom and right, at 315°.

Carrying currents toward you, the wires to the right also produce fields at P of 7.07 μT, in the following directions:

 C: downward and to the right, at 315°
 D: downward and to the left, at 225°.

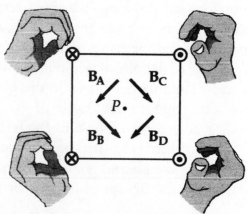

The total field is then $4(7.07\ \mu\text{T})\sin 45.0° = 20.0\ \mu\text{T}$ toward the bottom of the page. $\quad \lozenge$

---

25. A packed bundle of 100 long, straight, insulated wires forms a cylinder of radius $R = 0.500$ cm. (a) If each wire carries 2.00 A, what are the magnitude and direction of the magnetic force per unit length acting on a wire located 0.200 cm from the center of the bundle? (b) **What if?** Would a wire on the outer edge of the bundle experience a force greater or smaller than the value calculated in part (a)?

Solution

Conceptualize: The force **on** one wire comes from its interaction with the magnetic field created **by** the other ninety-nine wires. According to Ampere's law, at a distance r from the center, only the wires enclosed within a radius r contribute to this net magnetic field; the other wires outside the radius produce magnetic field vectors in opposite directions that cancel out at r. Therefore, the magnetic field (and also the force on a given wire at radius r) will be greater for larger radii within the bundle, and will decrease for distances beyond the radius of the bundle, as shown in the graph to the right. Applying $\mathbf{F} = I\boldsymbol{\ell} \times \mathbf{B}$, the magnetic force on a single wire will be directed toward the center of the bundle, so that all the wires tend to attract each other.

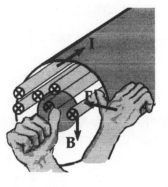

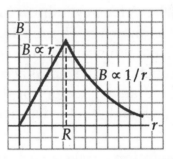

Categorize: We model the current carried by the ninety-nine wires as uniformly spread across the whole cylinder. Using Ampere's law, we can find the magnetic field at any radius, so that the magnetic force $\mathbf{F} = I\boldsymbol{\ell} \times \mathbf{B}$ on a single wire can then be calculated.

Analyze:

(a) Ampere's law is used to derive Equation 30.15, which we can use to find the magnetic field at $r = 0.200$ cm from the center of the cable:

$$B = \frac{\mu_0 I_0 r}{2\pi R^2} = \frac{(4\pi \times 10^{-7} \text{ T·m / A})(99)(2.00 \text{ A})(0.200 \times 10^{-2} \text{ m})}{2\pi (0.500 \times 10^{-2} \text{ m})^2} = 3.17 \times 10^{-3} \text{ T}$$

This field points tangent to a circle of radius 0.200 cm and exerts a force $\mathbf{F} = I\boldsymbol{\ell} \times \mathbf{B}$ toward the center of the bundle, on the single hundredth wire:

$$\frac{F}{\ell} = IB\sin\theta = (2.00 \text{ A})(3.17 \times 10^{-3} \text{ T})(\sin 90°) = 6.34 \text{ mN / m} \qquad \diamond$$

(b) As is shown above in the graph taken from the text, the magnetic field increases linearly as a function of r until it reaches a maximum at the outer surface of the cable. Therefore, the force on a single wire at the outer radius $r = 5.00$ cm would be greater than at $r = 2.00$ cm by a factor of $5/2$. $\qquad \diamond$

Finalize: We did not estimate the expected magnitude of the force, but 200 amperes is a lot of current. It would be interesting to see if the magnetic force that pulls together the individual wires in the bundle is enough to hold them against their own weight: If we assume that the insulation accounts for about half the volume of the bundle, then a single copper wire in this bundle would have a cross sectional area of about

$$(1/2)(0.01)\pi(0.500 \text{ cm})^2 = 4\times10^{-7} \text{ m}^2$$

with a weight per unit length of

$$\rho gA = (8\,920 \text{ kg}/\text{m}^3)(9.8 \text{ N}/\text{kg})(4\times10^{-7} \text{ m}^2) = 0.03 \text{ N}/\text{m}$$

Therefore, the outer wires experience an inward magnetic force that is about half the magnitude of their own weight. If placed on a table, this bundle of wires would form a loosely held mound without the outer sheathing to hold them together.

---

29. A long cylindrical conductor of radius R carries a current I as shown in Figure P30.29. The current density J, however, is not uniform over the cross section of the conductor but is a function of the radius according to $J = br$, where b is a constant. Find an expression for the magnetic field B (a) at a distance $r_1 < R$ and (b) at a distance $r_2 > R$, measured from the axis.

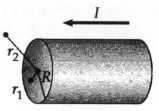

Figure P30.29

Solution   Take a circle of radius r_1 or r_2 to apply $\oint \mathbf{B}\cdot d\mathbf{s} = \mu_0 I$, where for nonuniform current density $I = \int \mathbf{J}\cdot d\mathbf{A}$. In this case $\mathbf{B}$ is parallel to $d\mathbf{s}$ and $\mathbf{J}$ is parallel to $d\mathbf{A}$, so Ampère's law gives

$$\oint B\,ds = \mu_0 \int J\,d\text{A}$$

(a)   For $r_1 < R$, $2\pi r_1 B = \mu_0 \int_0^{r_1} br(2\pi r\,dr)$ and $B = \frac{1}{3}\left(\mu_0 b r_1^2\right)$ (inside)   ◊

(b)   For $r_2 > R$, $2\pi r_2 B = \mu_0 \int_0^{R} br(2\pi r\,dr)$ and $B = \dfrac{\mu_0 b R^3}{3r_2}$ (outside)   ◊

---

31. What current is required in the windings of a long solenoid that has 1 000 turns uniformly distributed over a length of 0.400 m, to produce at the center of the solenoid a magnetic field of magnitude 1.00×10^{-4} T?

Solution   The magnetic field at the center of a solenoid is $B = \mu_0 \dfrac{N}{\ell}I$.

So $I = \dfrac{B\ell}{\mu_0 N} = \dfrac{(1.00\times10^{-4} \text{ T}\cdot\text{A})(0.400 \text{ m})}{(4\pi\times10^{-7} \text{ T}\cdot\text{m})(1000)} = 31.8 \text{ mA}$ ◊

---

35. A cube of edge length $\ell = 2.50$ cm is positioned as shown in Figure P30.35. A uniform magnetic field given by $\mathbf{B} = (5\hat{\mathbf{i}} + 4\hat{\mathbf{j}} + 3\hat{\mathbf{k}})$ T exists throughout the region. (a) Calculate the flux through the shaded face. (b) What is the total flux through the six faces?

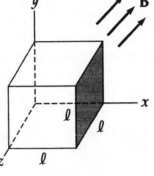

Figure P30.35

Solution

The flux is defined as $\quad\quad \Phi_B = \mathbf{B} \cdot \mathbf{A}$

(a) $\quad\quad\quad\quad\quad\quad \Phi_B = B_x A_x + B_y A_y + B_z A_z$

Here, $\quad\quad\quad\quad\quad\quad A_y = A_z = 0$

so $\quad\quad\quad\quad\quad\quad \Phi_B = (5.00 \text{ T})(0.0250 \text{ m})^2 = 3.12 \times 10^{-3}$ T·m^2 ◊

(b) For a closed surface, $\quad \oint \mathbf{B} \cdot d\mathbf{A} = 0 \quad\quad$ so $\quad\quad \Phi_B = 0$ ◊

---

37. A 0.100-A current is charging a capacitor that has square plates 5.00 cm on each side. The plate separation is 4.00 mm. Find (a) the time rate of change of electric flux between the plates and (b) the displacement current between the plates.

Solution The electric field in the space between the plates is $\quad E = \dfrac{\sigma}{\epsilon_0} = \dfrac{Q}{\epsilon_0 A}$

The flux of this field is $\quad\quad\quad\quad \Phi_E = \mathbf{E} \cdot \mathbf{A} = \left(\dfrac{Q}{\epsilon_0 A}\right) A \cos 0° = \dfrac{Q}{\epsilon_0}$

(a) The rate of change of flux is $\quad \dfrac{d\Phi_E}{dt} = \dfrac{d}{dt}\dfrac{Q}{\epsilon_0} = \dfrac{1}{\epsilon_0}\dfrac{dQ}{dt} = \dfrac{I}{\epsilon_0}$

$$\dfrac{d\Phi_E}{dt} = \left(\dfrac{0.100 \text{ A}}{8.85 \times 10^{-12} \text{ C}^2/\text{N} \cdot \text{m}^2}\right)(1 \text{ C}/\text{A} \cdot \text{s}) = 1.13 \times 10^{10} \text{ N} \cdot \text{m}^2/\text{C} \cdot \text{s}$$ ◊

(b) The displacement current is defined as

$$I_d = \epsilon_0 \dfrac{d\Phi}{dt} = (8.85 \times 10^{-12} \text{ C}^2/\text{N} \cdot \text{m}^2)(1.13 \times 10^{10} \text{ N} \cdot \text{m}^2/\text{C} \cdot \text{s}) = 0.100 \text{ A}$$ ◊

---

41. A toroid with a mean radius of 20.0 cm and 630 turns (see Figure 30.30) is filled with powdered steel whose magnetic susceptibility χ is 100. The current in the windings is 3.00 A. Find B (assumed uniform) inside the toroid.

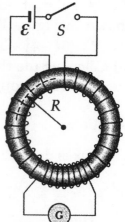

Solution

If the coil had a vacuum inside, the magnetic field would be given by the equation

$$B = \frac{\mu_0 N I}{2\pi r}$$

That is, the magnetic field strength would be

Figure 30.30

$$H = \frac{B}{\mu_0} = \frac{N I}{2\pi r}$$

With the steel inside, $B = \mu_0(1 + \chi)H = \mu_0(1 + \chi)\dfrac{N I}{2\pi r}$

$$B = \frac{(4\pi \times 10^{-7} \text{ T·m / A})(101)(630)(3.00 \text{ A})}{2\pi(0.200 \text{ m})} = 0.191 \text{ T}$$ ◊

---

47. The magnetic moment of the Earth is approximately 8.00×10^{22} A·m². (a) If this were caused by the complete magnetization of a huge iron deposit, how many unpaired electrons would this correspond to? (b) At two unpaired electrons per iron atom, how many kilograms of iron would this correspond to? (Iron has a density of 7 900 kg/m³, and approximately 8.50×10^{28} iron atoms/m³.)

Solution

Conceptualize: We know that most of the Earth is not iron, so if the situation described provides an accurate model, then the iron deposit must certainly be less than the mass of the Earth ($M_{\text{Earth}} = 5.98 \times 10^{24}$ kg). One mole of iron has a mass of 55.8 g and contributes $2(6.02 \times 10^{23})$ unpaired electrons, so we should expect the total unpaired electrons to be less than 10^{50}.

Categorize: The Bohr magneton μ_B is the measured value for the magnetic moment of a single unpaired electron. Therefore, we can find the number of unpaired electrons by dividing the magnetic moment of the Earth by μ_B. We can then use the density of iron to find the mass of the iron atoms that each contribute two electrons.

Analyze:

(a)
$$\mu_B = \left(9.27\times10^{-24}\ \text{J}/\text{T}\right)\left(\frac{1\ \text{N}\cdot\text{m}}{\text{J}}\right)\left(\frac{1\ \text{T}}{\text{N}\cdot\text{s}/\text{C}\cdot\text{m}}\right)\left(\frac{1\ \text{A}}{\text{C}/\text{s}}\right) = 9.27\times10^{-24}\ \text{A}\cdot\text{m}^2$$

The number of unpaired electrons is

$$N = \frac{8.00\times10^{22}\ \text{A}\cdot\text{m}^2}{9.27\times10^{-24}\ \text{A}\cdot\text{m}^2} = 8.63\times10^{45}\ \text{e}^-$$ ◊

(b) Each iron atom has two unpaired electrons, so the number of iron atoms required is

$$\tfrac{1}{2}N = \tfrac{1}{2}(8.63\times10^{45}) = 4.31\times10^{45}\ \text{iron atoms}$$

Thus, $$M_{Fe} = \frac{(4.31\times10^{45}\ \text{atoms})(7\ 900\ \text{kg}/\text{m}^3)}{8.50\times10^{28}\ \text{atoms}/\text{m}^3} = 4.01\times10^{20}\ \text{kg}$$ ◊

Finalize: The calculated answers seem reasonable based on the limits we expected. From the data in this problem, the iron deposit required to produce the magnetic moment would only be about 1/15 000 the mass of the Earth and would form a sphere 500 km in diameter. Although this is certainly a large amount of iron, it is much smaller than the inner core of the Earth, which is estimated to have a diameter of about 3000 km.

---

49. A very long, thin strip of metal of width w carries a current I along its length as shown in Figure P30.49. Find the magnetic field at the point P in the diagram. The point P is in the plane of the strip at distance b away from it.

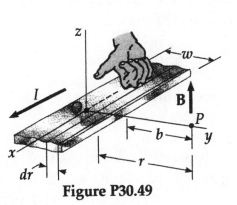

Figure P30.49

Solution Consider a long filament of the strip, which has width dr and is a distance r from point P. The magnetic field at a distance r from a long conductor is

$$B = \frac{\mu_0 I}{2\pi r}$$

Thus, the field due to the thin filament is $d\mathbf{B} = \dfrac{\mu_0\,dI}{2\pi r}\hat{k}$ where $dI = I\left(\dfrac{dr}{w}\right)$

so $$\mathbf{B} = \int_b^{b+w} \frac{\mu_0}{2\pi r}\left(I\frac{dr}{w}\right)\hat{k} = \frac{\mu_0 I}{2\pi w}\int_b^{b+w}\frac{dr}{r}\hat{k} = \frac{\mu_0 I}{2\pi w}\ln\left(1+\frac{w}{b}\right)\hat{k}$$ ◊

Note: Compare Problem 53 and its solution with Problem 54 and its solution. They contain the same steps of physical analysis and reasoning, but in problem 54 the numerical "plug and chug" steps are not included.

53. A nonconducting ring of radius 10.0 cm is uniformly charged with a total positive charge 10.0 μC. The ring rotates at a constant angular speed 20.0 rad/s about an axis through its center, perpendicular to the plane of the ring. What is the magnitude of the magnetic field on the axis of the ring 5.00 cm from its center?

54. A nonconducting ring of radius R is uniformly charged with a total positive charge q. The ring rotates at a constant angular speed ω about an axis through its center, perpendicular to the plane of the ring. What is the magnitude of the magnetic field on the axis of the ring a distance $R/2$ from its center?

Solution   The 'static' charge on the ring constitutes an electric current as the ring rotates.

The time required for the 10.0-μC charge to go past a point is the period of rotation,

$$T = \frac{\theta}{\omega} = \frac{2\pi \text{ rad}}{20 \text{ rad/s}} = 0.314 \text{ s}$$

The time required for the charge q to go past a point is the period of rotation,

$$T = \frac{\theta}{\omega} = \frac{2\pi}{\omega}$$

The current is $I = q/T$:

$$I = \frac{10.0 \times 10^{-6} \text{ C}}{0.314 \text{ s}} = 3.18 \times 10^{-5} \text{ A}$$

The current is $I = q/T$:

$$I = \frac{q}{2\pi/\omega} = \frac{q\omega}{2\pi}$$

The current has the shape of a flat compact circular coil with one turn, so the magnetic field it creates is

$$B = \frac{\mu_0 I R^2}{2(x^2 + R^2)^{3/2}}$$

$$B = \frac{\mu_0 I R^2}{2(x^2 + R^2)^{3/2}}$$

Substituting numeric values for these variables,

In this case, the distance x is equal to $R/2$, so

$$B = \frac{\left(4\pi \times 10^{-7} \, \frac{\text{T·m}}{\text{A}}\right)(3.18 \times 10^{-5} \text{ A})(0.100 \text{ m})^2}{2\left((0.0500 \text{ m})^2 + (0.100 \text{ m})^2\right)^{3/2}}$$

$$B = \frac{\mu_0 q\omega R^2}{4\pi(R^2/4 + R^2)^{3/2}}$$

$$B = \frac{4.00 \times 10^{-13} \text{ T·m}^3}{2(1.40 \times 10^{-3} \text{ m}^3)} = 1.43 \times 10^{-10} \text{ T} \qquad \lozenge$$

$$B = \frac{\mu_0 q\omega R^2}{4\pi(5R^2/4)^{3/2}} = \frac{2\mu_0 q\omega}{5\sqrt{5}\pi R} \qquad \lozenge$$

On the side of the ring from which it is seen as turning counterclockwise, the magnetic field is along the axis away from its center. The ring also creates an electric field in the same direction here. On the other side of the ring, the magnetic and electric fields are in opposite directions.

67. A wire is formed into the shape of a square of edge length L (Fig. P30.67). Show that when the current in the loop is I, the magnetic field at point P, a distance x from the center of the square along its axis is

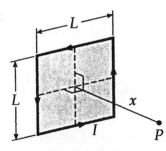

$$B = \frac{\mu_0 I L^2}{2\pi(x^2 + L^2/4)\sqrt{x^2 + L^2/2}}$$

Figure P30.67

Solution Consider the top side of the square. The distance from its center to point P is

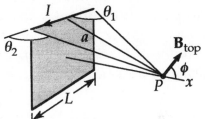

$$a = \sqrt{x^2 + (L/2)^2}$$

and Equation 30.4 describes the field it creates at point P. The distance from one of the corners of the square to P is

$$\sqrt{x^2 + (L/2)^2 + (L/2)^2} = \sqrt{x^2 + L^2/2}$$

 Thus $\quad \cos\theta_1 = \dfrac{L}{2\sqrt{x^2 + L^2/2}} \qquad$ and $\qquad \cos\theta_2 = -\cos\theta_1 = -\dfrac{L}{2\sqrt{x^2 + L^2/2}}$

Then, $\quad B_{top} = \dfrac{\mu_0 I}{4\pi a}(\cos\theta_1 - \cos\theta_2)$

$$B_{top} = \frac{\mu_0 I}{4\pi\sqrt{x^2 + L^2/4}}\left(\frac{L}{2\sqrt{x^2 + L^2/2}} + \frac{L}{2\sqrt{x^2 + L^2/2}}\right)$$

or $\quad B_{top} = \dfrac{\mu_0 I L}{4\pi\sqrt{x^2 + L^2/4}\sqrt{x^2 + L^2/2}}$

The component of this field along the direction of x is $B_{top}\cos\phi = B_{top}\dfrac{L/2}{a}$

Each of the four sides of the square produces this same size field along the direction of x, with the other components adding to zero. We assemble our results, and find that the net magnetic field points away from the center of the square, with magnitude

$$B = 4B_{top}\frac{L}{2a} = \frac{\mu_0 I L^2}{2\pi\left(x^2 + L^2/4\right)\sqrt{x^2 + L^2/2}} \qquad \Diamond$$

Chapter 31

FARADAY'S LAW

EQUATIONS AND CONCEPTS

Faraday's law of induction states that the average emf induced in a circuit is proportional to the rate of change of magnetic flux through the circuit. *The minus sign is included to indicate the polarity of the induced emf, which can be found by use of Lenz's law.*

$$\varepsilon = -\frac{d\Phi_B}{dt} \tag{31.1}$$

The **total magnetic flux** through a plane area, A, placed in a uniform magnetic field depends on the angle between the direction of the magnetic field and the normal to the surface area. *The maximum flux through the area occurs when the magnetic field is perpendicular to the plane of the surface area.* The unit of magnetic flux is the weber, Wb.

$$\Phi_B \equiv B_\perp A = BA\cos\theta$$

$$\Phi_{B,\,max} = BA$$

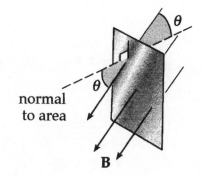

normal
to area

B

The **induced emf** can be expressed in terms of magnetic field (**B**), surface area (**A**) and the angle θ between **B** and the normal to the surface. From Equation 31.3, an emf will be induced in a circuit when any one or more of the following change with time:

- magnitude of **B**
- area enclosed by the circuit
- angle θ between **B** and the normal

$$\varepsilon = -\frac{d}{dt}(BA\cos\theta) \tag{31.3}$$

For a **non-uniform magnetic field,** the magnetic flux through an area is the integral of the normal component of the magnetic field over the area.

$$\Phi_B = \int \mathbf{B} \cdot d\mathbf{A}$$

Lenz's law states that the polarity of the induced emf (and the direction of the associated current in a closed circuit) would produce a current whose magnetic field opposes the change in the flux through the circuit. *That is, the induced current tends to maintain the original flux through the circuit.*

A motional emf is induced in a conductor moving through a magnetic field. The induced emf is given by Equation 31.5 when a conductor of length ℓ moves with speed v perpendicular to a constant magnetic field **B**. *Recall the significance of the negative sign as required by Lenz's law.*

$$\mathcal{E} = -B\ell v \qquad (31.5)$$

An **induced current** will exist in a conductor moving in a magnetic field if it is part of a complete circuit. *Confirm that Lenz's law correctly predicts the direction of the induced current as shown in the figure.*

$$I = \frac{|\mathcal{E}|}{R} = \frac{B\ell v}{R} \qquad (31.6)$$

Faraday's law in a more general form can be written as the integral of the electric field around a closed path. In this form the electric field is a non-conservative field that is generated by a changing magnetic field.

$$\oint \mathbf{E} \cdot d\mathbf{s} = -\frac{d\Phi_B}{dt} \qquad (31.9)$$

$$\oint \mathbf{E} \cdot d\mathbf{s} = -\frac{d}{dt} \int \mathbf{B} \cdot d\mathbf{A}$$

A **sinusoidally varying emf** is produced when a conducting loop of N turns and cross-sectional area, A, rotates with a constant angular velocity in a magnetic field. *For a given loop, the maximum value of the induced emf will be proportional to the angular velocity of the loop.*

$$\mathcal{E} = NAB\omega\sin\omega t \qquad (31.10)$$

$$\mathcal{E}_{max} = NAB\omega \qquad (31.11)$$

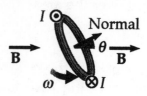

Maxwell's equations as applied to free space (i.e. in the absence of any dielectric or magnetic material) are as follows:

Gauss's law

$$\oint_S \mathbf{E}\cdot d\mathbf{A} = \frac{Q}{\epsilon_0} \qquad (31.12)$$

Gauss's law in magnetism

$$\oint_S \mathbf{B}\cdot d\mathbf{A} = 0 \qquad (31.13)$$

Faraday's law

$$\oint \mathbf{E}\cdot d\mathbf{s} = -\frac{d\Phi_B}{dt} \qquad (31.14)$$

Ampère-Maxwell law

$$\oint \mathbf{B}\cdot d\mathbf{s} = \mu_0 I + \epsilon_0\mu_0\frac{d\Phi_E}{dt} \qquad (31.15)$$

The **Lorentz force** is the combined effect of the electric force and the magnetic force exerted on a particle of charge q at some point in space.

$$\mathbf{F} = q\mathbf{E} + q\mathbf{v}\times\mathbf{B} \qquad (31.16)$$

SUGGESTIONS, SKILLS, AND STRATEGIES

CALCULATING AN INDUCED EMF

It is important to distinguish clearly between the **instantaneous value** of emf induced in a circuit and the **average value** of the emf induced in the circuit over a finite time interval.

To calculate the average induced emf, it is often useful to write Equation 31.2 as

$$\varepsilon_{avg} = -N\left(\frac{d\Phi_B}{dt}\right)_{avg} = -N\frac{\Delta\Phi_B}{\Delta t}$$

or

$$\varepsilon_{avg} = -N\left(\frac{\Phi_{B,f} - \Phi_{B,i}}{\Delta t}\right)$$

where the subscripts i and f refer to the magnetic flux through the circuit at the beginning and end of the time interval Δt. For a circuit or coil in a single plane, $\Phi_B = BA\cos\theta$, where θ is the angle between the vector normal to the plane of the circuit (conducting coil) and the direction of the magnetic field.

Equation 31.3 can be used to calculate the **instantaneous value of an induced emf.** For a multi-turn loop, the induced emf is

$$\varepsilon = -N\frac{d}{dt}(BA\cos\theta)$$

where in a particular case B, A, θ, or any combination of those parameters can be time dependent while the others remain constant. The expression resulting from the differentiation is then evaluated using the known or calculated values of B, A, and θ and their time derivatives.

Use the right-hand rule when applying Lenz's law to determine the direction of an induced current. **Remember, the induced current in a circuit will have a direction which tends to maintain the flux through the circuit.**

AN EXAMPLE OF LENZ'S LAW

Consider two single-turn, concentric coils **lying in the plane of the paper** as shown in the figure. The outside coil is part of a circuit containing a resistor (R), a battery ($\mathcal{E}$) and switch (S). The inner coil is not part of the circuit. When the switch is moved from **"open"** to **"closed"** the direction of the induced current in the inner coil can be predicted by Lenz's law.

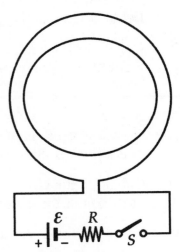

Consider the following steps:

(1) When the switch is in the **"open"** position as shown, there will be no current in the circuit.

(2) When the switch is moved to the **"closed"** position, there will be a clockwise current in the outside coil.

(3) Magnetic field lines due to current in the outside coil will be directed into the page through the area enclosed by the coil. **Use the right-hand rule (first application) to confirm this.** Magnetic flux into the page will penetrate the entire area enclosed by the outside coil **including the area of the inner coil.**

(4) By Faraday's law, the increasing flux produces an induced emf (and current) in the inner coil.

(5) Lenz's law requires that the induced current have a direction which will tend to maintain the initial flux condition **(which in this case was zero).** By using the **right-hand rule (second application),** you should be able to determine that the direction of the induced current in the inner coil must be counter-clockwise, contributing to a flux out of the page. Try this!

As a second example you should follow steps similar to those above to predict the direction of the induced current in the inner coil when the switch is moved from **"closed"** to **"open".**

REVIEW CHECKLIST

You should be able to:

▷ Calculate the emf (or current) induced in a circuit when the magnetic flux through the circuit is changing in time. The variation in flux might be due to a change in (a) the area of the circuit, (b) the magnitude of the magnetic field, (c) the direction of the magnetic field, or (d) the orientation/location of the circuit in the magnetic field. (Section 31.1)

▷ Calculate the emf induced between the ends of a conducting bar as it moves through a region where there is a constant magnetic field (motional emf). (Section 31.2)

▷ Apply Lenz's law to determine the direction of an induced emf or current. You should also understand that Lenz's law is a consequence of the law of conservation of energy. (Section 31.3)

▷ Calculate the electric field at various points in a charge-free region when the time variation of the magnetic field over the region is specified. (Section 31.4)

▷ Calculate the maximum and instantaneous values of the sinusoidal emf generated in a conducting loop rotating in a constant magnetic field. (Section 31.5)

ANSWERS TO SELECTED QUESTIONS

5. The bar in Figure Q31.5 moves on rails to the right with a velocity **v**, and the uniform, constant magnetic field is directed out of the page. Why is the induced current clockwise? If the bar were moving to the left, what would be the direction of the induced current?

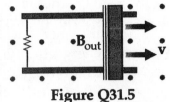

Figure Q31.5

Answer

The external magnetic field is out of the paper; as the area A enclosed by the loop increases, the external flux increases according to $\Phi_B = BA\cos\theta = BA$.

As predicted by Lenz's law, due to the increase in flux through the loop, free electrons will produce a current to create a magnetic flux inside the loop to oppose the change in flux.

In this case, the magnetic field due to the current must point into the paper in order to oppose the increasing magnetic flux of the external field coming out of the paper. By the right-hand rule (with your thumb pointing in the direction of the current along each wire), the current must be clockwise.

If the bar were moving toward the left, the area would decrease, and the flux would decrease. The flux due to **B** pointing out of the paper, then, would be decreasing. In order to cancel this change, the current would have to create a magnetic field inside the loop pointing out of the paper. This time, by the right-hand rule, we see that the current must be counter-clockwise.

□   □   □   □

9. How is energy produced in dams that is then transferred out by electrical transmission? (That is, how is the energy of motion of the water converted to energy that is transmitted by AC electricity?)

Answer

As the water falls, it gains kinetic energy. It is then forced to pass through a water wheel, transferring some of its energy to the rotor of a large AC electric generator.

The rotor of the generator is supplied with a small amount of DC current, which powers electromagnets in the rotor. Because the rotor is spinning, the electromagnets then create a magnetic flux that changes with time, according to the equation $\Phi_B = BA\cos\omega t$.

Coils of wire that are placed near the rotor then experience an induced emf according to the equation $\mathcal{E} = -Nd\Phi_B/dt$.

Finally, a small amount of this electricity is used to supply the rotor with its DC current; the rest is sent out over power lines to supply customers with electricity.

In terms of energy, the hydroelectric generating station is a nonisolated system in steady state. It takes in mechanical energy and puts out a precisely equal quantity of energy, nearly all of it by electrical transmission.

☐   ☐   ☐   ☐

SOLUTIONS TO SELECTED PROBLEMS

5.   A strong electromagnet produces a uniform magnetic field of 1.60 T over a cross-sectional area of 0.200 m². We place a coil having 200 turns and a total resistance of 20.0 Ω around the electromagnet. We then smoothly reduce the current in the electromagnet until it reaches zero in 20.0 ms. What is the current induced in the coil?

Solution

Conceptualize: A strong magnetic field turned off in a short time (20.0 ms) will produce a large emf, maybe on the order of 1 kV. With only 20.0 Ω of resistance in the coil, the induced current produced by this emf will probably be larger than 10 A but less than 1 000 A.

Categorize: According to Faraday's law, if the magnetic field is reduced uniformly, then a constant emf will be produced. The definition of resistance can be applied to find the induced current from the emf.

Analyze: The induced voltage is

$$\mathcal{E} = -N\frac{d(\mathbf{B}\cdot\mathbf{A})}{dt} = -N\left(\frac{0 - B_iA\cos\theta}{\Delta t}\right)$$

$$\mathcal{E} = \frac{(+200)(1.60\text{ T})(0.200\text{ m}^2)(\cos 0°)}{20.0\times 10^{-3}\text{ s}}\left(\frac{1\text{ N}\cdot\text{s}/\text{C}\cdot\text{m}}{\text{T}}\right)\left(\frac{1\text{ V}\cdot\text{C}}{\text{N}\cdot\text{m}}\right) = 3\,200\text{ V}$$

(**Note:** The unit conversions come from $\mathbf{F} = q\mathbf{v}\times\mathbf{B}$ and $U = qV$.)

$$I = \frac{\mathcal{E}}{R} = \frac{3\,200\text{ V}}{20.0\,\Omega} = 160\text{ A} \qquad \diamond$$

Finalize: This is a large current, as we expected. The positive sign means that the current in the coil is in the same direction as the current in the electromagnet.

7. An aluminum ring of radius 5.00 cm and resistance $3.00 \times 10^{-4}\ \Omega$ is placed on top of a long air-core solenoid with 1 000 turns per meter and radius 3.00 cm, as shown in Figure P31.7. Over the area of the end of the solenoid, assume that the axial component of the field produced by the solenoid is half as strong as at the center of the solenoid. Assume the solenoid produces negligible field outside its cross-sectional area. The current in the solenoid is increasing at a rate of 270 A/s. (a) What is the induced current in the ring? At the center of the ring, what are (b) the magnitude and (c) the direction of the magnetic field produced by the induced current in the ring?

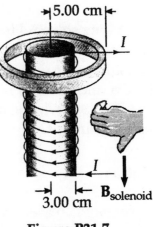

Figure P31.7

Solution

So as not to confuse variables, we define the radius of the ring to be r_{ring}, and its resistance to be R.

$$\varepsilon = -\frac{d}{dt}(BA\cos\theta) = -\frac{d}{dt}(0.500\mu_0 nIA\cos 0°) = -0.500\mu_0 nA\frac{dI}{dt}$$

Note that A must be interpreted as the area of the solenoid, where the field is strong:

$$\varepsilon = -0.500(4\pi\times10^{-7}\ \text{T·m}/\text{A})(1\,000\ \text{turns}/\text{m})\left[\pi(0.030\,0\ \text{m})^2\right](270\ \text{A}/\text{s})$$

$$\varepsilon = \left(-4.80\times10^{-4}\ \frac{\text{T·m}^2}{\text{s}}\right)\left(\frac{1\,\text{N·s}}{\text{C·m·T}}\right)\left(\frac{1\,\text{V·C}}{\text{N·m}}\right) = -4.80\times10^{-4}\ \text{V}$$

(a) $I_{ring} = \dfrac{|\varepsilon|}{R} = \dfrac{0.000\,480}{0.000\,300} = 1.60\ \text{A}$　　　　　◊

(b) $B_{ring} = \dfrac{\mu_0 I_{ring}}{2r_{ring}} = 2.01\times10^{-5}\ \text{T}$　　　　　◊

(c) The coil's field points downward, and is increasing, so B_{ring} points upward.　　◊

13. A long solenoid has 400 turns per meter and carries a current given by $I=(30.0\text{ A})(1-e^{-1.60t})$. Inside the solenoid and coaxial with it is a coil that has a radius of 6.00 cm and consists of a total of 250 turns of fine wire (Fig. P31.13). What emf is induced in the coil by the changing current?

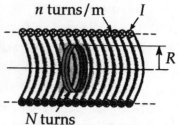

N turns

Figure P31.13

Solution

The solenoid creates a magnetic field

$$B = \mu_0 nI = (4\pi\times10^{-7}\text{ N}/\text{A}^2)(400\text{ turns}/\text{m})(30.0\text{ A})(1-e^{-1.60t})$$

$$B = (1.51\times10^{-2}\text{ N}/\text{m·A})(1-e^{-1.60t})$$

The magnetic flux through one turn of the flat coil is $\Phi_B=\int B\,dA\cos\theta$, but since dA refers to the area perpendicular to the flux, and the magnetic field is uniform over the area A of the flat coil,

$$\Phi_B = B\int dA = B(\pi R^2) = (1.51\times10^{-2}\text{ N}/\text{m·A})(1-e^{-1.60t})[\pi(0.060\,0\text{ m})^2]$$

$$\Phi_B = (1.71\times10^{-4}\text{ N·m}/\text{A})(1-e^{-1.60t})$$

The emf generated in the *N*-turn coil is $\mathcal{E} = -N\,d\Phi_B/dt$:

$$\mathcal{E} = -(250)\left(1.71\times10^{-4}\frac{\text{N·m}}{\text{A}}\right)\frac{d(1-e^{-1.60t})}{dt} = -\left(0.042\,6\frac{\text{N·m}}{\text{A}}\right)((1.60\text{ s}^{-1})e^{-1.60t})$$

$$\mathcal{E} = -(6.82\times10^{-2}\text{ N·m}/\text{C})e^{-1.60t} = -(6.82\times10^{-2}\text{ V})e^{-1.60t} \qquad \Diamond$$

The minus sign indicates that the emf will produce counterclockwise current in the smaller coil, opposite to the direction of the increasing current in the solenoid.

15. A coil formed by wrapping 50 turns of wire in the shape of a square is positioned in a magnetic field so that the normal to the plane of the coil makes an angle of 30.0° with the direction of the field. When the magnetic field is increased uniformly from 200 μT to 600 μT in 0.400 s, an emf of magnitude 80.0 mV is induced in the coil. What is the total length of the wire?

Solution

Conceptualize: If we assume that this square coil is some reasonable size between 1 cm and 1 m across, then the total length of wire would be between 2 m and 200 m.

Categorize: The changing magnetic field will produce an emf in the coil according to Faraday's law of induction. The constant area of the coil can be found from the change in flux required to produce the emf.

Analyze:

By Faraday's law,
$$\mathcal{E} = -N\frac{d\Phi_B}{dt} \quad \text{so} \quad \mathcal{E} = -N\frac{d}{dt}(BA\cos\theta) = -NA\cos\theta\frac{dB}{dt}$$

For magnitudes
$$|\bar{\mathcal{E}}| = NA\cos\theta\left(\frac{\Delta B}{\Delta t}\right) \quad \text{the area is} \quad A = \frac{|\bar{\mathcal{E}}|}{N\cos\theta\left(\frac{\Delta B}{\Delta t}\right)}$$

$$A = \frac{80.0\times10^{-3}\text{ V}}{50(\cos 30.0°)\left(\dfrac{600\times10^{-6}\text{ T} - 200\times10^{-6}\text{ T}}{0.400\text{ s}}\right)} = 1.85\text{ m}^2$$

Each side of the coil has length $d = \sqrt{A}$, so the total length of the wire is

$$L = N(4d) = 4N\sqrt{A} = (4)(50)\sqrt{1.85\text{ m}^2} = 272\text{ m} \qquad \lozenge$$

Finalize: The total length of wire is slightly longer than we predicted. With $d = 1.36$ m, a normal person could easily step through this large coil! As a bit of foreshadowing to a future chapter on AC circuits, an even bigger coil with more turns could be hidden in the ground below high-power transmission lines so that a significant amount of power could be "stolen" from the electric utility. There is a story of one man who did this and was arrested when investigators finally found the reason for a large power loss in the transmission lines!

---

21. Figure P31.20 shows a top view of a bar that can slide without friction. The resistor is 6.00 Ω and a 2.50-T magnetic field is directed perpendicularly downward, into the paper. Let $\ell = 1.20$ m. (a) Calculate the applied force required to move the bar to the right at a constant speed of 2.00 m/s. (b) At what rate is energy delivered to the resistor?

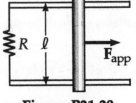

Figure P31.20

Solution

(a) We use the rigid body in equilibrium model. At constant speed, the net force on the moving bar equals zero,

or

$$|F_{app}| = I|\ell \times B|$$

where the current in the bar is $I = \mathcal{E}/R$

and $\mathcal{E} = B\ell v$

Therefore, $F_{app} = \left(\dfrac{B\ell v}{R}\right)\ell B = \dfrac{B^2\ell^2 v}{R}$

$$F_{app} = \frac{(2.50\text{ T})^2(1.20\text{ m})^2(2.00\text{ m}/\text{s})}{6.00\ \Omega} = 3.00\text{ N} \qquad \Diamond$$

(b) $\mathcal{P} = F_{app}v = (3.00\text{ N})(2.00\text{ m}/\text{s}) = 6.00\text{ W}$ $\qquad \Diamond$

In terms of energy, the circuit is a nonisolated system in steady state. The energy, six joules every second, delivered to the circuit by work done by the applied force is delivered to the resistor by electrical transmission. The energy can leave the resistor as energy transferred by heat into the surrounding air. When we studied an electric circuit consisting of a resistor connected across a battery, we did not consider the details of the chemical reaction in the battery and the battery's loss of chemical energy. The circuit in this problem is equally simple electrically. The outside agent pulling the bar through the magnetic field takes the place of the battery. This is the first circuit in which we can account completely for the energy transformations.

---

27. A helicopter has blades of length 3.00 m, extending out from a central hub and rotating at 2.00 rev/s. If the vertical component of the Earth's magnetic field is 50.0 μT, what is the emf induced between the blade tip and the center hub?

Solution   Following Example 31.4,

$$\mathcal{E} = \tfrac{1}{2}B\omega\ell^2 = \tfrac{1}{2}(5.00\times10^{-5}\text{ T})\left(\frac{2.00\text{ rev}}{\text{s}}\right)\left(\frac{2\pi\text{ rad}}{1.00\text{ rev}}\right)(3.00\text{ m})^2 = 2.83\text{ mV} \qquad \Diamond$$

---

29. A rectangular coil with resistance R has N turns, each of length ℓ and width w as shown in Figure P31.29. The coil moves into a uniform magnetic field **B** with constant velocity **v**. What are the magnitude and direction of the total magnetic force on the coil (a) as it enters the magnetic field, (b) as it moves within the field, and (c) as it leaves the field?

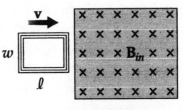

Figure P31.29

Solution

(a) Call x the distance that the leading edge has penetrated into the strong field region. The flux (Bwx away from you) through the coil increases in time so the voltage, of

$$|\varepsilon| = N\frac{d}{dt}Bwx = NBw\frac{dx}{dt} = NBwv$$

is induced in the coil, tending to produce counterclockwise current

$$|I| = \frac{|\varepsilon|}{R} = \frac{NBwv}{R}$$

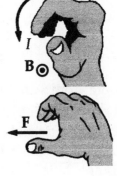

The current upward in the leading edge experiences a force

$$\mathbf{F} = NI\mathbf{L}\times\mathbf{B} = N\left(\frac{NBwv}{R}\right)w\hat{\jmath}\times B(-\hat{k}) = \left(\frac{N^2B^2w^2v}{R}\right)(-\hat{\imath}) \qquad \Diamond$$

This retarding force models eddy-current damping.

(b) The flux through the coil is constant when it is wholly within the high-field region. The induced emf, induced current, and magnetic force are zero. $\qquad \Diamond$

(c) As the coil leaves the field, the away-from-you flux it encloses decreases. The coil carries clockwise current to make some away-from-you field of its own. Again,

$$|I| = \frac{NBwv}{R}$$

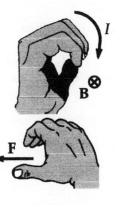

Now the trailing edge carries upward current to experience a force of

$$\mathbf{F} = \left(\frac{N^2B^2w^2v}{R}\right) \text{ to the left} \qquad \Diamond$$

33. A magnetic field directed into the page changes with time according to $B = (0.030\,0t^2 + 1.40)$ T, where t is in seconds. The field has a circular cross section of radius $R = 2.50$ cm (Fig. P31.32). What are the magnitude and direction of the electric field at point P_1 when $t = 3.00$ s and $r_1 = 0.020\,0$ m?

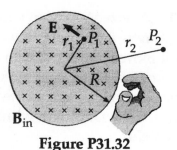

Figure P31.32

Solution $\qquad \oint \mathbf{E} \cdot d\mathbf{s} = -\dfrac{d\Phi_B}{dt}$

Consider a circular integration path of radius r_1

$$E(2\pi r_1) = -\frac{d}{dt}(BA) = -A\left(\frac{dB}{dt}\right)$$

$$|E| = \frac{A}{2\pi r_1}\frac{d}{dt}(0.030\,0t^2 + 1.40)\,\text{T} = \frac{\pi r_1^2}{2\pi r_1}(0.060\,0t) = \frac{r_1}{2}(0.060\,0t)$$

At $t = 3.00$ s, $\qquad E = \frac{1}{2}(0.020\,0 \text{ m})(0.060\,0 \text{ T/sec})(3.00 \text{ sec}) = 1.80 \times 10^{-3}$ N/C $\qquad \Diamond$

If there were a circle of wire of radius r_1, it would enclose increasing magnetic flux due to a magnetic field away from you. It would carry counterclockwise current to make its own magnetic field toward you, to oppose the change. Even without the wire and current, the counterclockwise electric field that would cause the current is lurking. At point P_1, it is upward and to the left, perpendicular to r_1. $\qquad \Diamond$

---

35. A coil of area 0.100 m² is rotating at 60.0 rev/s with the axis of rotation perpendicular to a 0.200-T magnetic field. (a) If the coil has 1 000 turns, what is the maximum emf generated in it? (b) What is the orientation of the coil with respect to the magnetic field when the maximum induced voltage occurs?

Solution For a coil rotating in a magnetic field,

(a) By Equation 31.10, $\qquad \mathcal{E} = NBA\omega \sin\theta$

so $\qquad \mathcal{E}_{max} = NBA\omega$

With these data, $\qquad \mathcal{E}_{max} = (1\,000)(0.200 \text{ T})(0.100 \text{ m}^2)(60.0 \text{ rev/s})(2\pi \text{ rad/rev})$

$$\mathcal{E}_{max} = 7\,540 \text{ V} \qquad \Diamond$$

(b) $|\mathcal{E}|$ approaches $\mathcal{E}_{max}$ when $|\sin\theta|$ approaches 1 or $\theta = \pm\pi/2$.

Therefore, at maximum emf, the plane of the coil is parallel to the field. $\qquad \Diamond$

43. A conducting rectangular loop of mass M, resistance R, and dimensions w by ℓ falls from rest into a magnetic field **B** as shown in Figure P31.43. During the time interval before the top edge of the loop reaches the field, the loop approaches a terminal speed v_T. (a) Show that $v_T = MgR/B^2w^2$. (b) Why is v_T proportional to R? (c) Why is it inversely proportional to B^2?

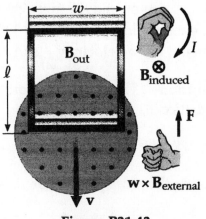

Figure P31.43

Solution Let y represent the average vertical height of the strong-field region above the bottom edge of the loop.

As the loop falls, y increases; the loop encloses increasing flux toward you and has induced in it an emf to produce a current to make its own magnetic field away from you. This current is to the left in the bottom side of the loop, and feels an upward force in the external field.

(a) Symbolically, the flux is

$$\Phi = BA\cos\theta = Bwy\cos 0°$$

The emf is

$$\mathcal{E} = -N\frac{d}{dt}(Bwy) = -Bw\frac{dy}{dt} = -Bwv$$

The magnitude of the current is

$$I = \frac{|\mathcal{E}|}{R} = \frac{Bwv}{R}$$

and the force is

$$\mathbf{F}_B = I\mathbf{w}\times\mathbf{B} = \left(\frac{Bwv}{R}\right)wB\sin 90° = \frac{B^2w^2v}{R}$$

When the loop is moving at terminal speed, we may model it as a rigid body in equilibrium:

$\sum F_y = 0$ becomes

$$\frac{+B^2w^2v_T}{R} - Mg = 0$$

Thus,

$$v_T = \frac{MgR}{B^2w^2} \qquad \diamond$$

(b) The emf is directly proportional to v, but the current is inversely proportional to R. A large R means a small current at a given speed, so the loop must travel faster to get $F_B = $ weight. $\qquad \diamond$

(c) At a given speed, the current is directly proportional to the magnetic field. But the force is proportional to the product of the current and the field. For a small B, the speed must increase to compensate for both the small B and also the current, so

$v_T \propto B^{-2}$. $\qquad \diamond$

45. A proton moves through a uniform electric field $\mathbf{E} = 50.0\hat{\jmath}$ V/m and a uniform magnetic field $\mathbf{B} = (0.200\hat{\imath} + 0.300\hat{\jmath} + 0.400\hat{k})$ T. Determine the acceleration of the proton when it has a velocity $\mathbf{v} = 200\hat{\imath}$ m/s.

Solution The combined force on the proton is the Lorentz force, as described by Equation 31.16:

$$\mathbf{F} = m\mathbf{a} = q\mathbf{E} + q\mathbf{v} \times \mathbf{B} \qquad \text{so that} \qquad \mathbf{a} = \frac{e}{m}[\mathbf{E} + \mathbf{v} \times \mathbf{B}]$$

Taking the cross product of $\mathbf{v}$ and $\mathbf{B}$,

$$\mathbf{v} \times \mathbf{B} = \begin{vmatrix} \hat{\imath} & \hat{\jmath} & \hat{k} \\ 200 & 0 & 0 \\ 0.200 & 0.300 & 0.400 \end{vmatrix} = -200(0.400)\hat{\jmath} + 200(0.300)\hat{k}$$

$$\mathbf{a} = \left(\frac{1.60 \times 10^{-19}}{1.67 \times 10^{-27}}\right)[50.0\hat{\jmath} - 80.0\hat{\jmath} + 60.0\hat{k}] = 9.58 \times 10^{7}[-30\hat{\jmath} + 60\hat{k}]$$

$$\mathbf{a} = (2.87 \times 10^{9})(-\hat{\jmath} + 2\hat{k}) \text{ m/s}^2 = (-2.87 \times 10^{9}\hat{\jmath} + 5.75 \times 10^{9}\hat{k}) \text{ m/s}^2 \qquad \Diamond$$

55. The plane of a square loop of wire with edge length $a = 0.200$ m is perpendicular to the Earth's magnetic field at a point where $B = 15.0$ μT, as shown in Figure P31.55. The total resistance of the loop and the wires connecting it to a sensitive ammeter is 0.500 Ω. If the loop is suddenly collapsed by horizontal forces as shown, what total charge passes through the ammeter?

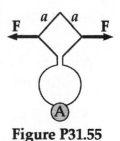

Figure P31.55

Solution

Conceptualize: For the situation described, the maximum current is probably less than 1 mA. So if the loop is closed in 0.1 s, then the total charge would be

$$Q = I\Delta t = (1 \text{ mA})(0.1 \text{ s}) = 100 \text{ } \mu\text{C}$$

Categorize: We do not know how quickly the loop is collapsed, but we can find the total charge by integrating the change in magnetic flux due to the change in area of the loop $(a^2 \rightarrow 0)$.

Analyze:

$$Q = \int I\, dt = \int \frac{\mathcal{E}\, dt}{R} = \frac{1}{R} \int -\left(\frac{d\Phi_B}{dt}\right) dt = -\frac{1}{R} \int d\Phi_B = -\frac{1}{R} \int d(BA) = -\frac{B}{R} \int_{A_1 = a^2}^{A_2 = 0} dA$$

$$Q = -\frac{B}{R} A \bigg]_{A_1 = a^2}^{A_2 = 0} = \frac{Ba^2}{R} = \frac{(15.0 \times 10^{-6}\ \text{T})(0.200\ \text{m})^2}{0.500\ \Omega} = 1.20 \times 10^{-6}\ \text{C} \qquad \lozenge$$

Finalize: The total charge is less than the maximum charge we predicted, so the answer seems reasonable. It is interesting that this charge can be calculated without knowing either the current or the time to collapse the loop.

Note: We ignored the internal resistance of the ammeter. D'Arsonval galvanometers typically have an internal resistance of 50 to 100 Ω, significantly more than the resistance of the wires given in the problem. A proper solution that includes R_G would reduce the total charge by about 2 orders of magnitude ($Q \sim 0.01\ \mu\text{C}$).

---

61. A rectangular coil of 60 turns, dimensions 0.100 m by 0.200 m and total resistance 10.0 Ω, rotates with angular speed 30.0 rad/s about the y axis in a region where a 1.00-T magnetic field is directed along the x axis. The rotation is initiated so that the plane of the coil is perpendicular to the direction of **B** at $t = 0$. Calculate (a) the maximum induced emf in the coil, (b) the maximum rate of change of magnetic flux through the coil, (c) the induced emf at $t = 0.0500$ s, and (d) the torque exerted by the magnetic field on the coil at the instant when the emf is a maximum.

Solution

Let θ represent the angle between the perpendicular to the coil and the magnetic field. Then $\theta = 0$ at $t = 0$ and $\theta = \omega t$ at all later times.

(a) $\quad \mathcal{E} = -N \dfrac{d}{dt}(BA\cos\theta) = -NBA\dfrac{d}{dt}(\cos\omega t) = +NBA\omega\sin\omega t$

$\quad \mathcal{E}_{\text{max}} = NBA\omega = (60)(1.00\ \text{T})(0.020\,0\ \text{m}^2)(30.0\ \text{rad}/\text{s}) = 36.0\ \text{V} \qquad \lozenge$

(b) $\quad \dfrac{d\Phi_B}{dt} = \dfrac{d}{dt}(BA\cos\theta) = BA\omega\sin\omega t$

$\quad \left(\dfrac{d\Phi_B}{dt}\right)_{\text{max}} = BA\omega = (1.00\ \text{T})(0.020\,0\ \text{m}^2)(30.0\ \text{rad}/\text{s}) = 0.600\ \text{T}\cdot\text{m}^2/\text{s} \qquad \lozenge$

(c) $\mathcal{E} = NBA\omega \sin\omega t = (36.0 \text{ V})\left(\sin\left[(30.0 \text{ rad}/\text{s})(0.050\,0\text{ s})\right]\right)$

$\mathcal{E} = (36.0 \text{ V})(\sin(1.50 \text{ rad})) = (36.0 \text{ V})(\sin 85.9°) = 35.9 \text{ V}$ ◊

(d) The emf is maximum when $\theta = 90°$ and $\boldsymbol{\tau} = \boldsymbol{\mu} \times \mathbf{B}$, so

$$\tau_{max} = \mu B \sin 90° = NIAB = N\mathcal{E}_{max}\frac{AB}{R}$$

and $\tau_{max} = (60)(36.0 \text{ V})\dfrac{(0.020\,0\text{ m}^2)(1.00\text{ T})}{10.0\ \Omega} = 4.32 \text{ N}\cdot\text{m}$ ◊

---

65. The magnetic flux through a metal ring varies with time t according to $\Phi_B = 3(at^3 - bt^2)$ T·m^2, with $a = 2.00$ s^{-3} and $b = 6.00$ s^{-2}. The resistance of the ring is 3.00 Ω. Determine the maximum current induced in the ring during the interval from $t = 0$ to $t = 2.00$ s.

Solution

Substituting the given values, $\Phi_B = (6.00t^3 - 18.0t^2)$ T·m^2

Therefore, the emf induced is $\mathcal{E} = -\dfrac{d\Phi_B}{dt} = -18.0t^2 + 36.0t$

The maximum $\mathcal{E}$ occurs when $\dfrac{d\mathcal{E}}{dt} = -36.0t + 36.0 = 0$, which gives $t = 1.00$ s.

Thus, maximum current (at $t = 1.00$ s) is $I_{max} = \dfrac{\mathcal{E}}{R} = \dfrac{(-18.0 + 36.0)\text{ V}}{3.00\ \Omega} = 6.00 \text{ A}$ ◊

---

71. A long, straight wire carries a current that is given by $I = I_{max}\sin(\omega t + \phi)$ and lies in the plane of a rectangular coil of N turns of wire, as shown in Figure P31.9. The quantities I_{max}, ω, and ϕ are all constants. Determine the emf induced in the coil by the magnetic field created by the current in the straight wire. Assume $I_{max} = 50.0$ A, $\omega = 200\pi$ s^{-1}, $N = 100$, $h = w = 5.00$ cm, and $L = 20.0$ cm.

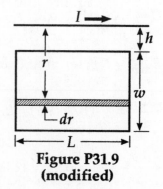

Figure P31.9
(modified)

Solution The coil is the boundary of a rectangular area. The magnetic field produced by the current in the straight wire is perpendicular to the plane of the area at all points. The magnitude of the field is

$$B = \frac{\mu_0 I}{2\pi r}$$

Thus the flux is

$$N\Phi_B = \frac{\mu_0 NL}{2\pi} I \int_h^{h+w} \frac{dr}{r} = \frac{\mu_0 NL}{2\pi} I_{max} \ln\left(\frac{h+w}{h}\right)\sin(\omega t + \phi)$$

Finally, the induced emf is in absolute value

$$\mathcal{E} = -N\frac{d\Phi_B}{dt} = -\frac{\mu_0 NL}{2\pi} I_{max}\omega \ln\left(\frac{h+w}{h}\right)\cos(\omega t + \phi)$$

$$\mathcal{E} = -(4\pi\times10^{-7}\ \text{T}\cdot\text{m}/\text{A})\frac{(100)(0.200\ \text{m})}{2\pi}(50.0\ \text{A})(200\pi\ \text{s}^{-1})\ln\left(\frac{10.0\ \text{cm}}{5.00\ \text{cm}}\right)\cos(\omega t + \phi)$$

$$\mathcal{E} = -(87.1\ \text{mV})\cos(200\pi t\ \text{rad}/\text{s} + \phi) \qquad \diamond$$

Related Comment: The factor $\sin(\omega t + \phi)$ in the expression for the current in the straight wire does not change appreciably when ωt changes by 0.1 rad or less. Thus, the current does not change appreciably during a time interval

$$\Delta t < \frac{0.100}{200\pi\ \text{s}^{-1}} = 1.59\times10^{-4}\ \text{s}$$

We define a critical length,

$$c\Delta t = (3.00\times10^8\ \text{m}/\text{s})(1.59\times10^{-4}\ \text{s}) = 4.77\times10^4\ \text{m}$$

equal to the distance to which field changes could be propagated during an interval of 1.59×10^{-4} s. This length is so much larger than any dimension of the loop or its distance from the wire that, although we consider the straight wire to be infinitely long, we can also safely ignore the field propagation effects in the vicinity of the loop. Moreover, the phase angle can be considered to be constant along the wire in the vicinity of the loop. If the angular frequency ω were much larger, say $200\pi\times10^5\ \text{s}^{-1}$, the corresponding critical length would be only 48 cm. In this situation, propagation effects would be important and the above expression of $\mathcal{E}$ would require modification. As a "rule of thumb," we can consider field propagation effects for circuits of laboratory size to be negligible for frequencies, $f = \omega/2\pi$, that are less than about 10^6 Hz.

Chapter 32
INDUCTANCE

EQUATIONS AND CONCEPTS

A self-induced emf is present in a circuit element (coil, solenoid, toroid, coaxial cable, etc.) when the current in the circuit changes in time. *The inductance, L, of a given device depends on its physical characteristics and is a measure of the opposition of the device to a change in the current. An **inductor** is a circuit element which has a large inductance.*

$$\varepsilon_L = -L\frac{dI}{dt} \qquad (32.1)$$

The SI unit of inductance is the henry, H. A rate of change of current of 1 ampere per second in an inductor of 1 henry will produce a self-induced emf of 1 volt.

$$1\,H = 1\frac{V \cdot s}{A} = 1\,\Omega \cdot s$$

The inductance of a circuit element can be expressed as the ratio of magnetic flux to current (Equation 32.2) or the ratio of induced emf to rate of change of current (Equation 32.3).

$$L = \frac{N\Phi_B}{I} \qquad (32.2)$$

$$L = -\frac{\varepsilon_L}{dI / dt} \qquad (32.3)$$

The inductance of a uniformly wound solenoid is proportional to the square of the number of turns per unit of length, n. In Equation 32.5, A is the cross-sectional area of a solenoid of length ℓ.

$$L = \mu_0 n^2 A\ell \qquad (32.5)$$

A series *RL* circuit is shown in the figure at right. The circuit elements are a battery, resistor, inductor and switch. *The inductor opposes any change in the current in the circuit.*

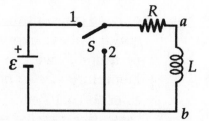

150

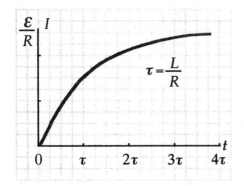

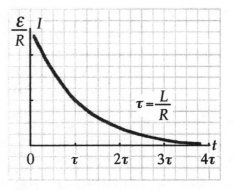

Increase of current (from zero) in the *RL* circuit when the switch is moved to position 1.

Exponential decay of current (from maximum) in the *RL* circuit when the switch is moved from position 1 to position 2.

Current increases in an *RL* circuit (from $I=0$) in a characteristic fashion when the switch in the circuit as shown is moved to position 1. This is shown in the graph above left and described by Equation 32.7. *The maximum current is achieved in a time that is long compared to the time constant, τ.*

$$I = \frac{\mathcal{E}}{R}(1 - e^{-t/\tau}) \qquad (32.7)$$

The time constant of the circuit is the time required for the current to reach 63% of its final value.

$$\tau = L/R \qquad (32.8)$$

Current will decay exponentially (from $I_0 = \mathcal{E}/R$) when the switch in the circuit is moved from position 1 to position 2. The curve of the decay is shown in the graph above right and is described by Equation 32.10.

$$I = \frac{\mathcal{E}}{R}e^{-t/\tau} = I_0 e^{-t/\tau} \qquad (32.10)$$

$$\text{where } \frac{\mathcal{E}}{R} = I_0$$

The stored energy U in the magnetic field of an inductor is proportional to the square of the current in the inductor.

$$U = \frac{1}{2}LI^2 \qquad (32.12)$$

The **energy density** u_B is the energy per unit volume stored in a magnetic field. *This expression is valid for any region of space in which a magnetic field exists.*

$$u_B = \frac{U}{A\ell} = \frac{B^2}{2\mu_0}$$

(32.14)

Mutual inductance, M, is characteristic of a system of two nearby circuits (e.g. coils). The value of the mutual inductance in a particular case depends on the geometry of both circuits and on their relation with respect to each other. *M is a shared property of the pair of two circuits.* In the equations shown, M_{12} is the mutual inductance of coil 2 with respect to coil 1 and Φ_{12} is the magnetic flux through coil 2 due to the current in coil 1. Corresponding notations apply for M_{21}. The unit of mutual inductance is the henry, H.

$$M_{12} \equiv \frac{N_2 \Phi_{12}}{I_1}$$

(32.15)

$$M_{21} = \frac{N_1 \Phi_{21}}{I_2}$$

$$M_{12} = M_{21}$$

The **emf induced in one coil** is always proportional the to rate at which the current in the other coil is changing.

$$\mathcal{E}_2 = -M_{12} \frac{dI_1}{dt}$$

(32.16)

$$\mathcal{E}_1 = -M_{21} \frac{dI_2}{dt}$$

(32.17)

A **series LC circuit** is shown in the figure at right. The charged capacitor can be connected by switch S to the inductor and the circuit is assumed to have zero resistance. When the switch S is closed, the circuit exhibits a continual transfer of energy between the electric field of the capacitor and the magnetic field of the inductor.

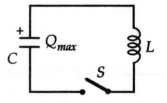

The **natural frequency** of transfer of energy between the capacitor and the inductor depends on the values of inductance and capacitance in the circuit.

$$\omega = \frac{1}{\sqrt{LC}}$$

(32.22)

The **charge on the capacitor and the current in the inductor** vary sinusoidally in time and are 90° out of phase with each other.

Q varies between $+Q_{max}$ and $-Q_{max}$

$$Q = Q_{max} \cos \omega t \quad (32.24)$$

I varies between $+I_{max}$ and $-I_{max}$

$$I = -I_{max} \sin \omega t \quad (32.25)$$

Equations 32.24 and 32.25 are valid for $Q = Q_{max}$ when $t = 0$.

The **total energy in an LC circuit** is shared between the electric field of the capacitor and the magnetic field of the inductor. *If the system has zero resistance and electromagnetic radiation is ignored, the total energy of the system remains constant.*

$$U = U_C + U_L$$

$$U = \frac{Q_{max}^2}{2C} \cos^2 \omega t + \frac{LI_{max}^2}{2} \sin^2 \omega t \quad (32.26)$$

SUGGESTIONS, SKILLS, AND STRATEGIES

Equation 32.2, $L = N\Phi_B / I$ and Equation 32.12, $U = \frac{1}{2}LI^2$, provide two different approaches for the calculation of the inductance L for a particular device.

In order to use Equation 32.2 to calculate L, take the following steps:

- Assume a current I to exist in the conductor for which you wish to calculate L (coil, solenoid, coaxial cable, or other device).
- Calculate the magnetic flux through the appropriate cross section using $\Phi_B = \int \mathbf{B} \cdot d\mathbf{A}$. Remember that in many cases, $\mathbf{B}$ will not be uniform over the area.
- Calculate L directly from the defining Equation 32.2.

In order to use Equation 32.12 to calculate L, take the following steps:

- Assume a current I in the conductor.
- Find an expression for $\mathbf{B}$ for the magnetic field produced by I.
- Use Equation 32.14, $u_B = B^2/2\mu_0$, and integrate this value of u_B over the appropriate volume to find the total energy stored in the magnetic field of the inductor $U_L = \int u \, dV$.
- Substitute this value of U_L into Equation 32.12 and solve for L.

REVIEW CHECKLIST

You should be able to:

▷ Calculate the inductance of a device of suitable geometry. (Section 32.1)

▷ Calculate the magnitude and direction of the self-induced emf in a circuit containing one or more inductive elements when the current changes with time. (Section 32.1)

▷ Determine the current, rate of change of current, time constant, and self induced emf in an LR circuit. (Section 32.2)

▷ Calculate the total magnetic energy stored in a magnetic field. You should be able to perform this calculation if (1) you are given the values of the inductance of the device with which the field is associated and the current in the circuit, or (2) given the value of the magnetic field throughout the region of space in which the magnetic field exists. In the latter case, you must integrate the expression for the energy density u_B over an appropriate volume. (Section 32.3)

▷ Determine the mutual inductance of two coils and calculate the emf induced by mutual inductance in one coil due to a time-varying current in a nearby coil. (Section 32.4)

▷ Calculate the natural frequency of oscillation of an LC circuit. Calculate the value of total energy and determine charge and current as a function of time. (Section 32.5)

ANSWERS TO SELECTED QUESTIONS

2. The current in a circuit containing a coil, resistor, and battery has reached a constant value. Does the coil have an inductance? Does the coil affect the value of the current?

Answer   The coil has an inductance regardless of the nature of the current in the circuit. Inductance depends only on the coil geometry and its construction. Since the current is constant, the self-induced emf in the coil is zero, and the coil does not affect the steady-state current. (We assume the resistance of the coil is negligible.)

□   □   □   □

9. If the current in an inductor is doubled, by what factor does the stored energy change?

Answer   The energy stored in an inductor carrying a current I is given by $U = \frac{1}{2}LI^2$. Therefore, doubling the current will quadruple the energy stored in the inductor.

□   □   □   □

14. In the *LC* circuit shown in Figure 32.16, the charge on the capacitor is sometimes zero, but at such instants the current in the circuit is not zero. How is this possible?

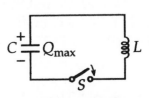

Answer When the capacitor is fully discharged, the current in the circuit is a maximum. The inductance of the coil is making the current continue to flow. At this time the magnetic field of the coil contains all the energy that was originally stored in the charged capacitor.

Figure 32.16

□   □   □   □

SOLUTIONS TO SELECTED PROBLEMS

3. A 2.00-H inductor carries a steady current of 0.500 A. When the switch in the circuit is opened, the current is effectively zero after 10.0 ms. What is the average induced emf in the inductor during this time?

Solution $\varepsilon_L = -L\dfrac{dI}{dt}$

$$\varepsilon_{L,\text{ave}} = -L\left(\frac{I_f - I_0}{t}\right) = (-2.00\text{ H})\left(\frac{0 - 0.500\text{ A}}{1.00 \times 10^{-2}\text{ s}}\right)\left(1\frac{\text{V·s/A}}{\text{H}}\right) = +100\text{ V} \qquad ◊$$

5. A 10.0-mH inductor carries a current $I = I_{\text{max}} \sin \omega t$, with $I_{\text{max}} = 5.00$ A and $\omega/2\pi = 60.0$ Hz. What is the back emf as a function of time?

Solution $\varepsilon_{\text{back}} = -\varepsilon_L = L\dfrac{dI}{dt} = L\dfrac{d}{dt}(I_{\text{max}} \sin \omega t)$

$$\varepsilon_{\text{back}} = L\omega I_{\text{max}} \cos \omega t = (0.0100\text{ H})(120\pi\text{ s}^{-1})(5.00\text{ A})\cos(120\pi t)$$

$$\varepsilon_{\text{back}} = (18.8\text{ V})\cos(377 t) \qquad ◊$$

Chapter 32

7. An inductor in the form of a solenoid contains 420 turns, is 16.0 cm in length, and has a cross-sectional area of 3.00 cm^2. What uniform rate of decrease of current through the inductor induces an emf of 175 μV?

Solution The inductance is $$L = \frac{\mu_0 N^2 A}{\ell}$$

and the emf as a function of time is $$\mathcal{E}_L = -L\frac{dI}{dt}$$

Therefore, rearranging terms, $$\frac{dI}{dt} = \frac{-\mathcal{E}_L}{L} = \frac{-\mathcal{E}_L \ell}{\mu_0 N^2 A}$$

Substituting the given values, we find that

$$\frac{dI}{dt} = \frac{(-175\times10^{-6}\text{ V})(0.160\text{ m})}{(4\pi\times10^{-7}\text{ N/A}^2)(420)^2(3.00\times10^{-4}\text{ m}^2)} = -0.421\text{ A/s} \qquad \Diamond$$

15. A 12.0-V battery is connected into a series circuit containing a 10.0-Ω resistor and a 2.00-H inductor. How long will it take the current to reach (a) 50.0% and (b) 90.0% of its final value?

Solution

Conceptualize: The time constant for this circuit is $\tau = L/R = 0.2$ s, which means that in 0.2 s, the current will reach $1/e = 63\%$ of its final value, as shown in the graph to the right. We can see from this graph that the time to reach 50% of I_{max} should be slightly less than the time constant, perhaps about 0.15 s, and the time to reach $0.9I_{max}$ should be about $2.5\tau = 0.5$ s.

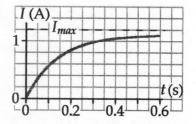

Categorize: The exact times can be found from the equation that describes the rising current in the above graph and gives the current as a function of time for a known emf, resistance, and time constant.

Analyze: At time t after connecting the circuit, $$I(t) = \frac{\mathcal{E}(1-e^{-t/\tau})}{R}$$

where, after a long time, the current $$I_{max} = \frac{\mathcal{E}(1-e^{-\infty})}{R} = \frac{\mathcal{E}}{R}$$

156

(a) At 50% of this maximum value, $\qquad I(t) = 0.500\, I_{max}$

So

$$0.500\frac{\mathcal{E}}{R} = \frac{\mathcal{E}\left(1 - e^{t/0.200\,s}\right)}{R}$$

This then yields $\qquad 0.500 = 1 - e^{-t/0.200\,s}$

Isolating the constants on the right, $\qquad \ln\left(e^{-t/2.00\,s}\right) = \ln(0.500)$

and solving for t, $\qquad -\dfrac{t}{0.200\,s} = -0.693$

Thus, to reach 50% of I_{max} takes: $\qquad t = 0.139\,s$ ◊

(b) Similarly, to reach 90% of I_{max}, $\qquad 0.900 = 1 - e^{-t/\tau}$

so $t = -\tau\ln(1 - 0.900)$ or $\qquad t = -(0.200\,s)\ln(0.100) = 0.461\,s$ ◊

Finalize: The calculated times agree reasonably well with our predictions. We must be careful to avoid confusing the equation for the rising current with the similar equation for the falling current. Checking our answers against predictions is a safe way to prevent such mistakes.

---

19. For the *RL* circuit shown in Figure P32.17, let the inductance be 3.00 H, the resistance 8.00 Ω, and the battery emf 36.0 V. (a) Calculate the ratio of the potential difference across the resistor to that across the inductor when the current is 2.00 A. (b) Calculate the voltage across the inductor when the current is 4.50 A.

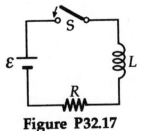

Figure P32.17

Solution

Conceptualize: The voltage across the resistor is proportional to the current, $\Delta V_R = IR$, while the voltage across the inductor is proportional to the **rate of change** in the current, $\mathcal{E}_L = -L\,dI/dt$. When the switch is first closed, the voltage across the inductor will be large as it opposes the sudden change in current. As the current approaches its steady state value, the voltage across the resistor increases and the inductor's emf decreases.

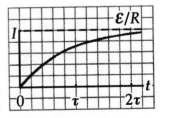

The maximum current will be $\mathcal{E}/R = 4.50\,A$, so when $I = 2.00\,A$, the resistor and inductor will share similar voltages at this mid-range current, but when $I = 4.50\,A$, the entire circuit voltage will be across the resistor, and the voltage across the inductor will be zero.

Categorize: We can use the definition of resistance to calculate the voltage across the resistor for each current. We will find the voltage across the inductor by using Kirchhoff's loop rule.

Analyze:

(a)   When $I = 2.00$ A, the voltage across the resistor is

$$\Delta V_R = IR = (2.00 \text{ A})(8.00 \text{ }\Omega) = 16.0 \text{ V}$$

Kirchhoff's loop rule tells us that the sum of the changes in potential around the loop must be zero:

$$\mathcal{E} - \Delta V_R - \mathcal{E}_L = 36.0 \text{ V} - 16.0 \text{ V} - \mathcal{E}_L = 0$$

so $\mathcal{E}_L = 20.0$ V   and $\dfrac{\Delta V_R}{\mathcal{E}_L} = \dfrac{16.0 \text{ V}}{20.0 \text{ V}} = 0.800$ ◊

(b)   Similarly, for $I = 4.50$ A, $\Delta V_R = IR = (4.50 \text{ A})(8.00 \text{ }\Omega) = 36.0 \text{ V}$

$$\mathcal{E} - \Delta V_R - \mathcal{E}_L = 36.0 \text{ V} - 36.0 \text{ V} - \mathcal{E}_L = 0$$

So $\mathcal{E}_L = 0$ ◊

Finalize: We see that when $I = 2.00$ A, $\Delta V_R < \mathcal{E}_L$, but they are similar in magnitude as expected. Also as predicted, the voltage across the inductor goes to zero when the current reaches its maximum value. A worthwhile exercise would be to consider the ratio of these voltages for several different times after the switch is reopened.

---

27.   A 140-mH inductor and a 4.90-Ω resistor are connected with a switch to a 6.00-V battery as shown in Figure P32.27. (a) If the switch is thrown to the left (connecting the battery), how much time elapses before the current reaches 220 mA? (b) What is the current in the inductor 10.0 s after the switch is closed? (c) Now the switch is quickly thrown from A to B. How much time elapses before the current falls to 160 mA?

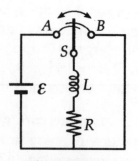

Figure P32.27

Solution The general equation for increasing current in an *LR* circuit is obtained by combining Equations 32.7 and 32.8:

(a)
$$I = \frac{\mathcal{E}(1 - e^{-Rt/L})}{R}$$

or
$$0.220 \text{ A} = \frac{6.00 \text{ V}}{4.90 \text{ }\Omega}\left(1 - e^{-(4.90 \text{ }\Omega)t/0.140 \text{ H}}\right)$$

$$0.180 = 1 - e^{-(35.0 \text{ s}^{-1})t} \qquad \text{or} \qquad e^{(35.0 \text{ s}^{-1})t} = 1.22$$

Thus,
$$t = \frac{\ln 1.22}{35.0 \text{ s}^{-1}} = 5.66 \text{ ms} \qquad\qquad \Diamond$$

(b) Again referring to the general equation,

$$I = \frac{6.00 \text{ V}}{4.90 \text{ }\Omega}\left(1 - e^{(-35.0 \text{ s}^{-1})(10.0 \text{ s})}\right) = 1.22 \text{ A} \qquad\qquad \Diamond$$

(c)
$$0.160 \text{ A} = (1.22 \text{ A})e^{(-4.90 \text{ }\Omega)t/0.140 \text{ H}}$$

$$7.65 = e^{(35.0 \text{ s}^{-1})t} \qquad \text{and} \qquad t = \frac{\ln 7.65}{35.0 \text{ s}^{-1}} = 58.1 \text{ ms} \qquad\qquad \Diamond$$

---

31. An air-core solenoid with 68 turns is 8.00 cm long and has a diameter of 1.20 cm. How much energy is stored in its magnetic field when it carries a current of 0.770 A?

Solution For a solenoid of length ℓ, $L = \dfrac{\mu_0 N^2 A}{\ell}$

Thus, since $U_B = \dfrac{1}{2}LI^2 = \dfrac{\mu_0 N^2 A I^2}{2\ell}$

$$U_B = \frac{(4\pi \times 10^{-7} \text{ N}/\text{A}^2)(68)^2 \pi (6.00 \times 10^{-3} \text{ m})^2 (0.770 \text{ A})^2}{2(0.0800 \text{ m})} = 2.44 \times 10^{-6} \text{ J} \qquad \Diamond$$

---

33. On a clear day at a certain location, a 100-V/m vertical electric field exists near the Earth's surface. At the same place, the Earth's magnetic field has a magnitude of 0.500×10^{-4} T. Compute the energy densities of the two fields.

Solution $u_E = \dfrac{\epsilon_0 E^2}{2} = \dfrac{(8.85 \times 10^{-12} \text{ C}^2/\text{N} \cdot \text{m}^2)(100 \text{ N}/\text{C})^2(1 \text{ J}/\text{N} \cdot \text{m})}{2} = 44.2 \text{ nJ}/\text{m}^3$ ◊

$$u_B = \dfrac{B^2}{2\mu_0} = \dfrac{(5.00 \times 10^{-5} \text{ T})^2}{2(4\pi \times 10^{-7} \text{ T} \cdot \text{m}/\text{A})} = 995 \times 10^{-6} \text{ T} \cdot \text{A}/\text{m}$$

$$u_B = (995 \times 10^{-6} \text{ T} \cdot \text{A}/\text{m})(1 \text{ N} \cdot \text{s}/\text{T} \cdot \text{C} \cdot \text{m})(1 \text{ J}/\text{N} \cdot \text{m}) = 995 \text{ } \mu\text{J}/\text{m}^3 \qquad ◊$$

Magnetic energy density is 22 500 times greater than that in the electric field.

---

43. Two solenoids A and B, spaced close to each other and sharing the same cylindrical axis, have 400 and 700 turns, respectively. A current of 3.50 A in coil A produces an average flux of 300 μWb through each turn of A and a flux of 90.0 μWb through each turn of B. (a) Calculate the mutual inductance of the two solenoids. (b) What is the self-inductance of A? (c) What emf is induced in B when the current in A increases at the rate of 0.500 A/s?

Solution

(a) $M_{12} = \dfrac{N_2 \Phi_{12}}{I_1} = \dfrac{(700)(90 \times 10^{-6} \text{ Wb})}{3.50 \text{ A}} = 18.0 \text{ mH}$ ◊

(b) $L = \dfrac{N\Phi_B}{I} = \dfrac{(400)(300 \times 10^{-6} \text{ Wb})}{3.50 \text{ A}} = 34.3 \text{ mH}$ ◊

(c) $\mathcal{E}_2 = -M_{12} \dfrac{dI_1}{dt} = -(18.0 \times 10^{-3} \text{ H})(0.500 \text{ A}/\text{s}) = -9.00 \text{ mV}$ ◊

---

49. A fixed inductance $L = 1.05$ μH is used in series with a variable capacitor in the tuning section of a radiotelephone on a ship. What capacitance tunes the circuit to the signal from a transmitter broadcasting at 6.30 MHz?

Solution

Conceptualize: It is difficult to predict a value for the capacitance without doing the calculations, but we might expect a typical value in the μF or pF range.

Categorize: We want the resonance frequency of the circuit to match the broadcasting frequency, and for a simple LC circuit, the resonance frequency only depends on the magnitudes of the inductance and capacitance.

Analyze: The resonance frequency is $f_0 = \dfrac{1}{2\pi\sqrt{LC}}$

Thus, $C = \dfrac{1}{(2\pi f_0)^2 L} = \dfrac{1}{\left[2\pi(6.30\times10^6 \text{ Hz})\right]^2(1.05\times10^{-6}\text{ H})} = 608 \text{ pF}$ ◊

Finalize: This is indeed a typical capacitance, so our calculation appears reasonable. You probably would not hear any familiar music on this broadcast frequency. The frequency range for FM radio broadcasting is 88.0–108.0 MHz, and AM radio is 535–1605 kHz. The 6.30 MHz frequency falls in the Maritime Mobile SSB Radiotelephone range, so you might hear a ship captain instead of Top 40 tunes! This and other information about the radio frequency spectrum can be found on the National Telecommunications and Information Administration (NTIA) website, which at the time of this printing was at http://www.ntia.doc.gov/osmhome/allochrt.html

---

53. An LC circuit like that in Figure 32.16 consists of a 3.30-H inductor and an 840-pF capacitor, initially carrying a 105-μC charge. The switch is open for $t<0$ and then closed at $t=0$. Compute the following quantities at $t=2.00$ ms: (a) the energy stored in the capacitor; (b) the energy stored in the inductor; (c) the total energy in the circuit.

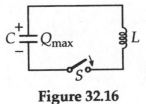

Figure 32.16

Solution At $t=0$ the capacitor charge is at its maximum value, so $\phi=0$ in

$$Q = Q_{max}\cos(\omega t + \phi) = Q_{max}\cos\left(\dfrac{t}{\sqrt{LC}}\right)$$

$$Q = (105\times10^{-6} \text{ C})\cos\left(\dfrac{2.00\times10^{-3} \text{ s}}{\sqrt{(3.30 \text{ H})(840\times10^{-12}\text{ F})}}\right)$$

$$Q = (105\times10^{-6} \text{ C})(\cos 38.0 \text{ rad}) = 1.01\times10^{-4}\text{C}$$

(a) $U_C = \dfrac{Q^2}{2C} = \dfrac{(1.01\times10^{-4}\text{ C})^2}{2(840\times10^{-12}\text{ F})} = 6.03 \text{ J}$ ◊

(c) The constant total energy is that originally of the capacitor:

$$U = \dfrac{Q_{max}^2}{2C} = \dfrac{(1.05\times10^{-4}\text{ C})^2}{2(840\times10^{-12}\text{ F})} = 6.56 \text{ J}$$ ◊

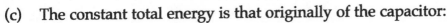

(b) $U_L = 6.56\ \text{J} - 6.03\ \text{J} = 0.529\ \text{J}$ ◊

We could also find this from

$$\tfrac{1}{2}LI^2 = \tfrac{1}{2}L\left(\frac{d}{dt}Q_{max}\cos\omega t\right)^2 = \tfrac{1}{2}LQ_{max}^2\omega^2\sin^2\omega t$$

55. Consider an LC circuit in which $L = 500$ mH and $C = 0.100\ \mu$F. (a) What is the resonance frequency ω_0? (b) If a resistance of 1.00 kΩ is introduced into this circuit, what is the frequency of the (damped) oscillations? (c) What is the percent difference between the two frequencies?

Solution

(a) $\omega_0 = \dfrac{1}{\sqrt{LC}} = \dfrac{1}{\sqrt{(0.50\ \text{H})(1.00\times10^{-7}\ \text{F})}} = 4.47\times10^3\ \text{rad/s}$ ◊

(b) $\omega_d = \sqrt{\dfrac{1}{LC} - \left(\dfrac{R}{2L}\right)^2} = \sqrt{\left(\dfrac{1}{(0.500\ \text{H})(1.00\times10^{-7}\ \text{F})}\right) - \left(\dfrac{1.00\times10^3\ \Omega}{2(0.500\ \text{H})}\right)^2}$

   $\omega_d = 4.36\times10^3\ \text{rad/s}$ ◊

(c) $\dfrac{\Delta\omega}{\omega_0} = \dfrac{4.47 - 4.36}{4.47} = 0.0253 = 2.53\%$ ◊

   Thus, the damped frequency is 2.53% lower than the undamped frequency.

69. At $t = 0$, the open switch in Figure P32.69 is closed. By using Kirchhoff's rules for the instantaneous currents and voltages in this two-loop circuit, show that the current in the inductor at time $t > 0$ is $I(t) = (\mathcal{E}/R_1)[1 - e^{-(R'/L)t}]$ where $R' = R_1R_2/(R_1 + R_2)$.

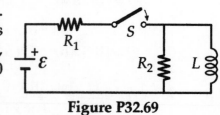

Figure P32.69

Solution

Call I the downward current through the inductor and I_2 the downward current through R_2. Then $I + I_2$ is the current in R_1.

Left-hand loop:
$$\mathcal{E} - (I + I_2)R_1 - I_2R_2 = 0$$

Outside loop:
$$\mathcal{E} - (I + I_2)R_1 - L\frac{dI}{dt} = 0$$

Eliminate I_2, obtaining
$$I\underbrace{\left(\frac{R_1R_2}{R_1 + R_2}\right)}_{R'} + L\frac{dI}{dt} = \underbrace{\left(\frac{R_2}{R_1 + R_2}\right)\mathcal{E}}_{\mathcal{E}'}$$

Thus,
$$\mathcal{E}' - IR' - L\frac{dI}{dt} = 0$$

This is of the same form as Equation 32.6, so the reasoning on page 1007 of the text shows that the solution is the same form as Equation 32.7,

$$I = \frac{\mathcal{E}'}{R'}\left[1 - e^{-R't/L}\right] \qquad \text{with} \qquad \frac{\mathcal{E}'}{R'} = \frac{\mathcal{E}R_2/(R_1 + R_2)}{R_1R_2/(R_1 + R_2)} = \frac{\mathcal{E}}{R_1}:$$

$$I(t) = \frac{\mathcal{E}}{R_1}\left[1 - e^{-(R't/L)}\right] \qquad\qquad \lozenge$$

---

71. In Figure P32.71, the switch is closed at $t<0$, and steady-state conditions are established. The switch is opened at $t=0$. (a) Find the initial voltage $\mathcal{E}_0$ across L just after $t=0$. Which end of the coil is at the higher potential: a or b? (b) Make freehand graphs of the currents in R_1 and in R_2 as a function of time, treating the steady-state directions as positive. Show values before and after $t=0$. (c) How long after $t=0$ does the current in R_2 have the value 2.00 mA?

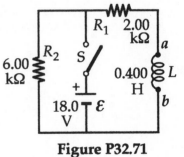

Figure P32.71

Solution Before time $t=0$, the current downward in R_2 is equal to 18.0 V/6.00 kΩ=3.00 mA, and the current clockwise in R_1 and the coil is equal to 18.0 V/2.00 kΩ=9.00 mA.

(a) The inductance prevents instantaneous change of the current in the coil. Just after $t=0$, the current in the outer loop is 9.00 mA clockwise but decreasing. Noting that $\mathcal{E}_0 = V_b - V_a$, the outer loop gives (clockwise):

$$V_b - (9.00 \text{ mA})(6.00 \text{ k}\Omega) - (9.00 \text{ mA})(2.00 \text{ k}\Omega) = V_a$$

$$\mathcal{E}_0 = V_b - V_a = 72.0 \text{ V} \qquad\qquad \lozenge$$

Thus, point b is at the higher potential. The coil creates an emf four times larger than that of the battery. $\qquad\qquad \lozenge$

(b) The currents in R_1 and R_2 are shown below. Just after $t=0$, the current in R_1 decreases from an initial value of 9.00 mA according to $I = I_0 e^{-Rt/L}$. Taking the downward direction as positive in each resistor, the current decreases from +9.00 mA (downward) to zero in R_1. In R_2 the current goes from +3.00 mA (downward) to −9.00 mA (upward) and then decreases in magnitude to zero.   ◊

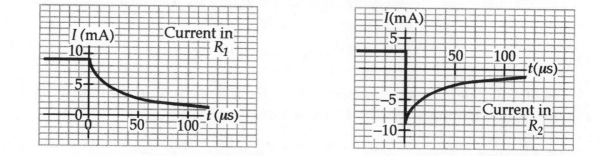

(c) $I = \dfrac{\varepsilon}{R} e^{-Rt/L}$: $2.00 \times 10^{-3}\,\text{A} = \dfrac{18.0\,\text{V}}{2.00\,\text{k}\Omega} e^{(-8.00\,\text{k}\Omega)t/0.400\,\text{H}}$

Resistance R_1 establishes the original value of the current in the outer loop, but the series combination of R_1 and R_2 establishes the decay constant.

$$t = \dfrac{0.400\,\text{H}}{8.00\,\text{k}\Omega} \ln 4.50 = 75.2\ \mu s$$ ◊

---

73. To prevent damage from arcing in an electric motor, a discharge resistor is sometimes placed in parallel with the armature. If the motor is suddenly unplugged while running, this resistor limits the voltage that appears across the armature coils. Consider a 12.0-V DC motor with an armature that has a resistance of 7.50 Ω and an inductance of 450 mH. Assume the back emf in the armature coils is 10.0 V when the motor is running at normal speed. (The equivalent circuit for the armature is shown in Figure P32.73.) Calculate the maximum resistance R that limits the voltage across the armature to 80.0 V when the motor is unplugged.

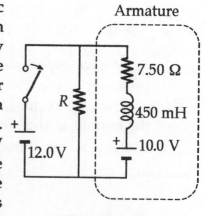

Figure P32.73

Solution

Conceptualize: We should expect R to be significantly greater than the resistance of the armature coil, for otherwise a large portion of the source current would be diverted through R and much of the total power would be wasted on heating this discharge resistor.

Categorize: When the motor is unplugged, the 10-V back emf will still exist for a short while because the motor's inertia will tend to keep it spinning. Now the circuit is reduced to a simple series loop with an emf, inductor, and two resistors. The current that was flowing through the armature coil must now flow through the discharge resistor, which will create a voltage across R that we wish to limit to 80 V. As time passes, the current will be reduced by the opposing back emf, and as the motor slows down, the back emf will be reduced to zero, and the current will stop.

Analyze:

The steady-state coil current when the switch is closed is found from applying Kirchhoff's loop rule to the outer loop:

$$+12.0\,\text{V} - I(7.50\,\Omega) - 10.0\,\text{V} = 0$$

so

$$I = \frac{2.00\,\text{V}}{7.50\,\Omega} = 0.267\,\text{A}$$

We then require that

$$\Delta V_R = 80.0\,\text{V} = (0.267\,\text{A})R$$

so

$$R = \frac{\Delta V_R}{I} = \frac{80.0\,\text{V}}{0.267\,\text{A}} = 300\,\Omega \qquad \Diamond$$

Finalize: As we expected, this discharge resistance is considerably greater than the coil's resistance. Note that while the motor is running, the discharge resistor turns $\mathcal{P} = (12\,\text{V})^2/300\,\Omega = 0.48$ joule in every second, from electrically transmitted energy into internal energy. The power wasted is 0.48 W. The source delivers power at the rate of about $\mathcal{P} = I\Delta V = [0.267\,\text{A} + (12\,\text{V}/300\,\Omega)](12\,\text{V}) = 3.68\,\text{W}$, so the discharge resistor wastes about 13% of the total power. To give a sense of perspective, this 4-W motor could lift a 40-N weight at the rate of 0.1 m/s.

Chapter 33
ALTERNATING-CURRENT CIRCUITS

EQUATIONS AND CONCEPTS

A **series alternating current circuit** with a sinusoidal source of emf is shown in the figure to the right. The rectangle ▭ used here represents the circuit element(s) which, in a particular case, may be a resistor R, a capacitor C, an inductor L, or some combination of the above.

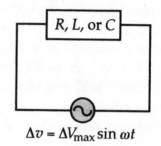

$$\Delta v = \Delta V_{max} \sin \omega t$$

The **applied sinusoidal voltage** of angular frequency ω has a maximum value of ΔV_{max}.

$$\Delta v = \Delta V_{max} \sin \omega t$$

When a **resistor** (R) is the only circuit element, the current and voltage across the resistor are **in phase**, and I_{max} is the maximum current.

$$i_R = \frac{\Delta v_R}{R} = I_{max} \sin \omega t \qquad (33.2)$$

$$\Delta v_R = I_{max} R \sin \omega t \qquad (33.3)$$

When an **inductor** (L) is the only circuit element, the **current lags the voltage** across the inductor by 90°.

$$\Delta v_L = L\frac{di}{dt} = \Delta V_{max} \sin \omega t \qquad (33.6)$$

$$i_L = \frac{\Delta V_{max}}{\omega L} \sin\left(\omega t - \frac{\pi}{2}\right) \qquad (33.8)$$

When a **capacitor** (C) is the only circuit element, **the current leads the voltage across the capacitor by 90°.**

$$\Delta v_C = \Delta V_{max} \sin \omega t \qquad (33.13)$$

$$i_C = \omega C \Delta V_{max} \sin\left(\omega t + \frac{\pi}{2}\right) \qquad (33.16)$$

The maximum value of the current (or current amplitude) through each element is proportional to the amplitude of the AC voltage across the element. In the case of an inductor and a capacitor, the maximum value of the current depends also on the angular frequency of the source of emf.

Resistor:

$$I_{max} = \frac{\Delta V_{max}}{R}$$

Inductor:

$$I_{max} = \frac{\Delta V_{max}}{\omega L} = \frac{\Delta V_{max}}{X_L} \qquad (33.9; 33.11)$$

Capacitor:

$$I_{max} = \frac{\Delta V_{max}}{(1/\omega C)} = \frac{\Delta V_{max}}{X_C} \qquad (33.17; 33.19)$$

The *RLC* circuit, as shown, includes a resistor, inductor, capacitor, and a sinusoidally varying voltage source.

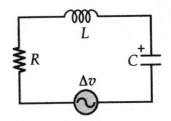

Instantaneous current has the same amplitude and phase at every point in the series circuit. *The phase angle ϕ (defined in Eq. 33.27, following page) is the angle between the current and the applied voltage.*

$$i = I_{max} \sin(\omega t - \phi)$$

The **phase relationship** between the current and instantaneous voltage across R, L, and C can be shown relative to the **common current phase.**

$$\Delta v_R = I_{max} R \sin \omega t = \Delta V_R \sin \omega t \qquad (33.21)$$

$$\Delta v_L = I_{max} X_L \sin\left(\omega t + \frac{\pi}{2}\right) = \Delta V_L \cos \omega t \qquad (33.22)$$

$$\Delta v_C = I_{max} X_C \sin\left(\omega t - \frac{\pi}{2}\right) = -\Delta V_C \cos \omega t \qquad (33.23)$$

The **maximum voltage** across each circuit element can be written in the form of Ohm's law.

$$\Delta V_R = I_{max}R$$

$$\Delta V_L = I_{max}X_L$$

$$\Delta V_C = I_{max}X_C$$

The **inductive reactance** (X_L) and the **capacitive reactance** (X_C) are frequency dependent. *The inductive reactance increases with increasing frequency, while the capacitive reactance decreases with increasing frequency.*

$$X_L = \omega L \tag{33.10}$$

$$X_C = \frac{1}{\omega C} \quad \text{where } \omega = 2\pi f \tag{33.18}$$

The **maximum current** in a series AC circuit depends on the angular frequency ω of the source of emf, as well as the values of ΔV_{max}, R, X_L, and X_C.

$$I_{max} = \frac{\Delta V_{max}}{\sqrt{R^2 + (X_L - X_C)^2}}$$

Impedance, Z, is a parameter of the circuit defined by Equation 33.25. Equation 33.26 relates ΔV_{max} and I_{max} in the form of Ohm's law. The SI unit of impedance is the ohm (Ω).

$$Z \equiv \sqrt{R^2 + (X_L - X_C)^2} \tag{33.25}$$

$$\Delta V_{max} = I_{max}Z \tag{33.26}$$

The **phase angle**, ϕ, can be determined from the impedance triangle of the circuit. It is a measure of the phase difference between the applied voltage and the current in the circuit.

$$\phi = \tan^{-1}\left(\frac{X_L - X_C}{R}\right) \tag{33.27}$$

The **average power** delivered by a generator (source of emf) to an *RLC* series circuit is directly proportional to $\cos \phi$, the **power factor** of the circuit. *There is zero power loss in ideal inductors and capacitors; the average power delivered by the source is converted to internal energy in the resistor.*

$$\mathcal{P}_{av} = I_{rms} \Delta V_{rms} \cos \phi \qquad (33.31)$$

$$\mathcal{P}_{av} = I_{rms}^2 R \qquad (33.32)$$

Root - mean - square (rms) values of current and voltage are those values to which measuring instruments usually respond. *You should notice that ΔV_{rms} and I_{rms} are in the same ratio as ΔV_{max} and I_{max}. Compare this equation to Equation 33.26.*

$$\Delta V_{rms} = \frac{\Delta V_{max}}{\sqrt{2}} \qquad (33.5)$$

$$I_{rms} = \frac{I_{max}}{\sqrt{2}} \qquad (33.4)$$

$$I_{rms} = \Delta V_{rms} / Z \qquad (33.33)$$

The **resonance frequency**, ω_0, is that frequency for which $X_L = X_C$ (and $Z = R$). *At this frequency the current has its maximum value and is in phase with the applied voltage.*

$$\omega_0 = \frac{1}{\sqrt{LC}} \qquad (33.35)$$

A **transformer** consists of a primary coil of N_1 turns and a secondary coil of N_2 turns wound on a common core. In Equation 33.41:

ΔV_1 = voltage across the primary

ΔV_2 = voltage across the secondary

I_1 = current in the primary

I_2 = current in the secondary

$$\Delta V_2 = \frac{N_2}{N_1} \Delta V_1 \qquad (33.41)$$

$$I_1 \Delta V_1 = I_2 \Delta V_2 \qquad (33.42)$$

In the ideal transformer, the ratio of voltages is equal to the ratio of turns, and the ratio of currents is equal to the inverse of the ratio of turns.

SUGGESTIONS, SKILLS, AND STRATEGIES

The **phasor diagram** is a very useful technique to use in the analysis of RLC circuits. In such a diagram, each of the rotating quantities ΔV_R, ΔV_L, ΔV_C, and I_{max} is represented by a separate phasor (rotating vector). A 'phasor' diagram which describes the AC circuit of figure (a) below is shown in figure (b). Each phasor has a length which is proportional to the magnitude of the voltage or current which it represents and rotates counterclockwise about the common origin with an angular frequency which equals the angular frequency of the alternating source, ω. The direction of the phasor which represents the current in the circuit is used as the **reference direction** to establish the correct phase differences among the phasors, which represent the voltage drops across the resistor, inductor, and capacitor. The **instantaneous values** Δv_R, Δv_L, Δv_C, and i are given by the **projection onto the vertical axis of the corresponding phasor**.

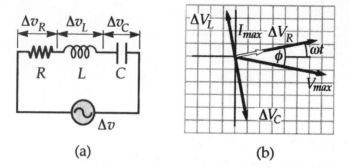

(a)   (b)

Consider the phasor diagram in figure (b) above, where the maximum voltage across the resistor, ΔV_R, is greater than the maximum voltage across the inductor, ΔV_L. At the instant shown, the **instantaneous** value of the voltage across the inductor is greater than that across the resistor. Also notice that as time increases and the phasors rotate counterclockwise, maintaining their constant relative phase, ΔV_R, ΔV_L, and ΔV_C (the voltage amplitudes) will remain constant in magnitude but the instantaneous values Δv_R, Δv_L, and Δv_C will vary sinusoidally with time. For the case shown in figure (b), the phase angle ϕ is negative (this is because $X_C > X_L$ and therefore $\Delta V_C > \Delta V_L$); hence, the current in the circuit leads the applied voltage in phase.

The maximum voltage across each element in the circuit is the product of I_{max} and the resistance or reactance of that component.

The following procedures are recommended when solving alternating current problems:

- The first step in analyzing alternating current circuits is to calculate as many of the unknown quantities such as X_L and X_C as possible. (Note that when calculating X_C, the capacitance should be expressed in farads, rather than, say, microfarads).

- Apply the equation $\Delta V = IZ$ to that portion of the circuit of interest. That is, if you want to know the voltage drop across the combination of an inductor and a resistor, the equation reduces to $\Delta V = I\sqrt{R^2 + X_L^2}$.

REVIEW CHECKLIST

You should be able to:

▷ Calculate instantaneous values of current and voltage for an AC source applied to a resistor, inductor, or capacitor. (Sections 33.1, 33.2, 33.3 and 33.4)

▷ Calculate (when values of resistance, inductance, capacitance, and the characteristics of the generator are known): (i) the instantaneous and rms voltage drop across each component, (ii) the instantaneous and rms current in the circuit, (iii) the phase angle by which the current leads or lags the voltage, (iv) the power expended in the circuit, and (v) the resonance frequency of the circuit. (Sections 33.5, 33.6 and 33.7)

▷ Construct and use phasor diagrams for the description and analysis of AC circuits. (Sections 33.2, 33.3, 33.4 and 33.5)

▷ Sketch circuit diagrams for high-pass and low-pass filter circuits; calculate the ratio of output to input voltage for given circuit parameters. (Section 33.8)

▷ Make calculations of primary to secondary voltage and current ratios for an ideal transformer. (Section 33.9)

ANSWERS TO SELECTED QUESTIONS

9. Does the phase angle depend on frequency? What is the phase angle when the inductive reactance equals the capacitive reactance?

Answer Yes. Since the phase angle is a function of the reactance, which depends on frequency, it must be frequency dependent. The phase angle is zero when the inductive reactance equals the capacitive reactance.

□ □ □ □

18. Will a transformer operate if a battery is used for the input voltage across the primary? Explain.

Answer No. A voltage can only be induced in the secondary coil if the flux through the core changes in time.

□ □ □ □

SOLUTIONS TO SELECTED PROBLEMS

5. The current in the circuit shown in Figure 33.2 equals 60.0% of the peak current at $t = 7.00$ ms. What is the smallest frequency of the source that gives this current?

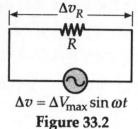

$\Delta v = \Delta V_{max} \sin \omega t$

Figure 33.2

Solution $i = \dfrac{\Delta v}{R} = \left(\dfrac{\Delta V_{max}}{R}\right)\sin \omega t$

$$\frac{0.600\Delta V_{max}}{R} = \left(\frac{\Delta V_{max}}{R}\right)\sin[\omega(7.00 \text{ ms})]$$

$$0.600 = \sin[\omega(7.00 \text{ ms})]$$

To find the lowest frequency we choose the smallest angle satisfying this relation:

$$\omega(7.00 \text{ ms}) = 36.9° = 0.644 \text{ rad}$$

Thus, $\omega = 91.9 \text{ rad/s}$ and $f = \dfrac{\omega}{2\pi} = 14.6 \text{ cycle/s}$ ◊

9. In a purely inductive AC circuit, as shown in Figure 33.6, $\Delta V_{max} = 100$ V. (a) The maximum current is 7.50 A at 50.0 Hz. Calculate the inductance L. (b) **What if?** At what angular frequency ω is the maximum current 2.50 A?

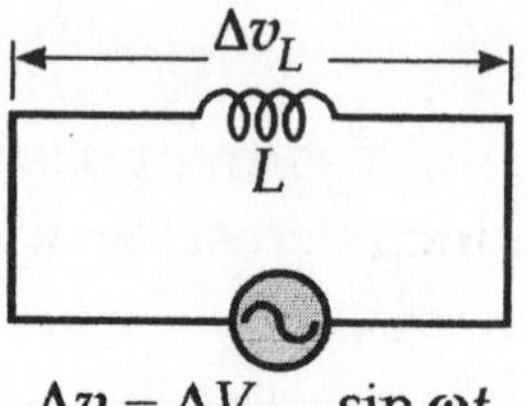

$\Delta v = \Delta V_{max} \sin \omega t$

Figure 33.6

Solution

$$I_{max} = \Delta V_{max} / X_L$$

(a) $X_L = \dfrac{\Delta V_{max}}{I_{max}} = \dfrac{100 \text{ V}}{7.50 \text{ A}} = 13.3\ \Omega = \omega L:$ $L = \dfrac{13.3\ \Omega}{\omega} = \dfrac{13.3\ \Omega}{2\pi(50.0 \text{ s}^{-1})} = 42.4 \text{ mH}$ ◊

(b) $X_L = \dfrac{\Delta V_{max}}{I_{max}} = \dfrac{100 \text{ V}}{2.50 \text{ A}} = 40.0\ \Omega = \omega L:$ $\omega = \dfrac{40.0\ \Omega}{42.4 \text{ mH}} = 942 \text{ rad/s}$ ◊

A frequency 3 times higher makes the inductive reactance 3 times larger.

11. For the circuit shown in Figure 33.6, $\Delta V_{max} = 80.0$ V, $\omega = 65.0\pi$ rad/s, and $L = 70.0$ mH. Calculate the current in the inductor at $t = 15.5$ ms.

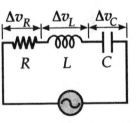

$\Delta v = \Delta V_{max} \sin \omega t$

Figure 33.6

Solution $X_L = \omega L = (65.0\pi \text{ s}^{-1})(70.0 \times 10^{-3} \text{ V·s / A}) = 14.3 \, \Omega$

$$I_{max} = \frac{\Delta V_{max}}{X_L} = \frac{80.0 \text{ V}}{14.3 \, \Omega} = 5.60 \text{ A}$$

$$I = -I_{max} \cos \omega t = -(5.60 \text{ A})\cos\left[(65.0\pi \text{ s}^{-1})(0.0155 \text{ s})\right]$$

$$I = -(5.60 \text{ A})\cos(3.17 \text{ rad}) = +5.60 \text{ A} \qquad \lozenge$$

---

17. What maximum current is delivered by an AC source with $\Delta V_{max} = 48.0$ V and $f = 90.0$ Hz when connected across a 3.70-μF capacitor?

Solution $I_{max} = \dfrac{\Delta V_{max}}{X_C} = \Delta V_{max}\omega C = \Delta V_{max}(2\pi f C)$

$$I_{max} = (48.0 \text{ V})(2\pi)(90.0 \text{ Hz})(3.70 \times 10^{-6} \text{ F}) = 0.100 \text{ A} = 100 \text{ mA} \qquad \lozenge$$

---

19. An inductor ($L = 400$ mH), a capacitor ($C = 4.43 \, \mu$F), and a resistor ($R = 500 \, \Omega$) are connected in series. A 50.0-Hz AC source produces a peak current of 250 mA in the circuit. (a) Calculate the required peak voltage ΔV_{max}. (b) Determine the phase angle by which the current leads or lags the applied voltage.

Solution We first find the impedance of the capacitor and the inductor:

(a) $X_L = \omega L = 2\pi(50.0 \text{ Hz})(400 \times 10^{-3} \text{ H}) = 126 \, \Omega$

and $X_C = \dfrac{1}{\omega C} = \dfrac{1}{2\pi(50.0 \text{ Hz})(4.43 \times 10^{-6} \text{ F})} = 719 \, \Omega$

Then, we substitute these values into the equation for a series LRC circuit:

$$\Delta V_{max} = I_{max} Z = I_{max}\sqrt{R^2 + (X_L - X_C)^2}$$

Thus, $Z = \sqrt{(500 \, \Omega)^2 + (126 \, \Omega - 719 \, \Omega)^2} = 776 \, \Omega$

and $\Delta V_{max} = I_{max} Z = (0.250 \text{ A})(776 \, \Omega) = 194 \text{ V} \qquad \lozenge$

(b) $$\tan\phi = \frac{X_L - X_C}{R} \qquad \text{so} \qquad \phi = \tan^{-1}\left(\frac{126 - 719}{500}\right) = -49.9° \qquad \lozenge$$

The current **leads** the voltage by 49.9°.

---

23. An *RLC* circuit consists of a 150-Ω resistor, a 21.0-μF capacitor, and a 460-mH inductor, connected in series with a 120-V, 60.0-Hz power supply. (a) What is the phase angle between the current and the applied voltage? (b) Which reaches its maximum earlier, the current or the voltage?

Solution The reactance of the inductor is

$$X_L = \omega L = 2\pi f L = 2\pi (60.0 \text{ s}^{-1})(0.460 \text{ H}) = 173 \text{ }\Omega$$

The reactance of the capacitor is

$$X_C = \frac{1}{\omega C} = \frac{1}{2\pi f C} = \frac{1}{2\pi (60.0 \text{ s}^{-1})(21.0 \times 10^{-6} \text{ F})} = 126 \text{ }\Omega$$

(a) $$\tan\phi = \frac{X_L - X_C}{R} = \frac{173 \text{ }\Omega - 126 \text{ }\Omega}{150 \text{ }\Omega} = 0.314$$

so $\phi = 0.304 \text{ rad} = 17.4°$ $\qquad \lozenge$

(b) Since $X_L > X_C$, ϕ is positive, so Δv leads the current. $\qquad \lozenge$

---

31. An AC voltage of the form $\Delta v = (100 \text{ V})\sin(1\,000\,t)$ is applied to a series *RLC* circuit. Assume the resistance is 400 Ω, the capacitance is 5.00 μF, and the inductance is 0.500 H. Find the average power delivered to the circuit.

Solution

Conceptualize: Comparing $\Delta v = (100 \text{ V})\sin(1\,000t)$ with $\Delta v = \Delta V_{max}\sin \omega t$, we see that $\Delta V_{max} = 100 \text{ V}$ and $\omega = 1\,000 \text{ s}^{-1}$. Only the resistor takes electric energy out of the circuit, but the capacitor and inductor will impede the current flow and therefore reduce the voltage across the resistor. Because of this impedance, the average power dissipated by the resistor must be less than the power the source would deliver if it were connected directly across the resistor:

$$\mathcal{P}_{av} = \frac{(\Delta V_{max})^2}{2R} = \frac{(100 \text{ V})^2}{2(400 \text{ }\Omega)} = 12.5 \text{ W}$$

Categorize: The average power dissipated by the resistor can be found from $\mathcal{P}_{av} = I_{rms}^2 R$, where $I_{rms} = \Delta V_{rms}/Z$.

Analyze: $\Delta V_{rms} = \dfrac{100}{\sqrt{2}} = 70.7\text{ V}$

In order to calculate the impedance, we first need the capacitive and inductive reactances:

$$X_C = \frac{1}{\omega C} = \frac{1}{(1000\text{ s}^{-1})(5.00 \times 10^{-6}\text{ F})} = 200\ \Omega$$

$$X_L = \omega L = (1000\text{ s}^{-1})(0.500\text{ H}) = 500\ \Omega$$

Thus, $\quad Z = \sqrt{R^2 + (X_L - X_C)^2}: \qquad\qquad Z = \sqrt{(400\ \Omega)^2 + (500\ \Omega - 200\ \Omega)^2} = 500\ \Omega$

$$I_{rms} = \frac{\Delta V_{rms}}{Z} = \frac{70.7\text{ V}}{500\ \Omega} = 0.141\text{ A}: \quad \mathcal{P}_{av} = I_{rms}^2 R = (0.141\text{ A})^2 (400\ \Omega) = 8.00\text{ W} \quad \Diamond$$

Finalize: The power dissipated by the resistor is less than 12.5 W, so our answer appears to be reasonable. As with other *RLC* circuits, the power will be maximized at the resonance frequency where $X_L = X_C$ so that $Z = R$. Then the average power dissipated will simply be the 12.5 W we calculated first.

---

33. In a certain series *RLC* circuit, $I_{rms} = 9.00\text{ A}$, $\Delta V_{rms} = 180\text{ V}$, and the current leads the voltage by 37.0°. (a) What is the total resistance of the circuit? (b) Calculate the reactance of the circuit $(X_L - X_C)$.

Solution The power is $\quad \mathcal{P}_{av} = I_{rms} \Delta V_{rms} \cos\phi = I_{rms}^2 R$

(a) Therefore, $\qquad\qquad R = \dfrac{\Delta V_{rms} \cos\phi}{I_{rms}} = \dfrac{(180\text{ V})\cos(-37.0°)}{9.00\text{ A}} = 16.0\ \Omega \qquad \Diamond$

(b) $\quad \tan\phi = \dfrac{X_L - X_C}{R}: \qquad X_L - X_C = R\tan\phi = (16.0\ \Omega)\tan(-37.0°) = -12.0\ \Omega \qquad \Diamond$

---

37. An *RLC* circuit is used in a radio to tune into an FM station broadcasting at 99.7 MHz. The resistance in the circuit is 12.0 Ω, and the inductance is 1.40 μH. What capacitance should be used?

Solution The circuit is to be in resonance when $\omega L = \dfrac{1}{\omega C}$

$$C = \frac{1}{\omega^2 L} = \frac{1}{4\pi^2 f^2 L} = \frac{1}{4\pi^2 (99.7 \text{ MHz})^2 (1.40 \ \mu\text{V·s}/\text{A})} = 1.82 \text{ pF} \qquad \Diamond$$

45. A transformer has $N_1 = 350$ turns and $N_2 = 2\,000$ turns. If the input voltage is $\Delta v(t) = (170 \text{ V}) \cos \omega t$, what rms voltage is developed across the secondary coil?

Solution $\Delta V_{1,\text{rms}} = \dfrac{170 \text{ V}}{\sqrt{2}} = 120 \text{ V}$

$$\Delta V_{2,\text{rms}} = \frac{N_2}{N_1} \Delta V_{1,\text{rms}} = \frac{2\,000}{350} (120 \text{ V}) = 687 \text{ V} \qquad \Diamond$$

53. The *RC* high-pass filter shown in Figure 33.25 has a resistance $R = 0.500$ Ω. (a) What capacitance gives an output signal that has half the amplitude of a 300-Hz input signal? (b) What is the ratio $(\Delta V_{\text{out}}/\Delta V_{\text{in}})$ for a 600-Hz signal? You may use the result of Problem 51.

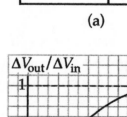

(a)

Solution

Conceptualize: It is difficult to estimate the capacitance required without actually calculating it, but we might expect a typical value in the μF to pF range. The nature of a high-pass filter is to yield a larger gain at higher frequencies, so if this circuit is designed to have a gain of 0.5 at 300 Hz, then it should have a higher gain at 600 Hz. We might guess this is near 1.0 based on the graph above.

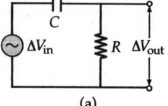

(b)

Figure 33.25

Categorize: The output voltage of this circuit is taken across the resistor, but the input sees the impedance of the resistor and the capacitor. Therefore, the gain will be the ratio of the resistance to the impedance.

Analyze: $\dfrac{\Delta V_{out}}{\Delta V_{in}} = \dfrac{R}{\sqrt{R^2 + (1/\omega C)^2}}$

(a) When $\Delta V_{out} / \Delta V_{in} = 0.500$, solving for C gives

$$C = \frac{1}{\omega R \sqrt{(\Delta V_{in}/\Delta V_{out})^2 - 1}} = \frac{1}{2\pi(300\ \text{Hz})(0.500\ \Omega)\sqrt{(2.00)^2 - 1}} = 613\ \mu\text{F} \qquad \lozenge$$

(b) At 600 Hz, we have $\omega = (2\pi\ \text{rad})(600\ \text{s}^{-1})$, so

$$\frac{\Delta V_{out}}{\Delta V_{in}} = \frac{0.500\ \Omega}{\sqrt{(0.500\ \Omega)^2 + \left(1/(1200\pi\ \text{rad}/\text{s})(613\ \mu\text{F})\right)^2}} = 0.756 \qquad \lozenge$$

Finalize: The capacitance value seems reasonable, but the gain is considerably less than we expected. Based on our calculation, we can modify the above graph to more transparently represent the characteristics of this high-pass filter. If this were an audio filter, it would reduce low frequency "humming" sounds while allowing high pitch sounds to pass through. A low pass filter would be needed to reduce high frequency "static" noise.

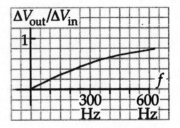

---

57. A series RLC circuit consists of an 8.00-Ω resistor, a 5.00-μF capacitor, and a 50.0-mH inductor. A variable frequency source applies emf 400 V (rms) across the combination. Determine the power delivered to the circuit when the frequency is equal to half the resonance frequency.

Solution

Conceptualize: Maximum power is delivered at the resonance frequency, and the power delivered at other frequencies depends on the quality factor, Q. For the relatively small resistance in this circuit, we could expect a high $Q = \omega_0 L / R$. So at half the resonant frequency, the power should be a small fraction of the maximum power,

$$\mathcal{P}_{av,\text{max}} = \Delta V_{rms}^2 / R = (400\ \text{V})^2 / 8\ \Omega = 20\ \text{kW}$$

$(Q \approx 12.5)$

Categorize: We must first calculate the resonance frequency in order to find half this frequency. Then the power delivered by the source must equal the power taken out by the resistor. This power can be found from $\mathcal{P}_{av} = I_{rms}^2 R$, where $I_{rms} = \Delta V_{rms}/Z$.

Analyze: The resonance frequency is

$$f_0 = \frac{1}{2\pi\sqrt{LC}} = \frac{1}{2\pi\sqrt{(0.050\ 0\ \text{H})(5.00\times10^{-6}\ \text{F})}} = 318\ \text{Hz}$$

The operating frequency is $f = f_0/2 = 159\ \text{Hz}$. We can calculate the impedance at this frequency:

$$X_L = 2\pi fL = 2\pi(159\ \text{Hz})(0.050\ 0\ \text{H}) = 50.0\ \Omega$$

$$X_C = \frac{1}{2\pi fC} = \frac{1}{2\pi(159\ \text{Hz})(5.00\times10^{-6}\ \text{F})} = 200\ \Omega$$

$$Z = \sqrt{R^2 + (X_L - X_C)^2} = \sqrt{8.00^2 + (50.0 - 200)^2}\ \Omega = 150\ \Omega$$

So, $\quad I_{rms} = \dfrac{\Delta V_{rms}}{Z} = \dfrac{400\ \text{V}}{150\ \Omega} = 2.66\ \text{A}$

The power delivered by the source is the power dissipated by the resistor:

$$\mathcal{P}_{av} = I_{rms}^2 R = (2.66\ \text{A})^2(8.00\ \Omega) = 56.7\ \text{W} \qquad\qquad \lozenge$$

Finalize: This power is only about 0.3% of the 20 kW peak power delivered at the resonance frequency. The significant reduction in power for frequencies away from resonance is a consequence of the relatively high Q-factor of about 12.5 for this circuit. A high Q is beneficial if, for example, you want to listen to your favorite radio station that broadcasts at 101.5 MHz, and you do not want to receive the signal from another local station that broadcasts at 101.9 MHz.

---

65. Consider a series *RLC* circuit having the following circuit parameters: $R = 200\ \Omega$, $L = 663\ \text{mH}$, and $C = 26.5\ \mu\text{F}$. The applied voltage has an amplitude of 50.0 V and a frequency of 60.0 Hz. Find the following amplitudes: (a) The current I_{max}, including its phase constant ϕ relative to the applied voltage Δv; (b) the voltage ΔV_R across the resistor and its phase relative to the current; (c) the voltage ΔV_C across the capacitor and its phase relative to the current; and (d) the voltage ΔV_L across the inductor and its phase relative to the current.

Solution   We identify that

$$R = 200\ \Omega,\ \ L = 663\ \text{mH},\ \ C = 26.5\ \mu\text{F},\ \ \omega = 377\ \text{rad/s},\ \ \text{and}\ \ \Delta V_{max} = 50.0\ \text{V}$$

So $\omega L = 250\ \Omega$,   and $1/\omega C = 100\ \Omega$

The impedance is

$$Z = \sqrt{R^2 + \left(\omega L - \frac{1}{\omega C}\right)^2} = \sqrt{(200\ \Omega)^2 + (250\ \Omega - 100\ \Omega)^2} = 250\ \Omega$$

(a) $I_{max} = \dfrac{\Delta V_{max}}{Z} = \dfrac{50.0\ \text{V}}{250\ \Omega} = 0.200\ \text{A}$ ◊

$\phi = \tan^{-1}\left(\dfrac{X_L - X_C}{R}\right) = 36.8°$ with Δv leading i ◊

(b) $\Delta V_R = I_{max}R = 40.0\ \text{V}$ at $\phi = 0°$ ◊

(c) $\Delta V_C = I_{max}X_C = (0.200\ \text{A})(100\ \Omega) = 20.0\ \text{V}$ at $\phi = -90.0°$ ◊

(d) $\Delta V_L = I_{max}X_L = (0.200\ \text{A})(250\ \Omega) = 50.0\ \text{V}$ at $\phi = 90.0°$ ◊

---

67.   Impedance matching. Example 28.2 showed that maximum power is transferred when the internal resistance of a DC source is equal to the resistance of the load. A transformer may be used to provide maximum power transfer between two AC circuits that have different impedances. (a) Show that the ratio of turns N_1/N_2 needed to meet this condition is

$$N_1/N_2 = \sqrt{Z_1/Z_2}$$

(b) Suppose you want to use a transformer as an impedance-matching device between an audio amplifier that has an output impedance of $8.00\ \text{k}\Omega$ and a speaker that has an input impedance of $8.00\ \Omega$. What should your N_1/N_2 ratio be?

Solution $\dfrac{N_1}{N_2} = \dfrac{\Delta V_1}{\Delta V_2}$

with $Z_1 = \dfrac{\Delta V_1}{I_1}$ and $Z_2 = \dfrac{\Delta V_2}{I_2}$

Thus, $\dfrac{N_1}{N_2} = \dfrac{Z_1 I_1}{Z_2 I_2}$

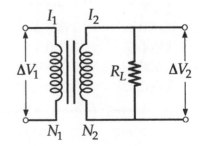

(a)   Since $\dfrac{I_1}{I_2} = \dfrac{N_2}{N_1}$ we find $\dfrac{N_1}{N_2} = \sqrt{\dfrac{Z_1}{Z_2}}$ ◊

(b) $\dfrac{N_1}{N_2} = \sqrt{\dfrac{8\,000\ \Omega}{8.00\ \Omega}} = 31.6$ ◊

72.   A series *RLC* circuit in which $R = 1.00\ \Omega$, $L = 1.00$ mH, and $C = 1.00$ nF is connected to an AC source delivering 1.00 V (rms). Make a precise graph of the power delivered to the circuit as a function of the frequency and verify that the full width of the resonance peak at half-maximum is $R/2\pi L$.

Solution

At resonance, $\omega = \dfrac{1}{\sqrt{LC}} = \dfrac{1}{\sqrt{(1.00 \times 10^{-3}\ \text{H})(1.00 \times 10^{-9}\ \text{F})}} = 1.00 \times 10^6\ \text{rad/s}$

At that point $Z = R = 1.00\ \Omega$

and $I = \dfrac{1.00\ \text{V}}{1.00\ \Omega} = 1.00\ \text{A}$

and the power is $I^2 R = (1.00\ \text{A})^2 (1.00\ \Omega) = 1.00\ \text{W}$

We compute the power at some other angular frequencies. Thus,

ω, 10^6 rad/s	ωL, Ω	$1/\omega C$, Ω	Z, Ω	I, A	$I^2 R = \mathcal{P}$, W
0.999 0	999	1 001	2.24	0.447	0.199 84
0.999 4	999.4	1 000.6	1.56	0.640	0.409 69
0.999 5	999.5	1 000.5	1.41	0.707	0.499 87
0.999 6	999.6	1 000.4	1.28	0.781	0.609 66
0.999 8	999.8	1 000.2	1.08	0.928	0.862 05
1	1 000	1 000	1	1	1
1.000 2	1 000.2	999.8	1.08	0.928	0.862 09
1.000 4	1 000.4	999.6	1.28	0.781	0.609 85
1.000 5	1 000.5	999.5	1.41	0.707	0.500 12
1.000 6	1 000.6	999.4	1.56	0.640	0.409 98
1.001	1 001	999	2.24	0.447	0.200 16

The angular frequencies giving half the maximum power are

$$0.999\,5\times10^6 \text{ rad/s}$$

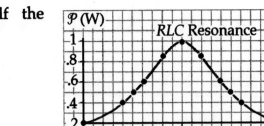

and $\qquad 1.000\,5\times10^6 \text{ rad/s,}$

so the full width at half the maximum is

$$\Delta\omega = (1.000\,5 - 0.999\,5)\times10^6 \text{ rad/s}$$

$$\Delta\omega = 1.00\times10^3 \text{ rad/s}$$

Since $\quad \Delta\omega = 2\pi\Delta f,$

$$\Delta f = 159 \text{ Hz} \qquad \text{and} \qquad \frac{R}{2\pi L} = \frac{1.00\,\Omega}{2\pi(1.00\times10^{-3}\text{ H})} = 159 \text{ Hz} \quad \text{(They agree.)} \quad \Diamond$$

Chapter 34

ELECTROMAGNETIC WAVES

EQUATIONS AND CONCEPTS

Maxwell's equations are the fundamental laws governing the behavior of electric and magnetic fields. Electromagnetic waves are a natural consequence of these laws. You should notice that the integrals in Equations 34.1 and 34.2 are **surface integrals** in which the normal components of electric and magnetic fields are integrated over a **closed surface**. Equations 34.3 and 34.4 involve line integrals in which the tangential components of electric and magnetic fields are integrated around a **closed path**.

$$\oint \mathbf{E} \cdot d\mathbf{A} = \frac{Q}{\epsilon_0} \tag{34.1}$$

$$\oint \mathbf{B} \cdot d\mathbf{A} = 0 \tag{34.2}$$

$$\oint \mathbf{E} \cdot d\mathbf{s} = -\frac{d\Phi_B}{dt} \tag{34.3}$$

$$\text{where } \Phi_B = \int \mathbf{B} \cdot d\mathbf{A}$$

$$\oint \mathbf{B} \cdot d\mathbf{s} = \mu_0 I + \mu_0 \epsilon_0 \frac{d\Phi_E}{dt} \tag{34.4}$$

$$\text{where } \Phi_E = \int \mathbf{E} \cdot d\mathbf{A}$$

The **wave equations for electromagnetic waves in free space** (where $Q=0$ and $I=0$) as stated here, represent linearly polarized waves traveling with a speed c. *Both* **E** *and* **B** *satisfy a differential equation which has the form of the general wave equation.*

$$\frac{\partial^2 E}{\partial x^2} = \mu_0 \epsilon_0 \frac{\partial^2 E}{\partial t^2} \tag{34.8}$$

$$\frac{\partial^2 B}{\partial x^2} = \mu_0 \epsilon_0 \frac{\partial^2 B}{\partial t^2} \tag{34.9}$$

The **speed of electromagnetic waves** in vacuum is the same as the speed of light in vacuum.

$$c = \frac{1}{\sqrt{\mu_0 \epsilon_0}} \tag{34.10}$$

The **electric and magnetic fields** of an electromagnetic wave of a particular frequency vary in position and time as sinusoidal transverse waves. *Their planes of vibration are perpendicular to each other and perpendicular to the direction of propagation.*

$$E = E_{max} \cos(kx - \omega t) \tag{34.11}$$

$$B = B_{max} \cos(kx - \omega t) \tag{34.12}$$

The **ratio of the magnitudes** of the electric and magnetic fields is constant and equal to the speed of light c.

$$\frac{E_{max}}{B_{max}} = \frac{E}{B} = c \qquad (34.14)$$

The **Poynting vector S** describes the energy flow associated with an electromagnetic wave. *The direction of S is along the direction of propagation and the magnitude of S is the rate at which electromagnetic energy crosses a unit surface area perpendicular to the direction of S.*

$$\mathbf{S} \equiv \frac{1}{\mu_0}\mathbf{E}\times\mathbf{B} \qquad (34.19)$$

The wave intensity is the time average of the magnitude of the Poynting vector. E_{max} and B_{max} are the maximum values of the field magnitudes.

$$I = S_{av} = \frac{E_{max}B_{max}}{2\mu_0} = \frac{E_{max}^2}{2\mu_0 c} = \frac{cB_{max}^2}{2\mu_0} \quad (34.21)$$

The **instantaneous energy densities** of the electric and magnetic fields are equal.

$$u_B = u_E = \frac{1}{2}\epsilon_0 E^2 = \frac{B^2}{2\mu_0}$$

The **total instantaneous energy density** u is proportional to E^2 and B^2.

$$u = u_E + u_B = \epsilon_0 E^2 = \frac{B^2}{\mu_0}$$

The **total average energy density** is proportional to E_{max}^2 and B_{max}^2. *The average energy density is also proportional to the wave intensity.*

$$u_{av} = \epsilon_0(E^2)_{av} = \frac{1}{2}\epsilon_0 E_{max}^2 = \frac{B_{max}^2}{2\mu_0} \quad (34.22)$$

The **intensity** of an electromagnetic wave equals the average energy density multiplied by the speed of light.

$$I = S_{av} = cu_{av} \qquad (34.23)$$

The **linear momentum** (p) **delivered to an absorbing surface** by an electromagnetic wave at normal incidence depends on the fraction of the total energy absorbed.

$$p = \frac{U}{c} \qquad \binom{\text{complete}}{\text{absorption}} \qquad (34.24)$$

$$p = \frac{2U}{c} \qquad \binom{\text{complete}}{\text{reflection}} \qquad (34.26)$$

Radiation pressure on an absorbing surface (at normal incidence) depends on the magnitude of the Poynting vector and the degree of absorption.

$$P = \frac{S}{c} \qquad \binom{\text{Perfectly}}{\text{absorbing surface}} \qquad (34.25)$$

$$P = \frac{2S}{c} \qquad \binom{\text{Perfectly}}{\text{reflecting surface}} \qquad (34.27)$$

All **electromagnetic waves are produced by accelerating charges.** Types of electromagnetic waves can be characterized by "typical" ranges of frequency or wavelength.

- **Radio waves** ($\sim10^4\,\text{m} > \lambda > \sim0.1\,\text{m}$) are the result of electric charges accelerating through a conducting wire (antenna).

- **Microwaves** ($\sim0.3\,\text{m} > \lambda > \sim10^{-4}\,\text{m}$) are generated by electronic devices.

- **Infrared waves** ($\sim10^{-3}\,\text{m} > \lambda > \sim7\times10^{-7}\,\text{m}$) are produced by high temperature objects and molecules.

- **Visible light** ($\sim7\times10^{-7}\,\text{m} > \lambda > \sim4\times10^{-7}\,\text{m}$) is produced by the rearrangement of electrons in atoms and molecules.

- **Ultraviolet (UV) light** ($\sim4\times10^{-7}\,\text{m} > \lambda > \sim6\times10^{-10}\,\text{m}$) is an important component of radiation from the Sun.

- **X-rays** ($\sim10^{-8}\,\text{m} > \lambda > \sim10^{-12}\,\text{m}$) are produced when high energy electrons bombard a metal target.

- **Gamma rays** ($\sim10^{-10}\,\text{m} > \lambda > \sim10^{-14}\,\text{m}$) are electromagnetic waves emitted by radioactive nuclei.

Remember, the wavelength ranges stated above are approximate. On the long wavelength end, radio waves can be arbitrarily long; and on the short wavelength end, gamma rays can be arbitrarily short. Regions of the electromagnetic spectrum overlap in wavelength; see Figure 34.12 of the text.

REVIEW CHECKLIST

You should be able to:

▷ State Maxwell's equations and describe the essential features of the apparatus and procedure used by Hertz in his experiments leading to the discovery and understanding of the source and nature of electromagnetic waves. (Section 34.1)

▷ Write the equation for the magnetic field of an electromagnetic wave when given the equation of the corresponding electric field. (Section 34.2)

▷ Calculate the values for the Poynting vector (magnitude), wave intensity, and instantaneous and average energy densities in a plane electromagnetic wave. (Section 34.3)

▷ Calculate the radiation pressure on a surface and the linear momentum delivered to a surface by an electromagnetic wave. (Section 34.4)

▷ Describe the production of electromagnetic waves and radiation of energy by a half-wave (or dipole) antenna. Use a diagram to show the relative directions for **E**, **B**, and **S** and their space and time dependencies. Account for the intensity of the radiated wave at points near the dipole and at distant points. (Section 34.5)

▷ Place the various types of electromagnetic waves in the correct sequence in the electromagnetic spectrum and describe the basis of production particular to each of the wave types. (Section 34.6)

ANSWERS TO SELECTED QUESTIONS

1. Radio stations often advertise "instant news." If they mean that you can hear the news the instant they speak it, is their claim true? About how long would it take for a message to travel across this country by radio waves, assuming that the waves could be detected at this range?

Answer Radio waves move at the speed of light. They can travel around the curved surface of the Earth, bouncing between the ground and the ionosphere, which has an altitude that is small when compared to the radius of the Earth. The distance across the lower forty-eight states is approximately $5\,000$ km, requiring a time of $(5\times10^6\ \text{m})/(3\times10^8\ \text{m/s}) \sim 10^{-2}$ s. To go halfway around the Earth takes only 0.07 s. In other words, a speech can be heard on the other side of the world before it is heard at the back of the room.

□   □   □   □

8. If you charge a comb by running it through your hair and then hold the comb next to a bar magnet, do the electric and magnetic fields produced constitute an electromagnetic wave?

Answer No. Charge on the comb creates an electric field and the bar magnet sets up a magnetic field. However, at any point these fields are constant in magnitude and direction. In an electromagnetic wave, the electric and magnetic fields must be changing in order to create each other.

□   □   □   □

18. What does a radio wave do to the charges in the receiving antenna to provide a signal for your car radio?

Answer Consider a typical metal rod antenna for a car radio. The rod detects the electric field portion of the carrier wave. Variations in the amplitude of the carrier wave cause the electrons in the rod to vibrate with amplitudes emulating those of the carrier wave. Likewise, for frequency modulation, the variations of the frequency of the carrier wave cause constant-amplitude vibrations of the electrons in the rod but at frequencies that imitate those of the carrier.

□   □   □   □

21. Suppose that a creature from another planet had eyes that were sensitive to infrared radiation. Describe what the alien would see if it looked around the room you are now in. In particular, what would be bright and what would be dim?

Answer Light bulbs and the toaster glow brightly in the infrared. Somewhat fainter are the back of the refrigerator and the back of the television set, while the TV screen is dark. The pipes under the sink show the same weak glow as the walls until you turn on the faucets. Then the pipe on the right gets darker while that on the left develops a rich gleam that quickly runs up along its length. The food on your plate shines; so does human skin, the same color for all races. Clothing is dark as a rule, but your seat glows like a monkey's rump when you get up from a chair, and you leave a patch of the same glow on your chair. Your face appears lit from within, like a jack-o-lantern; your nostrils and openings of your ear canals are bright; brighter still are the pupils of your eyes.

□   □   □   □

SOLUTIONS TO SELECTED PROBLEMS

5. Figure 34.3 shows a plane electromagnetic sinusoidal wave propagating in the x direction. Suppose that the wavelength is 50.0 m, and the electric field vibrates in the xy plane with an amplitude of 22.0 V/m. Calculate (a) the frequency of the wave and (b) the magnitude and direction of **B** when the electric field has its maximum value in the negative y direction. (c) Write an expression for **B** with the correct unit vector, with numerical values for B_{max}, k, and ω, and with its magnitude in the form $B = B_{max}\cos(kx - \omega t)$.

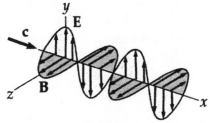

Figure 34.3

Solution

(a) $c = f\lambda$:

$$f = \frac{c}{\lambda} = \frac{3.00 \times 10^8 \text{ m/s}}{50.0 \text{ m}} = 6.00 \times 10^6 \text{ Hz} \qquad \Diamond$$

(b) $c = E/B$:

$$B = \frac{E}{c} = \frac{22.0 \text{ V/m}}{3.00 \times 10^8 \text{ m/s}} = 7.33 \times 10^{-8} \text{ T} = 73.3 \text{ nT} \qquad \Diamond$$

B is directed along **negative z direction** when **E** is in the negative y direction, so that $\mathbf{S} = \mathbf{E} \times \mathbf{B}/\mu_0$ will propagate in the direction of $(-\hat{\mathbf{j}}) \times (-\hat{\mathbf{k}}) = +\hat{\mathbf{i}}$.

(c) $B = B_{max}\cos(kx - \omega t)$: $k = \dfrac{2\pi}{\lambda} = \dfrac{2\pi}{50.0 \text{ m}} = 0.126 \text{ m}^{-1}$

$$\omega = 2\pi f = (2\pi \text{ rad})(6.00 \times 10^6 \text{ Hz}) = 3.77 \times 10^7 \text{ rad/s}$$

Thus,

$$\mathbf{B} = (73.3 \text{ nT})\cos\left[(0.126 \text{ rad/m})x - (3.77 \times 10^7 \text{ rad/s})t\right](-\hat{\mathbf{k}}) \Diamond$$

7. In SI units, the electric field in an electromagnetic wave is described by $E_y = 100\sin(1.00 \times 10^7 x - \omega t)$. Find (a) the amplitude of the corresponding magnetic field oscillations, (b) the wavelength λ, and (c) the frequency f.

Solution

(a) $\quad B_{max} = \dfrac{E_{max}}{c} = \dfrac{100\ \text{V/m}}{3.00\times10^{8}\ \text{m/s}} = 3.33\times10^{-7}\ \text{T}$ ◊

(b) We compare the given wave function with $y = A\sin(kx - \omega t)$

to see that $\qquad\qquad k = 1.00\times10^{7}\ \text{m}^{-1}$

and $\qquad\qquad \lambda = \dfrac{2\pi}{k} = \dfrac{2\pi}{1.00\times10^{7}\ \text{m}^{-1}} = 6.28\times10^{-7}\ \text{m}$ ◊

(c) $\qquad\qquad f = \dfrac{c}{\lambda} = \dfrac{3.00\times10^{8}\ \text{m/s}}{6.28\times10^{-7}\ \text{m}} = 4.77\times10^{14}\ \text{Hz}$ ◊

---

13. What is the average magnitude of the Poynting vector 5.00 miles from a radio transmitter broadcasting isotropically with an average power of 250 kW?

Solution

Conceptualize: As the distance from the source is increased, the power per unit area will decrease, so at a distance of 5 miles from the source, the power per unit area will be a small fraction of the Poynting vector near the source.

Categorize: The Poynting vector is the power per unit area, where A is the surface area of a sphere with a 5-mile radius.

Analyze:

The Poynting vector is $\qquad\qquad S_{av} = \dfrac{\mathcal{P}}{A} = \dfrac{\mathcal{P}}{4\pi r^{2}}$

In meters, $\qquad\qquad r = (5.00\ \text{mi})(1\,609\ \text{m/mi}) = 8\,045\ \text{m}$

and the magnitude is $\qquad\qquad S = \dfrac{250\times10^{3}\ \text{W}}{4\pi(8\,045)^{2}} = 3.07\times10^{-4}\ \text{W/m}^{2}$ ◊

Finalize: The magnitude of the Poynting vector ten meters from the source is 199 W/m^{2}, on the order of a million times larger than it is 5 miles away! It is surprising to realize how little power is actually received by a radio (at the 5-mile distance, the signal would only be about 30 nW, assuming a receiving area of about 1 cm^{2}).

---

15. A community plans to build a facility to convert solar radiation to electrical power. They require 1.00 MW of power, and the system to be installed has an efficiency of 30.0% (that is, 30.0% of the solar energy incident on the surface is converted to useful energy that can power the community). What must be the effective area of a perfectly absorbing surface used in such an installation, assuming sunlight has a constant intensity of 1 000 W/m^2?

Solution   At 30.0% efficiency, the power $\mathcal{P} = 0.300SA$:

$$A = \frac{\mathcal{P}}{0.300S} = \frac{1.00 \times 10^6 \text{ W}}{0.300(1\,000 \text{ W/m}^2)} = 3\,330 \text{ m}^2 \approx 0.8 \text{ acre} \qquad \Diamond$$

17. The filament of an incandescent lamp has a 150-Ω resistance and carries a direct current of 1.00 A. The filament is 8.00 cm long and 0.900 mm in radius. (a) Calculate the Poynting vector at the surface of the filament, associated with the static electric field producing the current and the current's static magnetic field. (b) Find the magnitude of the static electric and magnetic fields at the surface of the filament.

Solution   In this problem, the Poynting vector does not represent the light radiated or the energy convected away from the filament. Rather, it represents the energy flow in the static electric and magnetic fields created in the surrounding empty space, by current in the filament. The Poynting vector describes not only energy transfer by electromagnetic radiation but also energy transfer by electrical transmission.

The rate at which energy is delivered to the resistor is

$$\mathcal{P} = I^2 R = (1.00 \text{ A})^2 (150 \text{ }\Omega) = 150 \text{ W}$$

and the surface area $A = 2\pi r L = 2\pi (0.900 \times 10^{-3} \text{ m})(0.080\,0 \text{ m}) = 4.52 \times 10^{-4} \text{ m}^2$

(a)   The Poynting vector is directed radially inward:

$$S = \frac{\mathcal{P}}{A} = \frac{150 \text{ W}}{4.52 \times 10^{-4} \text{ m}^2} = 3.32 \times 10^5 \text{ W/m}^2 \qquad \Diamond$$

(b) $B = \mu_0 \dfrac{I}{2\pi r} = (4\pi \times 10^{-7} \text{ T} \cdot \text{m/A}) \dfrac{1.00 \text{ A}}{2\pi (0.900 \times 10^{-3} \text{ m})} = 2.22 \times 10^{-4} \text{ T}$ $\qquad \Diamond$

$$E = \frac{\Delta V}{\Delta x} = \frac{IR}{L} = \frac{150 \text{ V}}{0.080\,0 \text{ m}} = 1\,880 \text{ V/m} \qquad \Diamond$$

Note: We could also calculate the Poynting vector from $S = \dfrac{EB}{\mu_0} = 3.32 \times 10^5 \text{ W/m}^2$

27. A radio wave transmits 25.0 W/m² of power per unit area. A flat surface of area A is perpendicular to the direction of propagation of the wave. Calculate the radiation pressure on it, assuming the surface is a perfect absorber.

Solution For complete absorption, $P = \dfrac{S}{c} = \dfrac{25.0 \text{ W}/\text{m}^2}{3.00 \times 10^8 \text{ m}/\text{s}} = 8.33 \times 10^{-8} \text{ N}/\text{m}^2$ ◊

29. A 15.0-mW helium-neon laser ($\lambda = 632.8$ nm) emits a beam of circular cross section with a diameter of 2.00 mm. (a) Find the maximum electric field in the beam. (b) What total energy is contained in a 1.00-m length of the beam? (c) Find the momentum carried by a 1.00-m length of the beam.

Solution The intensity of the light is the average magnitude of the Poynting vector:

$$I = \frac{\mathcal{P}}{\pi r^2} = \frac{E_{max}^2}{2\mu_0 c}$$

(a) Therefore, the maximum electric field is $E_{max} = \sqrt{\dfrac{\mathcal{P}(2\mu_0 c)}{\pi r^2}} = 1.90 \times 10^3 \text{ N}/\text{C}$ ◊

(b) The power being 15.0 mW means that 15.0 mJ passes through a cross section of the beam in one second. This energy is uniformly spread through a beam length of 3.00×10^8 m, since that is how far the front end of the energy travels in one second. Thus, the energy in just a one-meter length is

$$\left(\frac{15.0 \times 10^{-3} \text{ J}/\text{s}}{3.00 \times 10^8 \text{ m}/\text{s}}\right)(1.00 \text{ m}) = 5.00 \times 10^{-11} \text{ J}$$ ◊

(c) The linear momentum carried by a 1.00-m length of the beam is the momentum that would be received by an absorbing surface, under complete absorption:

$$p = \frac{U}{c} = \frac{5.00 \times 10^{-11} \text{ J}}{3.00 \times 10^8 \text{ m}/\text{s}} = 1.67 \times 10^{-19} \text{ kg} \cdot \text{m}/\text{s}$$ ◊

35. Two radio-transmitting antennas are separated by half the broadcast wavelength and are driven in phase with each other. In which directions are (a) the strongest and (b) the weakest signals radiated?

Solution

Conceptualize: The strength of the radiated signal will be a function of the location around the two antennas and will depend on the interference of the waves.

Categorize: A diagram helps to visualize this situation. The two antennas are driven in phase, which means that they both create maximum electric field strength at the same time, as shown in the diagram. The radio EM waves travel radially outwards from the antennas, and the received signal will be the vector sum of the two waves.

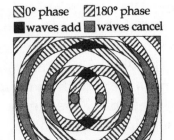

Analyze:

(a) Along the perpendicular bisector of the line joining the antennas, the distance is the same to both transmitting antennas. The transmitters oscillate in phase, so along this line the two signals will be received in phase, constructively interfering to produce a maximum signal strength that is twice the amplitude of one transmitter. ◊

(b) Along the extended line joining the sources, the wave from the more distant antenna must travel one-half wavelength farther, so the waves are received 180° out of phase. They interfere destructively to produce the weakest signal with zero amplitude. ◊

Finalize: Radio stations may use an antenna array to direct the radiated signal toward a highly-populated region and reduce the signal strength delivered to a sparsely-populated area.

---

41. What are the wavelengths of electromagnetic waves in free space that have frequencies of (a) 5.00×10^{19} Hz and (b) 4.00×10^9 Hz?

Solution

(a) $\lambda = \dfrac{c}{f} = \dfrac{3.00 \times 10^8 \text{ m/s}}{5.00 \times 10^{19} \text{ s}^{-1}} = 6.00$ pm ◊

This would be called an x-ray if it were emitted when an inner electron in an atom or an electron in a vacuum tube loses energy. It would be called a gamma ray if it were radiated by an atomic nucleus.

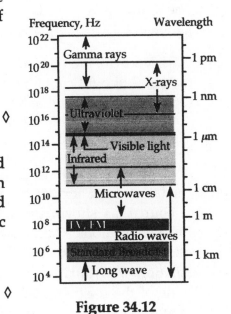

Figure 34.12

(b) $\lambda = \dfrac{c}{f} = \dfrac{3.00 \times 10^8 \text{ m/s}}{4.00 \times 10^9 \text{ s}^{-1}} = 7.50$ cm ◊

By Figure 34.12, this finger-length wave is called a radio wave or a microwave.

---

49. Review problem. In the absence of cable input or a satellite dish, a television set can use a dipole-receiving antenna for VHF channels and a loop antenna for UHF channels (Fig. Q34.12). The UHF antenna produces an emf from the changing magnetic flux through the loop. The TV station broadcasts a signal with a frequency f, and the signal has an electric-field amplitude E_{max} and a magnetic-field amplitude B_{max} at the location of the receiving antenna. (a) Using Faraday's law, derive an expression for the amplitude of the emf that appears in a single-turn circular loop antenna with a radius r, which is small compared with the wavelength of the wave. (b) If the electric field in the signal points vertically, what orientation of the loop gives the best reception?

Figure Q34.12

Solution We can approximate the magnetic field as uniform over the area of the loop while it oscillates in time as $B = B_{max}\cos\omega t$. The induced voltage is

$$\mathcal{E} = -\frac{d\Phi_B}{dt} = -\frac{d}{dt}(BA\cos\theta) = -A\frac{d}{dt}(B_{max}\cos\omega t\cos\theta)$$

$$\mathcal{E} = AB_{max}\omega(\sin\omega t\cos\theta)$$

$$\mathcal{E}(t) = 2\pi f B_{max}A\sin(2\pi ft)\cos\theta = 2\pi^2 r^2 f B_{max}\cos\theta\sin(2\pi ft)$$

(a) The amplitude of this emf is $\mathcal{E}_{max} = 2\pi^2 r^2 f B_{max}\cos\theta$, where θ is the angle between the magnetic field and the normal to the loop. ◊

(b) If **E** is vertical, then **B** is horizontal, so the plane of the loop should be vertical, and the plane should contain the line of sight to the transmitter. This will make $\theta = 0°$, so $\cos\theta$ takes on its maximum value. ◊

---

51. A dish antenna having a diameter of 20.0 m receives (at normal incidence) a radio signal from a distant source, as shown in Figure P34.51. The radio signal is a continuous sinusoidal wave with amplitude $E_{max} = 0.200\ \mu V/m$. Assume the antenna absorbs all the radiation that falls on the dish. (a) What is the amplitude of the magnetic field in this wave? (b) What is the intensity of the radiation received by this antenna? (c) What is the power received by the antenna? (d) What force is exerted by the radio waves on the antenna?

Figure P34.51

Solution

(a) $B_{max} = E_{max}/c = 6.67 \times 10^{-16}$ T ◊

(b) $S_{av} = E_{max}^2 / 2\mu_0 c = 5.31 \times 10^{-17}$ W/m^2 ◊

(c) $\mathcal{P}_{av} = S_{av}A = 1.67 \times 10^{-14}$ W ◊

(Do not confuse this power with the expression for pressure $P = S/c$, which we use to find the force.)

(d) $F = PA = (S_{av}/c)A = 5.56 \times 10^{-23}$ N (~3 000 Hydrogen atoms' weight!) ◊

---

53. In 1965, Arno Penzias and Robert Wilson discovered the cosmic microwave radiation left over from the Big Bang expansion of the Universe. Suppose the energy density of this background radiation is 4.00×10^{-14} J/m^3. Determine the corresponding electric field amplitude.

Solution

The energy density can be written as $u = \frac{1}{2}\epsilon_0 E_{max}^2$

so $E_{max} = \sqrt{\dfrac{2u}{\epsilon_0}} = \sqrt{\dfrac{2(4.00 \times 10^{-14} \text{ N/m}^2)}{8.85 \times 10^{-12} \text{ C}^2/\text{N} \cdot \text{m}^2}} = 95.1$ mV/m ◊

---

55. A linearly polarized microwave of wavelength 1.50 cm is directed along the positive x axis. The electric field vector has a maximum value of 175 V/m and vibrates in the xy plane. (a) Assume that the magnetic field component of the wave can be written in the form $B = B_{max}\sin(kx - \omega t)$ and give values for B_{max}, k, and ω. Also, determine in which plane the magnetic field vector vibrates. (b) Calculate the average value of the Poynting vector for this wave. (c) What radiation pressure would this wave exert if it were directed at normal incidence onto a perfectly reflecting sheet? (d) What acceleration would be imparted to a 500-g sheet (perfectly reflecting and at normal incidence) with dimensions of 1.00 m $\times$ 0.750 m?

Solution

(a) $B_{max} = \dfrac{E_{max}}{c} = \dfrac{175 \text{ V}/\text{m}}{3.00 \times 10^8 \text{ m}/\text{s}} = 5.83 \times 10^{-7}$ T $\quad\diamond$

The magnetic field is in the z direction so that $\mathbf{S} = (1/\mu_0)\mathbf{E} \times \mathbf{B}$ can be in the $\hat{\mathbf{j}} \times \hat{\mathbf{k}} = \hat{\mathbf{i}}$ direction.

$k = \dfrac{2\pi}{\lambda} = \dfrac{2\pi}{0.015 \text{ m}} = 419 \text{ m}^{-1} \quad\diamond$

$\omega = kc = (419 \text{ m}^{-1})(3.00 \times 10^8 \text{ m}/\text{s}) = 1.26 \times 10^{11} \text{ rad}/\text{s} \quad\diamond$

(b) $S_{av} = \dfrac{E_{max}B_{max}}{2\mu_0} = \dfrac{(175 \text{ V}/\text{m})(5.83 \times 10^{-7} \text{ T})}{2(4\pi \times 10^{-7} \text{ N}/\text{A}^2)} = 40.6 \text{ W}/\text{m}^2$

$\mathbf{S_{av}} = (40.6 \text{ W}/\text{m}^2)\hat{\mathbf{i}} \quad\diamond$

(c) For perfect reflection,

$P_r = \dfrac{2S}{c} = \dfrac{2(40.6 \text{ W}/\text{m}^2)}{3.00 \times 10^8 \text{ m}/\text{s}} = 2.71 \times 10^{-7} \text{ N}/\text{m}^2 \quad\diamond$

(d) We use the particle under a net force model:

$a = \dfrac{F}{m} = \dfrac{P_r A}{m} = \dfrac{(2.71 \times 10^{-7} \text{ N}/\text{m}^2)(0.750 \text{ m}^2)}{0.500 \text{ kg}} = 4.06 \times 10^{-7} \text{ m}/\text{s}^2$

$\mathbf{a} = (406 \text{ nm}/\text{s}^2)\hat{\mathbf{i}} \quad\diamond$

---

57. An astronaut, stranded in space 10.0 m from his spacecraft and at rest relative to it, has a mass (including equipment) of 110 kg. Because he has a 100-W light source that forms a directed beam, he considers using the beam as a photon rocket to propel himself continuously toward the spacecraft. (a) Calculate how long it takes him to reach the spacecraft by this method. (b) **What if?** Suppose, instead, that he decides to throw the light source away in a direction opposite the spacecraft. If the mass of the light source is 3.00 kg and, after being thrown, it moves at 12.0 m/s relative to the recoiling astronaut, how long does it take for the astronaut to reach the spacecraft?

Solution

Conceptualize: Based on our everyday experience, the force exerted by photons is too small to feel, so it may take a very long time (maybe days!) for the astronaut to travel 10 m with his "photon rocket." Using the momentum of the thrown light seems like a better solution, but it will still take a while (maybe a few minutes) for the astronaut to reach the spacecraft because his mass is so much larger than the mass of the light source.

Categorize: In part (a), the radiation pressure can be used to find the force that accelerates the astronaut toward the spacecraft. In part (b), the principle of conservation of momentum can be applied to find the time required to travel the 10 m.

Analyze:

(a) Light exerts on the astronaut a pressure $P = F/A = S/c$,

and a force of
$$F = \frac{SA}{c} = \frac{\mathcal{P}}{c} = \frac{100 \text{ J/s}}{3.00 \times 10^8 \text{ m/s}} = 3.33 \times 10^{-7} \text{ N}$$

By Newton's 2nd law,
$$a = \frac{F}{m} = \frac{3.33 \times 10^{-7} \text{ N}}{110 \text{ kg}} = 3.03 \times 10^{-9} \text{ m/s}^2$$

This acceleration is constant, so the distance traveled is $x = \frac{1}{2}at^2$, and the amount of time he travels is

$$t = \sqrt{\frac{2x}{a}} = \sqrt{\frac{2(10.0 \text{ m})}{3.03 \times 10^{-9} \text{ m/s}^2}} = 8.12 \times 10^4 \text{ s} = 22.6 \text{ h} \;\lozenge$$

(b) Because there are no external forces, the momentum of the astronaut before throwing the light is the same as afterwards when the now 107-kg astronaut is moving at speed v towards the spacecraft and the light is moving away from the spacecraft at $(12.0 \text{ m/s} - v)$.

Thus, $\mathbf{p}_i = \mathbf{p}_f$ gives
$$0 = (107 \text{ kg})v - (3.00 \text{ kg})(12.0 \text{ m/s} - v)$$
$$0 = (107 \text{ kg})v - (36.0 \text{ kg·m/s}) + (3.00 \text{ kg})v$$

$$v = \frac{36.0 \text{ kg·m/s}}{110 \text{ kg}} = 0.327 \text{ m/s}: \quad t = \frac{x}{v} = \frac{10.0 \text{ m}}{0.327 \text{ m/s}} = 30.6 \text{ s} \qquad\qquad \lozenge$$

Finalize: Throwing the light away is certainly a more expedient way to reach the spacecraft, but there is not much chance of retrieving the lamp unless it has a very long cord. How long would the cord need to be, and does its length depend on how hard the astronaut throws the lamp? (You should verify that the minimum cord length is 367 m, independent of the speed that the lamp is thrown.)

Chapter 35

THE NATURE OF LIGHT AND THE LAWS OF GEOMETRIC OPTICS

EQUATIONS AND CONCEPTS

Reflection and refraction can occur when a light ray is incident obliquely on a smooth, planar surface which forms the boundary between two transparent media of different optical densities. *As shown in the figure, the subscript 1 refers to parameters for light in the initial medium and subscript 2 refers to corresponding parameters in the new medium. A portion of the energy associated with each incoming ray will be reflected back into the original medium, while the remaining fraction will be transmitted into the second medium.*

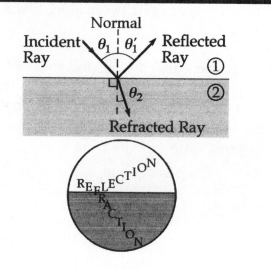

The **energy of a photon** is proportional to the frequency of the associated electromagnetic wave.

$$E = hf \tag{35.1}$$

Planck's constant, h, has SI units of J·s.

$$h = 6.63 \times 10^{-34} \text{ J·s}$$

The **law of reflection** states that the angle of incidence (the angle measured between the incident ray and the normal line) equals the angle of reflection (the angle measured between the reflected ray and the normal line). *The incident ray, reflected ray, and the normal line lie in the same plane.*

$$\theta_1' = \theta_1 \tag{35.2}$$

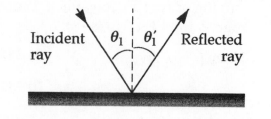

Snell's law of refraction can be expressed in terms of the speeds of light in the media on either side of the refracting surface (Equation 35.3); or in terms of the indices of refraction of the two media (Equation 35.8).

$$\frac{\sin \theta_2}{\sin \theta_1} = \frac{v_2}{v_1} = \text{constant} \tag{35.3}$$

Equation 35.8 is the most practical form of Snell's law. This equation involves a parameter called the index of refraction, the value of which is characteristic of a particular medium. The index of refraction is defined in Equation 35.4 and Equation 35.7.

$$n_1 \sin \theta_1 = n_2 \sin \theta_2 \qquad (35.8)$$

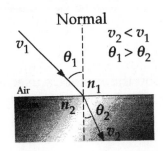

The **index of refraction** of a transparent medium equals the ratio of the speed of light in vacuum to the speed of light in the medium. The index of refraction of a given medium can be expressed as the ratio of the wavelength of light in vacuum to the wavelength in that medium. *The frequency of a wave is characteristic of the source; as light travels from one medium into another of different index of refraction, the frequency remains constant but the wavelength changes.*

$$n \equiv \frac{\text{Speed of light in vacuum}}{\text{Speed of light in a medium}} = \frac{c}{v} \qquad (35.4)$$

$$n = \frac{\lambda}{\lambda_n} \qquad (35.7)$$

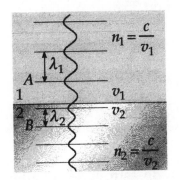

The **critical angle** is the minimum angle of incidence for which total internal reflection can occur.

$$\sin \theta_c = \frac{n_2}{n_1} \qquad (\text{for } n_1 > n_2) \qquad (35.10)$$

Total internal reflection occurs for angles of incidence equal to or greater than the critical angle. An incident ray will be totally internally reflected back into the first medium. *Total internal reflection is possible only when a light ray is directed from a medium of high index of refraction into a medium of lower index of refraction.*

REVIEW CHECKLIST

You should be able to:

▷ Describe the methods used by Roemer and Fizeau for the measurement of c and make calculations using sets of typical values for the quantities involved. (Section 35.2)

▷ Make calculations using the law of reflection and the law of refraction (Snell's law). (Sections 35.4 and 35.5)

▷ Calculate the angle of deviation and the angular dispersion in a prism. (Section 35.7)

▷ Understand the conditions under which total internal reflection can occur in a medium and determine the critical angle for a given pair of adjacent media. (Section 35.8)

ANSWERS TO SELECTED QUESTIONS

2. Why do astronomers looking at distant galaxies talk about looking backward in time?

Answer

Light travels through a vacuum at a speed of 300 000 km per second. Thus, an image we see from a distant star or galaxy must have been generated some time ago.

For example, the star Altair is 16 light-years away; if we look at an image of Altair today, we know only what was happening 16 years ago.

This may not initially seem significant, but astronomers who look at other galaxies can gain an idea of what galaxies looked like when they were significantly younger. Thus, it actually makes sense to speak of "looking backward in time."

□   □   □   □

9. As light travels from one medium to another, does the wavelength of the light change? Does the frequency change? Does the speed change? Explain.

Answer

You can think about this based upon the fact that the light wave must be continuous. Therefore, the frequency of the light must remain constant, since any one crest coming up to the interface must create one crest in the new medium. However, the speed of light does vary in different media, as described by Snell's law. In addition, since the speed of the light wave is equal to the light wave's frequency times its wavelength, and the frequency does not change, we can be sure that the wavelength must change. The speed decreases upon entering a new medium from vacuum; therefore the wavelength decreases as well.

□ □ □ □

15. Explain why a diamond sparkles more than a glass crystal of the same shape and size.

Answer

Diamond has a larger index of refraction than glass, and consequently has a smaller critical angle for internal reflection. A brilliant-cut diamond is shaped to admit light from above, reflect it totally at the converging facets on the underside of the jewel, and let the light escape only at the top. Glass will have less light internally reflected.

□ □ □ □

19. When two colors of light (X and Y) are sent through a glass prism, X is bent more than Y. Which color travels more slowly in the prism?

Answer

If the light slows down upon entering the prism, a light ray that is bent more suffers a greater loss of speed as it enters the new medium. Therefore, the X light rays travel more slowly.

SOLUTIONS TO SELECTED PROBLEMS

3. In an experiment to measure the speed of light using the apparatus of Fizeau (see Fig. 35.2), the distance between light source and mirror was 11.45 km and the wheel had 720 notches. The experimentally determined value of c was 2.998×10^8 m/s. Calculate the minimum angular speed of the wheel for this experiment.

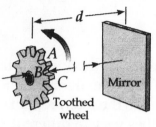

Figure 35.2

Solution The distance down to the mirror and back is

$$2 \times 11.45 \times 10^3 \text{ m} = 2.290 \times 10^4 \text{ m}$$

The round-trip time is expected to be $\quad t = \dfrac{x}{v} = \dfrac{2.290 \times 10^4 \text{ m}}{2.998 \times 10^8 \text{ m/s}} = 7.64 \times 10^{-5} \text{ s}$

In this time the wheel should turn by $\quad \frac{1}{2}(1/720)$ rev

to move a tooth into the place of a notch. Its angular speed should be

$$\omega = \frac{\theta}{t} = \frac{1}{2}\left(\frac{1 \text{ rev}}{720}\right)\left(\frac{1}{7.64 \times 10^{-5} \text{ s}}\right) = 9.09 \text{ rev/s} \quad \Diamond$$

Higher angular speeds must be available so that the experimenter can home in on the special dark setting from both sides. The demonstration is most convincing if the wheel turns at twice this angular speed, to replace a notch with another notch during the light's travel time, restoring the returning light to full brightness.

---

13. An underwater scuba diver sees the Sun at an apparent angle of 45.0° above the horizon. What is the elevation angle of the Sun above the horizon?

Solution

Conceptualize: The sunlight refracts as it enters the water from the air. Because the water has a higher index of refraction, the light slows down and bends toward the vertical line that is normal to the interface. Therefore, the elevation angle of the Sun above the water will be less than 45° as shown in the diagram to the right, even though it appears to the diver that the sun is 45° above the horizon.

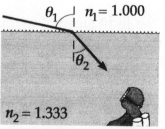

Categorize: We can use Snell's law of refraction to find the precise angle of incidence.

Analyze: Snell's law is

$$n_1 \sin \theta_1 = n_2 \sin \theta_2$$

which gives

$$\sin \theta_1 = 1.333 \sin 45.0°$$

$$\sin \theta_1 = (1.333)(0.707) = 0.943$$

The sunlight is at $\theta_1 = 70.5°$ to the vertical, so the Sun is 19.5° above the horizon. ◊

Finalize: The calculated result agrees with our prediction. When applying Snell's law, it is easy to mix up the index values and to confuse angles-with-the-normal and angles-with-the-surface. Making a sketch and a prediction as we did here helps avoid careless mistakes.

19. A ray of light strikes a flat block of glass ($n = 1.50$) of thickness 2.00 cm at an angle of 30.0° with the normal. Trace the light beam through the glass, and find the angles of incidence and refraction at each surface.

Solution   At entry, $n_1 \sin \theta_1 = n_2 \sin \theta_2$: $1.00 \sin 30° = 1.50 \sin \theta_2$

and

$$\theta_2 = \sin^{-1}\left(\frac{0.500}{1.50}\right) = 19.5°$$ ◊

To do geometric optics, you must remember some geometry. The surfaces of entry and exit are parallel so their normals are parallel. Then angle θ_2 of refraction at entry and the angle θ_3 of incidence at exit are alternate interior angles formed by the ray as a transversal cutting parallel lines.

So

$$\theta_3 = \theta_2 = 19.5°$$ ◊

At the exit, $n_2 \sin \theta_3 = n_1 \sin \theta_4$: $1.50 \sin 19.5° = 1.00 \sin \theta_4$

and

$$\theta_4 = 30.0°$$ ◊

The exiting ray in air is parallel to the original ray in air. Thus, a car windshield of uniform thickness will not distort, but shows the driver the actual direction to every object outside.

31. A prism that has an apex angle of 50.0° is made of cubic zirconia, with $n = 2.20$. What is its angle of minimum deviation?

Solution

From Equation 35.9, $\quad n = \dfrac{\sin\left(\dfrac{\Phi + \delta_{min}}{2}\right)}{\sin\left(\dfrac{\Phi}{2}\right)}$

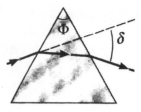

Solving for δ_{min}, $\quad \delta_{min} = 2\sin^{-1}\left(n\sin\dfrac{\Phi}{2}\right) - \Phi$

$$\delta_{min} = 2\sin^{-1}(2.20\sin 25.0°) - 50.0° = 86.8° \qquad \lozenge$$

---

33. A triangular glass prism with apex angle $\Phi = 60.0°$ has an index of refraction $n = 1.50$ (Fig. P35.33). What is the smallest angle of incidence θ_1 for which a light ray can emerge from the other side?

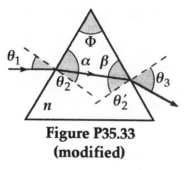

**Figure P35.33
(modified)**

34. A triangular glass prism with apex angle Φ has index of refraction n. (See Fig. P35.33.) What is the smallest angle of incidence θ_1 for which a light ray can emerge from the other side?

Solution Call the angles of incidence and refraction, at the surfaces of entry and exit, θ_1, θ_2, θ_2', and θ_3, in order as shown. The apex angle ϕ is the angle between the surfaces of entry and exit. The ray in the glass forms a triangle with these surfaces, in which the interior angles must add to $180°$. Thus, with

$\phi = 60.0°$,

$(90° - \theta_2) + 60° + (90° - \theta_2') = 180°$

so $\quad \theta_2 + \theta_2' = 60.0°$ **[1]**

which is a general rule for light going through prisms. At the first refraction, Snell's law gives

$\sin\theta_1 = 1.50\sin\theta_2$ **[2]**

At the second boundary, we want to almost reach the condition for total internal reflection:

$1.50\sin\theta_2' = 1.00\sin 90° = 1.00$

$\phi = \Phi$,

$(90° - \theta_2) + \Phi + (90° - \theta_2') = 180°$

so $\quad \theta_2 + \theta_2' = \Phi$ **[1]**

which is a general rule for light going through prisms. At the first interface between the air and prism,

$\sin\theta_1 = n\sin\theta_2$ **[2]**

At the second interface, we want to almost reach the condition for total internal reflection:

$n\sin\theta_2' = \sin 90° = 1$

or $\theta_2' = \sin^{-1}(1.00/1.50) = 41.8°$

Now by Equation [1] above,

$\theta_2 = 60.0° - 41.8° = 18.2°$

while by Equation [2], we find that

$\theta_1 = \sin^{-1}(1.50\sin 18.2°)$

So $\theta_1 = 27.9°$ ◊

Figure 35.27 in the text shows the effect nicely.

> **Note** that in the case of this problem, the numeric solution eliminated several of the analytic steps of the variable solution, and thus was much shorter. Whether a given problem is best solved analytically or numerically depends on the complexity of the problem.

or $\theta_2' = \sin^{-1}(1/n)$

Now by Equation [1] above,

$\theta_2 = \Phi - \theta_2' = \Phi - \sin^{-1}(1/n)$

while by Equation [2], we find that

$\theta_1 = \sin^{-1}\left[n\sin\left(\Phi - \sin^{-1}(1/n)\right)\right]$

Looking at the drawing of a right triangle, remember the identities that

$\sin(\alpha - \beta) = \sin\alpha\cos\beta - \cos\alpha\sin\beta$

$\cos\left[\sin^{-1}(1/n)\right] = \sqrt{n^2-1}/n$

θ_1 therefore simplifies to

$\theta_1 = \sin^{-1}\left[\left(\sqrt{n^2-1}\right)\sin\Phi - \cos\Phi\right]$ ◊

35. The index of refraction for violet light in silica flint glass is 1.66, and that for red light is 1.62. What is the angular dispersion of visible light passing through a prism of apex angle 60.0° if the angle of incidence is 50.0°? (See Fig. P35.35.)

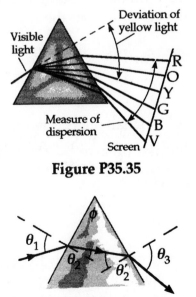

Figure P35.35

Solution Call the angles of incidence and refraction, at the surfaces of entry and exit, θ_1, θ_2, θ_2', and θ_3, in the order as shown. The apex angle ($\phi=60.0°$) is the angle between the surfaces of entry and exit. The ray in the glass forms a triangle with these surfaces, in which the interior angles must add to 180°.

Thus, $(90°-\theta_2)+\phi+(90°-\theta_2') = 180°$

and $\theta_2' = \phi - \theta_2$

This is a general rule for light going through prisms.

For the incoming ray, $\sin\theta_2 = \dfrac{\sin\theta_1}{n}$:

$(\theta_2)_{violet} = \sin^{-1}\left(\dfrac{\sin 50.0°}{1.66}\right) = 27.48°$

$(\theta_2)_{red} = \sin^{-1}\left(\dfrac{\sin 50.0°}{1.62}\right) = 28.22°$

For the outgoing ray, $\theta'_2 = 60° - \theta_2$ and $\sin\theta_3 = n\sin\theta'_2$

$(\theta_3)_{violet} = \sin^{-1}[1.66\sin 32.52°] = 63.17°$

$(\theta_3)_{red} = \sin^{-1}[1.62\sin 31.78°] = 58.56°$

The dispersion is the difference between these two angles:

$$\Delta\theta_3 = 63.17° - 58.56° = 4.61° \qquad \Diamond$$

---

39. Consider a common mirage formed by super-heated air just above a roadway. A truck driver whose eyes are 2.00 m above the road, where $n = 1.0003$, looks forward. She perceives the illusion of a patch of water ahead on the road, where her line of sight makes an angle of 1.20° below the horizontal. Find the index of refraction of the air just above the road surface. (*Suggestion:* Treat this as a problem in total internal reflection.)

Solution Think of the air as in two discrete layers, the first medium being cooler air with $n_1 = 1.0003$ and the second medium being hot air with a lower index, which reflects light from the sky by total internal reflection.

Use $n_1 \sin\theta_1 \ge n_2 \sin 90°$: $1.0003\sin 88.8° \ge n_2$ so $n_2 \le 1.00008$ $\qquad \Diamond$

---

49. A small underwater pool light is 1.00 m below the surface. The light emerging from the water forms a circle on the water surface. What is the diameter of this circle?

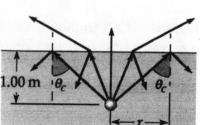

Solution

Conceptualize: Only the light that is directed upwards and hits the water's surface at less than the critical angle will be transmitted to the air so that someone outside can see it. The light that hits the surface farther from the center at an angle greater than θ_c will be totally reflected within the water, unable to be seen from the outside. From the diagram to the right, the diameter of this circle of light appears to be about 2 m.

Categorize: We can apply Snell's law to find the critical angle, and the diameter can then be found from the geometry.

Analyze: The critical angle is found when the refracted ray just grazes the surface ($\theta_2 = 90°$). The index of refraction of water is $n_1 = 1.333$, and $n_2 = 1.00$ for air, so

$$n_1 \sin\theta_c = n_2 \sin 90° \qquad \text{gives} \qquad \theta_c = \sin^{-1}(1/1.333) = \sin^{-1}(0.750) = 48.6°$$

The radius then satisfies
$$\tan\theta_c = \frac{r}{1.00 \text{ m}}$$

So the diameter is
$$d = 2r = 2(1.00 \text{ m})(\tan 48.6°) = 2.27 \text{ m} \qquad \Diamond$$

Finalize: Only the light rays within a 97.2° cone above the lamp escape the water and can be seen by an outside observer (Note: this angle does not depend on the depth of the light source). The path of a light ray is always reversible, so if a person were located beneath the water, they could see the whole hemisphere above the water surface within this cone; this is a good experiment to try the next time you go swimming!

---

53. A hiker stands on an isolated mountain peak near sunset and observes a rainbow caused by water droplets in the air 8.00 km away. The valley is 2.00 km below the mountain peak and entirely flat. What fraction of the complete circular arc of the rainbow is visible to the hiker? (See Fig. 35.24.)

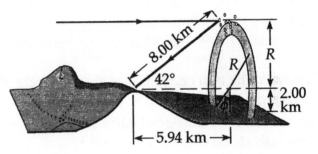

Solution Horizontal light rays from the setting Sun pass above the hiker. The light rays are twice refracted and once reflected, as in Figure 35.23. The most intense light reaching the hiker, that which represents the visible rainbow, is located between angles of 40.0° and 42.0° from the hiker's shadow. The hiker sees a greater percentage of the violet inner edge, so we consider the red outer edge. The radius R of the circle of droplets is

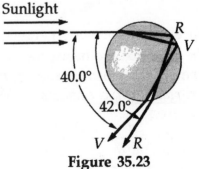

Figure 35.23

$$R = (8.00 \text{ km})\sin 42.0° = 5.35 \text{ km}$$

Then the angle ϕ between the vertical and the radius where the bow touches the ground, is given by

$$\cos\phi = \frac{2.00 \text{ km}}{R} = \frac{2.00 \text{ km}}{5.35 \text{ km}} = 0.374 \qquad \text{or} \qquad \phi = 68.1°$$

The angle filled by the visible bow is $360° - 2(68.1°) = 224°$, so the visible bow is

$$\frac{224°}{360°} = 62.2\% \text{ of a circle} \qquad \Diamond$$

This striking view motivated Charles Wilson's 1906 invention of the cloud chamber, a standard tool of nuclear physics. Look for a full circle of color around your shadow when you fly in an airplane.

55. A laser beam strikes one end of a slab of material, as shown in Figure P35.55. The index of refraction of the slab is 1.48. Determine the number of internal reflections of the beam before it emerges from the opposite end of the slab.

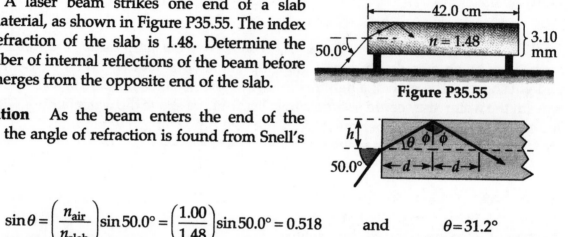

Figure P35.55

Solution As the beam enters the end of the slab, the angle of refraction is found from Snell's law:

$$\sin\theta = \left(\frac{n_{air}}{n_{slab}}\right)\sin 50.0° = \left(\frac{1.00}{1.48}\right)\sin 50.0° = 0.518 \qquad \text{and} \qquad \theta = 31.2°$$

Note that the two normal lines are perpendicular to each other. Using the right triangle having these lines as two of its sides, observe that the angle of incidence at the top of the slab is $\phi = 90.0° - \theta = 58.8°$. The critical angle as the light tries to go from the slab back into air is

$$\theta_c = \arcsin\left(\frac{n_{air}}{n_{slab}}\right) = \arcsin\left(\frac{1.00}{1.48}\right) = 42.5°$$

Since $\phi > \theta_c$, total internal reflection will indeed occur at the top and bottom of the slab. The distance the beam travels down the length of the slab for each reflection is $2d$, where d is the base of the right triangle shown in the sketch. Given that the altitude of the triangle, h, is one half the thickness of the slab,

$$d = \frac{h}{\tan\theta} = \frac{(3.10\text{ mm})/2}{\tan\theta} \qquad \text{and} \qquad 2d = \frac{3.10\times10^{-1}\text{ cm}}{\tan 31.2°} = 5.12\times10^{-1}\text{ cm}$$

The number of internal reflections made before reaching the opposite end of the slab is then

$$N = \frac{\text{length of slab}}{2d} = \frac{42.0\text{ cm}}{5.12\times10^{-1}\text{ cm}} = 82 \text{ reflections} \qquad \Diamond$$

59. The light beam in Figure P35.59 strikes surface 2 at the critical angle. Determine the angle of incidence θ_1.

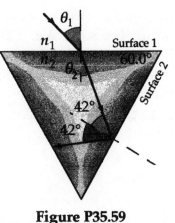

Figure P35.59

Solution

Conceptualize: From the diagram it appears that the angle of incidence is about 40°.

Categorize: We can find θ_1 by applying Snell's law at the first interface where the light is refracted. At surface 2, knowing that the 42.0° angle of reflection is the critical angle, we can work backwards to find θ_1.

Analyze: Define n_1 to be the index of refraction of the surrounding medium and n_2 to be that for the prism material. We can use the critical angle of 42.0° to find the ratio n_2/n_1:

$$n_2 \sin 42.0° = n_1 \sin 90.0°$$

So,
$$\frac{n_2}{n_1} = \frac{1}{\sin 42.0°} = 1.49$$

Call the angle of refraction θ_2 at the surface 1. The ray inside the prism forms a triangle with surfaces 1 and 2, so the sum of the interior angles of this triangle must be 180°.

Thus, $(90.0° - \theta_2) + 60.0° + (90.0° - 42.0°) = 180°$

Therefore, $\theta_2 = 18.0°$

Applying Snell's law at surface 1,

$$n_1 \sin \theta_1 = n_2 \sin 18.0°$$

$$\sin \theta_1 = (n_2/n_1) \sin \theta_2 = 1.49 \sin 18.0° \qquad \theta_1 = 27.5° \qquad \lozenge$$

Finalize: The result is a bit less than the 40.0° we expected, but this is probably because the figure is not drawn to scale. This problem was a bit tricky because it required four key concepts (refraction, reflection, critical angle, and geometry) in order to find the solution. One practical extension of this problem is to consider what would happen to the exiting light if the angle of incidence were varied slightly. Would all the light still be reflected off surface 2, or would some light be refracted and pass through this second surface? See Figure 35.27 in the textbook.

61. A light ray of wavelength 589 nm is incident at an angle θ on the top surface of a block of polystyrene, as shown in Figure P35.61. (a) Find the maximum value of θ for which the refracted ray undergoes total internal reflection at the left vertical face of the block. **What If?** Repeat the calculation for the case in which the polystyrene block is immersed in (b) water and (c) carbon disulfide.

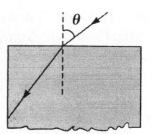

Figure P35.61

Solution

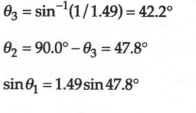

(a) The index of refraction (for 589 nm light) of each material is listed below:

Air:	1.00
Water:	1.33
Polystyrene:	1.49
Carbon disulfide:	1.63

For polystyrene **surrounded by air,**

the critical angle for total internal reflection is $\quad \theta_3 = \sin^{-1}(1/1.49) = 42.2°$

and then from geometry, $\quad \theta_2 = 90.0° - \theta_3 = 47.8°$

From Snell's law, $\quad \sin\theta_1 = 1.49\sin 47.8°$

This has no solution; thus, the real maximum value for θ_1 is 90.0°.

Total internal reflection always occurs. ◊

(b) For polystyrene **surrounded by water,** we have $\quad \theta_3 = \sin^{-1}\left(\dfrac{1.33}{1.49}\right) = 63.2°$

and $\quad \theta_2 = 26.8°$

From Snell's law, $n_1\sin\theta_1 = n_2\sin\theta_2$: $\quad 1.33\sin\theta_1 = 1.49\sin 26.8°$

and $\quad \theta_1 = 30.3°$ ◊

(c) **This is not possible** since the beam is initially traveling in a medium of lower index of refraction. ◊

63. A shallow glass dish is 4.00 cm wide at the bottom, as shown in Figure P35.63. When an observer's eye is placed as shown, the observer sees the edge of the bottom of the empty dish. When this dish is filled with water, the observer sees the center of the bottom of the dish. Find the height of the dish.

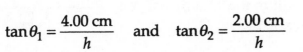

Figure P35.63

Solution

$$\tan\theta_1 = \frac{4.00 \text{ cm}}{h} \quad \text{and} \quad \tan\theta_2 = \frac{2.00 \text{ cm}}{h}$$

$$\tan^2\theta_1 = (2.00\tan\theta_2)^2 = 4.00\tan^2\theta_2$$

$$\frac{\sin^2\theta_1}{1-\sin^2\theta_1} = 4.00\left(\frac{\sin^2\theta_2}{1-\sin^2\theta_2}\right) \qquad [1]$$

Snell's law in this case is:

$$n_1\sin\theta_1 = n_2\sin\theta_2$$

$$\sin\theta_1 = 1.333\sin\theta_2$$

Squaring both sides,

$$\sin^2\theta_1 = 1.777\sin^2\theta_2 \qquad [2]$$

Substituting [2] into [1] yields

$$\frac{1.777\sin^2\theta_2}{1-1.777\sin^2\theta_2} = 4.00\left(\frac{\sin^2\theta_2}{1-\sin^2\theta_2}\right)$$

Defining

$$x = \sin^2\theta_2$$

and solving for x,

$$\frac{0.444}{1-1.777x} = \frac{1}{1-x}$$

$$0.444 - 0.444x = 1 - 1.777x$$

$$x = 0.417$$

From x we can solve for θ_2:

$$\theta_2 = \sin^{-1}\sqrt{0.417} = 40.2°$$

and

$$h = \frac{2.00 \text{ cm}}{\tan\theta_2} = \frac{2.00 \text{ cm}}{\tan 40.2°} = 2.36 \text{ cm} \qquad \Diamond$$

65. Derive the law of reflection (Eq. 35.2) from Fermat's principle of least time. (See the procedure outlined in Section 35.9 for the derivation of the law of refraction from Fermat's principle.)

Solution

To derive the law of **reflection**, locate point O so that the time of travel from point A to point B will be minimum. Let $c + d = R$, so that we can replace d with $R - c$. Since the light travels through the same medium for both paths, you only need to minimize the distance traveled, D.

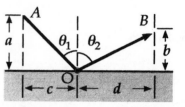

$$D = \sqrt{a^2 + c^2} + \sqrt{b^2 + (R - c)^2}$$

Require that $\dfrac{dD}{dc} = 0$:

$$\frac{2c}{\sqrt{a^2 + c^2}} - \frac{2(R - c)}{\sqrt{b^2 + (R - c)^2}} = 0$$

By inspecting the triangles of the diagram, it follows that

$$\sin\theta_1 = \frac{c}{\sqrt{a^2 + c^2}}$$

and

$$\sin\theta_2 = \frac{R - c}{\sqrt{b^2 + (R - c)^2}}$$

Substituting these values,

$$\sin\theta_1 = \sin\theta_2$$

Therefore,

$$\theta_1 = \theta_2 \qquad \Diamond$$

Chapter 36
IMAGE FORMATION

EQUATIONS AND CONCEPTS

In using the equations related to the image-forming properties of spherical mirrors, spherical refracting surfaces, and thin lenses, you must be very careful to use the correct algebraic sign for each physical quantity. *The sign conventions appropriate for the equation forms stated here are summarized in the Suggestions, Skills and Strategies section.*

Lateral magnification is defined as the ratio of image height to the object height. *This ratio always has a value of +1 for a plane mirror since the erect image is always the same size as the object.*

$$M = \frac{\text{image height}}{\text{object height}} = \frac{h'}{h} \qquad (36.1)$$

The **lateral magnification of a spherical mirror** can be stated either as a ratio of image size to object size or in terms of the ratio of image distance to object distance.

$$M = \frac{h'}{h} = -\frac{q}{p} \qquad (36.2)$$

The **focal length of a spherical mirror** is the distance from the vertex of the mirror to the focal point, *F*, located midway between the center of curvature and the vertex of the mirror. *For a concave mirror, the focal point is in front of the mirror and f is positive. For a convex mirror, the focal point is back of the mirror and f is negative.*

$$f = \frac{R}{2} \qquad (36.5)$$

The **mirror equation** is used to locate the position of an image formed by reflection of paraxial rays.

$$\frac{1}{p} + \frac{1}{q} = \frac{1}{f} \qquad (36.6)$$

211

A **single spherical refracting surface** of radius R which separates two media whose indices of refraction are n_1 and n_2 will form an image of an object. This equation is valid regardless of the relative values of the indices of refraction.

$$\frac{n_1}{p} + \frac{n_2}{q} = \frac{n_2 - n_1}{R}$$ (36.8)

A **flat refracting surface** ($R = \infty$) forms an image that is on the same side of the refracting surface as the object.

$$q = -\frac{n_2}{n_1} p$$ (36.9)

In the case of a **thin lens**, several equations can be used to relate different combinations of lens parameters and image characteristics.

Equation 36.14 is valid when the lens thickness is much less than R_1 and R_2. A given radius is positive if the center of curvature is behind the lens and negative if the center of curvature is in front of the lens. In the figure, R_1 is positive and R_2 is negative. *If the lens is surrounded by a medium other than air, the index of refraction given in Equations 36.14 and 36.15 must be the ratio of the index of refraction of the lens to that of the surrounding medium.*

$$\frac{1}{p} + \frac{1}{q} = (n-1)\left(\frac{1}{R_1} - \frac{1}{R_2}\right)$$ (36.14)

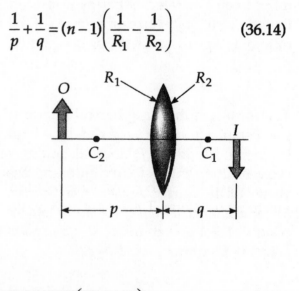

The **lens makers' equation** relates the focal length to the physical properties of the lens. *A hollow convex lens ("air lens"), if immersed in water, would have a negative focal length.*

$$\frac{1}{f} = (n-1)\left(\frac{1}{R_1} - \frac{1}{R_2}\right)$$ (36.15)

The **thin lens equation** can be used to find the image location when the focal length is known. Each lens has two focal points as illustrated in the figure below.

$$\frac{1}{p} + \frac{1}{q} = \frac{1}{f} \qquad (36.16)$$

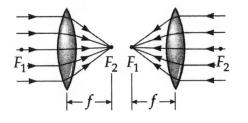 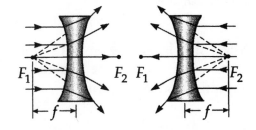

The **lateral magnification** of a thin lens is the same as that of a spherical mirror.

$$M = \frac{h'}{h} = -\frac{q}{p}$$

Two thin lenses in contact are equivalent to a single lens with a focal length given by Equation 36.17. *If the two lenses are of the same type (converging or diverging) the equivalent focal length will be lass than the focal length of either single lens.*

$$\frac{1}{f} = \frac{1}{f_1} + \frac{1}{f_2} \qquad (36.17)$$

Image forming properties of several optical instruments are described in the following paragraphs and equations.

Camera: The **light intensity** I incident on the film is inversely proportional to the square of a ratio of the focal length of the lens to its diameter. The *f*-number equals the ratio of the focal length to the lens diameter.

$$I \propto \frac{1}{(f/D)^2} \propto \frac{1}{(f\text{-number})^2} \qquad (36.19)$$

$$f\text{-number} \equiv \frac{f}{D} \qquad (36.18)$$

Eye: The **power of a lens** measured in diopters is the reciprocal of the focal length measured in meters (including the correct algebraic sign).

$$P = \frac{1}{f}$$

Simple magnifier: Angular magnification is the ratio of the angle subtended by an object using a lens to the angle subtended by the object at the near point without a lens.

$$m \equiv \frac{\theta}{\theta_0}$$ (36.20)

When the **image is at the near point** (25 cm), the angular magnification is maximum.

$$m_{max} = 1 + \frac{25 \text{ cm}}{f}$$ (36.22)

When the **image is at infinity** (most relaxed for the eye), the angular magnification is minimum.

$$m_{min} = \frac{25 \text{ cm}}{f}$$ (36.23)

Compound microscope: The overall magnification is the product of the lateral magnification (M_o) of the objective lens and the angular magnification (m_e) of the eyepiece. The two lenses are separated by a distance L.

$$M = M_o m_e = -\frac{L}{f_o}\left(\frac{25 \text{ cm}}{f_e}\right)$$ (36.24)

Astronomical telescope: The angular magnification is equal to the ratio of the objective focal length to the eyepiece focal length. *The two converging lenses are separated by a distance equal to the sum of their focal lengths.*

$$m = -\frac{f_o}{f_e}$$ (36.25)

SUGGESTIONS, SKILLS, AND STRATEGIES

A major portion of this chapter is devoted to the development and presentation of equations, which can be used to determine the location and nature of images formed by various optical components acting either singly or in combination. It is essential that these equations be used with the correct algebraic sign associated with each quantity involved. You must understand clearly the sign conventions for mirrors, refracting surfaces, and lenses. The following discussion represents a review of these sign conventions.

SIGN CONVENTIONS FOR SPHERICAL MIRRORS

Equations:
$$\frac{1}{p}+\frac{1}{q}=\frac{1}{f}=\frac{2}{R}$$
$$M=\frac{h'}{h}=-\frac{q}{p}$$

The front side of the mirror is the region on which light rays are incident and reflected.

> p is + if the object is in front of the mirror (real object).
> p is − if the object is in back of the mirror (virtual object).
>
> q is + if the image is in front of the mirror (real image).
> q is − if the image is in back of the mirror (virtual image).
>
> Both f and R are + if the center of curvature is in front (concave mirror).
> Both f and R are − if the center of curvature is in back (convex mirror).
>
> If M is positive, the image is upright.
> If M is negative, the image is inverted.

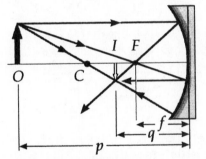

(a) Concave Mirror
$p > 2f$: $q+, f+, R+$
Image real, inverted, diminished

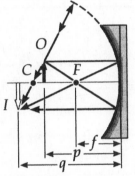

(b) Concave Mirror
$2f > p > f$: $q+, f+, R+$
Image real, inverted, enlarged

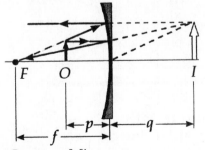

(c) Concave Mirror
$p < f$: $q-, f+, R+$
Image virtual, upright, enlarged

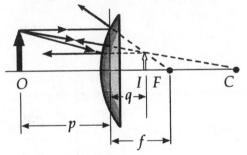

(d) Convex Mirror
$p+, q-, f-, R-$
Image virtual, upright, diminished
for any object distance

Figure 36.1 Figures describing sign conventions for mirrors.

SIGN CONVENTIONS FOR REFRACTING SURFACES

Equations:

$$\frac{n_1}{p} + \frac{n_2}{q} = \frac{n_2 - n_1}{R} \qquad M = \frac{h'}{h} = -\frac{n_1 q}{n_2 p}$$

In the following table, the **front** side of the surface is the side **from which the light is incident**.

> p is $+$ if the object is in front of the surface (real object).
> p is $-$ if the object is in back of the surface (virtual object).
>
> q is $+$ if the image is in back of the surface (real image).
> q is $-$ if the image is in front of the surface (virtual image).
>
> R is $+$ if the center of curvature is in back of the surface.
> R is $-$ if the center of curvature is in front of the surface.
>
> n_1 refers to the index of refraction of the first medium (before refraction).
> n_2 is the index of refraction of the second medium (after refraction).

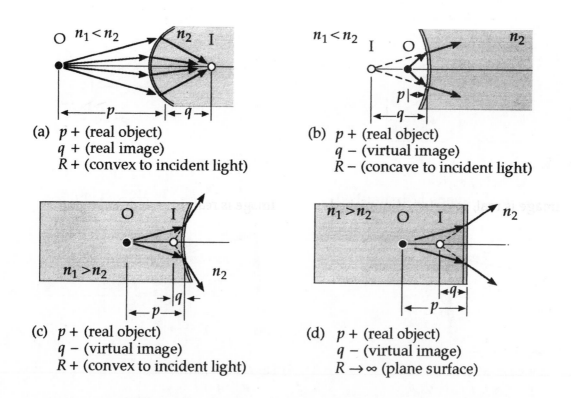

(a) $p +$ (real object)
     $q +$ (real image)
     $R +$ (convex to incident light)

(b) $p +$ (real object)
     $q -$ (virtual image)
     $R -$ (concave to incident light)

(c) $p +$ (real object)
     $q -$ (virtual image)
     $R +$ (convex to incident light)

(d) $p +$ (real object)
     $q -$ (virtual image)
     $R \rightarrow \infty$ (plane surface)

Figure 36.2   These figures describe sign conventions for refracting surfaces. The figures do not represent all possible cases; for example in figure (a) with p small enough that $q < 0$.

SIGN CONVENTIONS FOR THIN LENSES

Equations:
$$\frac{1}{p}+\frac{1}{q}=\frac{1}{f}=(n-1)\left(\frac{1}{R_1}-\frac{1}{R_2}\right) \qquad M=\frac{h'}{h}=-\frac{q}{p}$$

In the following table, the **front** of the lens is the **side from which the light is incident.**

> p is + if the object is in front of the lens.
> p is − if the object is in back of the lens.
>
> q is + if the image is in back of the lens.
> q is − if the image is in front of the lens.
>
> f is + if the lens is thickest at the center.
> f is − if the lens is thickest at the edges.
>
> R_1 and R_2 are + if the center of curvature is in back of the lens.
> R_1 and R_2 are − if the center of curvature is in front of the lens.

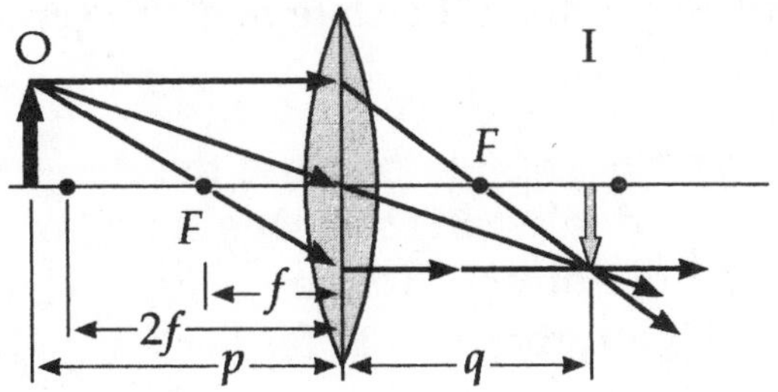

(a) Converging Lens ($f>0$)
$p>2f$: $p+$, $q+$
Image is real, inverted, diminished

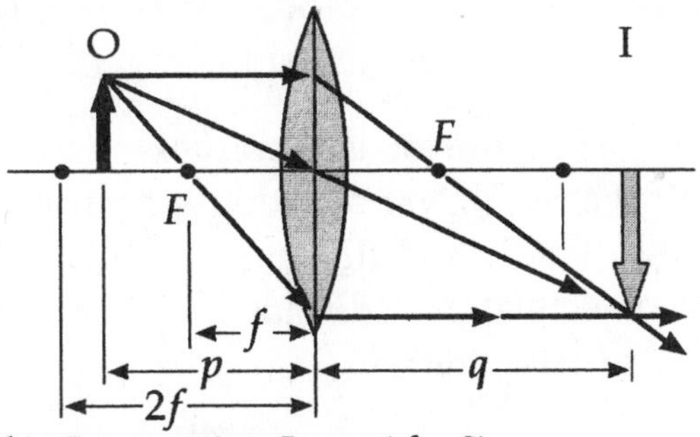

(b) Converging Lens ($f>0$)
$2f>p>f$: $p+$, $q+$
Image is real, inverted, enlarged

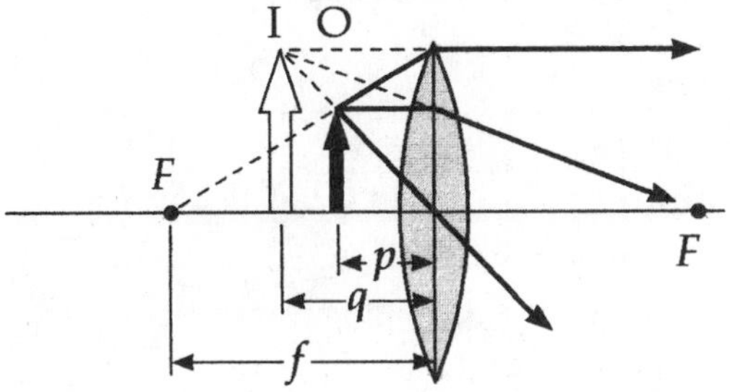

(c) Converging Lens ($f>0$)
$p<f$: $p+$, $q-$
Image is virtual, upright, enlarged

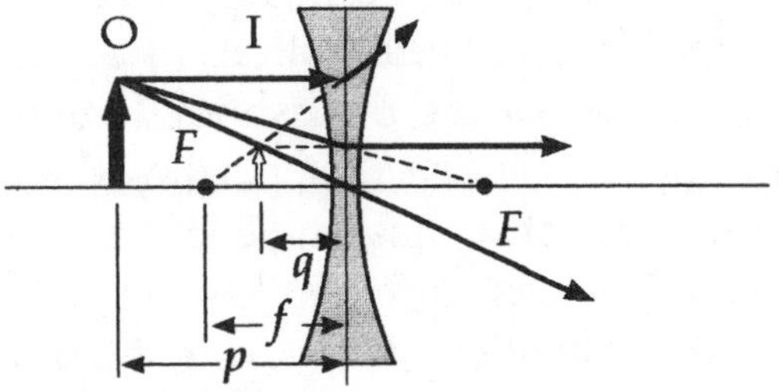

(d) Diverging Lens ($f<0$)
$p+$, $q-$
Image is virtual, upright, diminished

Figure 36.3 Figures describing the sign conventions for various thin lenses.

REVIEW CHECKLIST

You should be able to:

▷ Correctly identify the nature of the image (real or virtual, upright or inverted, enlarged or diminished) from the algebraic sign of the image distance and magnification.

▷ Correctly use the required equations and associated sign conventions to calculate the location of the image of a specified object as formed by a plane mirror, spherical mirror, plane refracting surface, spherical refracting surface, thin lens, or a combination of two or more of these devices. Determine the magnification and character of the image in each case. (Sections 36.1, 36.2, 36.3 and 36.4)

▷ Construct ray diagrams to determine the location and nature of the image of a given object when the geometrical characteristics of the optical device (lens or mirror) are known. (Sections 36.2 and 36.4)

▷ Make calculations of magnification for a simple magnifier, compound microscope, and refracting telescope. (Sections 36.8, 36.9 and 36.10)

ANSWERS TO SELECTED QUESTIONS

13. Why do some automobile mirrors have printed on them the statement "Objects in mirror are closer than they appear"? (See Fig. P36.19.)

Answer   The mirror is a convex mirror. This type is selected because the designers wish to give the driver a wide field of view, and a virtual, upright image for all object distances. The image distance is always less than the object distance. Nevertheless, because objects appear diminished, your brain can interpret them as farther away than the objects really are.

□   □   □   □

15. Explain why a fish in a spherical goldfish bowl appears larger than it really is.

Answer   As in the diagram, let the center of curvature C of the fishbowl and the bottom of the fish define the optical axis, intersecting the fishbowl at vertex V. A ray from the top of the fish that reaches the bowl along a radial line through C has angle of incidence zero and angle of refraction zero.   This ray exits the bowl unchanged in direction. A ray from the top of the fish to V is refracted to bend away from the normal.   Its extension back inside the fishbowl determines the location of the image and the characteristics of the image. It is upright, virtual, and enlarged.

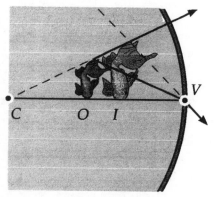

□   □   □   □

18. Lenses used in eyeglasses, whether converging or diverging, are always designed so that the middle of the lens curves away from the eye, like the center lenses of Figure 36.27a and b. Why?

Answer   With the meniscus design, when you direct your gaze near the outer circumference of the lens you receive a ray that has passed through glass with more nearly parallel surfaces of entry and exit. Thus, the lens minimally distorts the direction to the object you are looking at. If you wear glasses, you can demonstrate this by turning them around and looking through them the wrong way, maximizing the distortion.

☐   ☐   ☐   ☐

SOLUTIONS TO SELECTED PROBLEMS

3. Determine the minimum height of a vertical flat mirror in which a person 5′10″ in height can see his or her full image. (A ray diagram would be helpful.)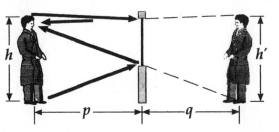

Solution

The flatness of the mirror is described by $R=\infty$, $f=\infty$, and $1/f=0$. By our general mirror equation,

$$\frac{1}{p}+\frac{1}{q}=\frac{1}{f} \qquad \text{or} \qquad q=-p$$

Thus, the image is as far behind the mirror as the person is in front. The magnification is then

$$M=\frac{-q}{p}=1=\frac{h'}{h} \qquad \text{so} \qquad h'=h=70.0 \text{ inches}$$

The required height of the mirror is defined by the triangle from the person's eyes to the top and bottom of the image, as shown. From the geometry of the triangle, we see that the mirror height must be:

$$h'\left(\frac{p}{p-q}\right)=h'\left(\frac{p}{2p}\right)=\frac{h'}{2}$$

Thus, the mirror must be at least 35.0 inches high.   ◊

9. A spherical convex mirror has a radius of curvature with a magnitude of 40.0 cm. Determine the position of the virtual image and the magnification for object distances of (a) 30.0 cm and (b) 60.0 cm. (c) Are the images upright or inverted?

Solution

The convex mirror is described by

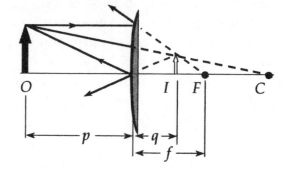

$$f = \frac{R}{2} = \frac{-40.0 \text{ cm}}{2} = -20.0 \text{ cm}$$

and $\dfrac{1}{p} + \dfrac{1}{q} = \dfrac{1}{f}$

(a) $\dfrac{1}{30.0 \text{ cm}} + \dfrac{1}{q} = \dfrac{1}{-20.0 \text{ cm}}$: $\qquad q = -12.0 \text{ cm}$ ◊

$$M = -\frac{q}{p} = -\left(\frac{-12.0 \text{ cm}}{30.0 \text{ cm}}\right) = +0.400$$ ◊

(c) As shown by the figure above, the image is behind the mirror, upright, virtual, and diminished. ◊

(b) $\dfrac{1}{60.0 \text{ cm}} + \dfrac{1}{q} = \dfrac{1}{-20.0 \text{ cm}}$: $\qquad q = -15.0 \text{ cm}$ ◊

$$M = \frac{-q}{p} = -\left(\frac{-15.0 \text{ cm}}{60.0 \text{ cm}}\right) = +0.250$$ ◊

(c) The principal ray diagram is an essential complement to the numerical description of the image. Add rays on this diagram for a 60-cm object distance.

Use your diagram to confirm that the image is behind the mirror, upright, virtual and diminished. ◊

11. A concave mirror has a radius of curvature of 60.0 cm. Calculate the image position and magnification of an object placed in front of the mirror at distances of (a) 90.0 cm and (b) 20.0 cm. (c) Draw ray diagrams to obtain the image characteristics in each case.

Solution

Conceptualize: It is always a good idea to first draw a ray diagram for any optics problem. This gives a qualitative sense of how the image appears relative to the object. From the ray diagrams below, we see that when the object is 90 cm from the mirror, the image will be real, inverted, diminished, and located about 45 cm in front of the mirror, midway between the center of curvature and the focal point. When the object is 20 cm from the mirror, the image will be virtual, upright, magnified, and located about 50 cm behind the mirror.

Categorize: The mirror equation can be used to find precise quantitative values.

Analyze:

(a) The mirror equation is applied using the sign conventions listed in **Suggestions, Skills, and Strategies:**

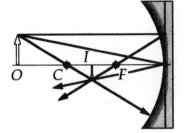

$$\frac{1}{p} + \frac{1}{q} = \frac{2}{R} \quad \text{or} \quad \frac{1}{90.0 \text{ cm}} + \frac{1}{q} = \frac{2}{60.0 \text{ cm}}$$

$$\frac{1}{q} = \frac{2}{60.0 \text{ cm}} - \frac{1}{90.0 \text{ cm}} = 0.0222 \text{ cm}^{-1}$$

$q = 45.0$ cm (real, in front of the mirror)

$$M = \frac{-q}{p} = -\frac{45.0 \text{ cm}}{90.0 \text{ cm}} = -0.500 \text{ (inverted)}$$

(b) We again use the mirror equation:

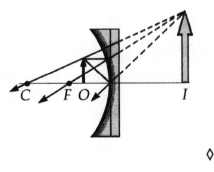

$$\frac{1}{p} + \frac{1}{q} = \frac{2}{R} \quad \text{or} \quad \frac{1}{20.0 \text{ cm}} + \frac{1}{q} = \frac{2}{60.0 \text{ cm}}$$

$$\frac{1}{q} = \frac{2}{60.0 \text{ cm}} - \frac{1}{20.0 \text{ cm}} = -0.0167 \text{ cm}^{-1}$$

$q = -60.0$ cm (virtual, behind the mirror)

$$M = -\frac{q}{p} = -\frac{-60.0 \text{ cm}}{20.0 \text{ cm}} = 3.00 \text{ (upright)}$$

Finalize: The calculated image characteristics agree well with our predictions. It is easy to miss a minus sign or to make a computational mistake when using the mirror-lens equation, so the qualitative values obtained from the ray diagrams are useful for a check on the reasonableness of the calculated values.

17. A spherical mirror is to be used to form, on a screen located 5.00 m from the object, an image five times the size of the object. (a) Describe the type of mirror required. (b) Where should the mirror be positioned relative to the object?

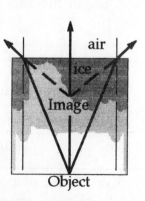

Solution

We are given that

$$q=(p+5.00 \text{ m}),\ M=-\frac{q}{p}\ \text{and}\ |M|=5.00$$

Since the image must be real, q must be positive, and $M=-5.00$ or $q=5.00p$

(b) Solving for the distance of the mirror from the object, $p+5.00 \text{ m}=5.00p$

$$p=1.25 \text{ m} \qquad \Diamond$$

(a) Applying Equation 36.16, $\dfrac{1}{f}=\dfrac{1}{p}+\dfrac{1}{q}:$ $\qquad \dfrac{1}{f}=\dfrac{1}{1.25 \text{ m}}+\dfrac{1}{6.25 \text{ m}}$

so that the focal length of the mirror is $\qquad f=1.04 \text{ m}$

Noting that the image is real, inverted and enlarged, we can say that the mirror must be concave, and must have a radius of curvature of $R=2f=2.08 \text{ m}$. $\qquad \Diamond$

---

21. A cubical block of ice 50.0 cm on a side is placed on a level floor over a speck of dust. Find the location of the image of the speck as viewed from above. The index of refraction of ice is 1.309.

Solution The upper surface of the block is a single refracting surface with zero curvature, and with infinite radius of curvature.

$$\frac{n_1}{p}+\frac{n_2}{q}=\frac{n_2-n_1}{R}\ \text{or}\ \frac{1.309}{50.0 \text{ cm}}+\frac{1}{q}=\frac{1.00-1.309}{\infty}=0$$

$$q=\frac{-50.0 \text{ cm}}{1.309}=-38.2 \text{ cm}$$

The speck of dust on the floor appears to be 38.2 cm below the upper surface of the ice. The image is virtual, upright, and actual size. $\qquad \Diamond$

---

23. A glass sphere ($n = 1.50$) with a radius of 15.0 cm has a tiny air bubble 5.00 cm above its center. The sphere is viewed looking down along the extended radius containing the bubble. What is the apparent depth of the bubble below the surface of the sphere?

Solution

From Equation 36.8,

$$\frac{n_1}{p} + \frac{n_2}{q} = \frac{n_2 - n_1}{R}$$

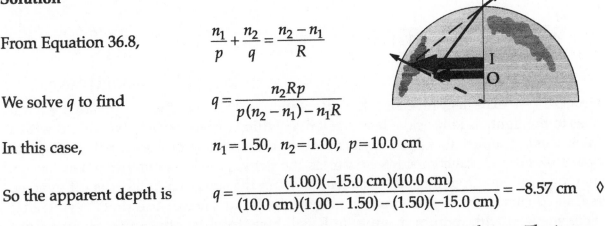

We solve q to find

$$q = \frac{n_2 R p}{p(n_2 - n_1) - n_1 R}$$

In this case,

$$n_1 = 1.50, \quad n_2 = 1.00, \quad p = 10.0 \text{ cm}$$

So the apparent depth is

$$q = \frac{(1.00)(-15.0 \text{ cm})(10.0 \text{ cm})}{(10.0 \text{ cm})(1.00 - 1.50) - (1.50)(-15.0 \text{ cm})} = -8.57 \text{ cm} \quad \Diamond$$

To sketch a ray diagram we use the wave under refraction model, as shown. The image is virtual, upright, and enlarged.

29. The left face of a biconvex lens has a radius of curvature of magnitude 12.0 cm, and the right face has a radius of curvature of magnitude 18.0 cm. The index of refraction of the glass is 1.44. (a) Calculate the focal length of the lens. (b) **What If?** Calculate the focal length the lens has after it is turned around to interchange the radii of curvature of the two faces.

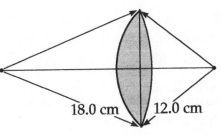

18.0 cm   12.0 cm

Solution

Conceptualize: Since this is a biconvex lens, the center is thicker than the edges, and the lens will tend to converge incident light rays. Therefore it has a positive focal length. Exchanging the radii of curvature amounts to turning the lens around so the light enters the opposite side first. However, this does not change the fact that the center of the lens is still thicker than the edges, so we should not expect the focal length of the lens to be different (assuming the thin-lens approximation is valid).

Categorize: The lens makers' equation can be used to find the focal length of this lens.

Analyze: The centers of curvature of the lens surfaces are on opposite sides, so the second surface has a negative radius:

(a) $\dfrac{1}{f} = (n-1)\left(\dfrac{1}{R_1} - \dfrac{1}{R_2}\right) = (1.44 - 1.00)\left(\dfrac{1}{12.0\ \text{cm}} - \dfrac{1}{-18.0\ \text{cm}}\right)$ so $f = 16.4\ \text{cm}$ ◊

(b) $\dfrac{1}{f} = 0.440\left(\dfrac{1}{18.0\ \text{cm}} - \dfrac{1}{-12.0\ \text{cm}}\right)$ so $f = 16.4\ \text{cm}$ ◊

Finalize: As expected, reversing the orientation of the lens does not change what it does to the light, as long as the lens is relatively thin (variations may be noticed with a thick lens). The fact that light rays can be traced forward or backward through an optical system is sometimes referred to as the **principle of reversibility**. We can see that the focal length of this biconvex lens is about the same magnitude as the average radius of curvature. A few approximations, useful as checks, are that a symmetric biconvex lens with radii of magnitude R will have focal length $f \approx R$; a plano-convex lens with radius R will have $f \approx R/2$; and a symmetric biconcave lens has $f \approx -R$. These approximations apply when the lens has $n \approx 1.5$, which is typical of many types of clear glass and plastic.

33. The nickel's image in Figure P36.33 has twice the diameter of the nickel and is 2.84 cm from the lens. Determine the focal length of the lens.

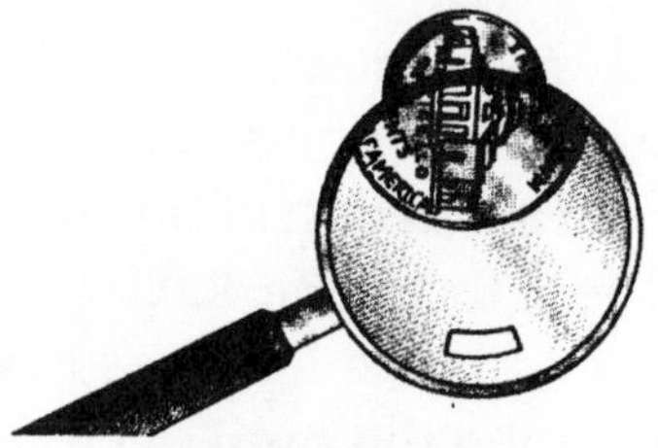

Solution   Looking through the lens, you see the image beyond the lens. Therefore, the image is virtual, with $q = -2.84$ cm.

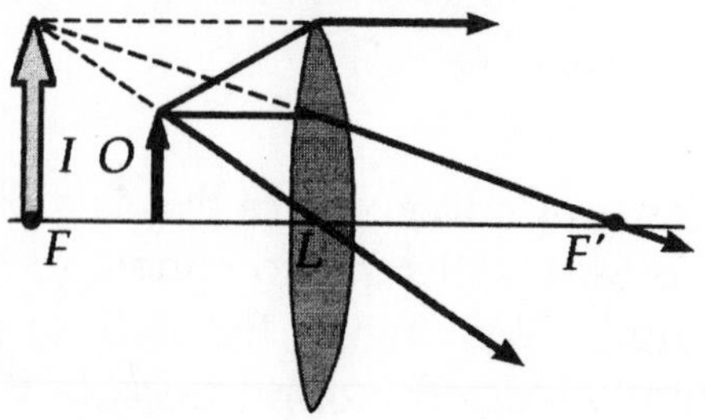

Figure P36.33

Now, $M = \dfrac{h'}{h} = 2 = -\dfrac{q}{p}$

so $p = -\dfrac{q}{2} = 1.42\ \text{cm}$

Thus, $f = \left(\dfrac{1}{p} + \dfrac{1}{q}\right)^{-1} = \left[\dfrac{1}{1.42\ \text{cm}} + \dfrac{1}{(-2.84\ \text{cm})}\right]^{-1} = 2.84\ \text{cm}$ ◊

37. An object is located 20.0 cm to the left of a diverging lens having a focal length $f = -32.0$ cm. Determine (a) the location and (b) the magnification of the image. (c) Construct a ray diagram for this arrangement.

Solution $\dfrac{1}{p} + \dfrac{1}{q} = \dfrac{1}{f}$ or $\dfrac{1}{20.0 \text{ cm}} + \dfrac{1}{q} = \dfrac{1}{-32.0 \text{ cm}}$

(a) So $q = -\left(\dfrac{1}{20.0 \text{ cm}} + \dfrac{1}{32.0 \text{ cm}} \right)^{-1} = -12.3$ cm ◊

(b) $M = -\dfrac{q}{p} = -\dfrac{(-12.3 \text{ cm})}{20.0 \text{ cm}} = 0.615$ ◊

(c) The image is virtual, upright, diminished; the ray diagram is given to the right. ◊

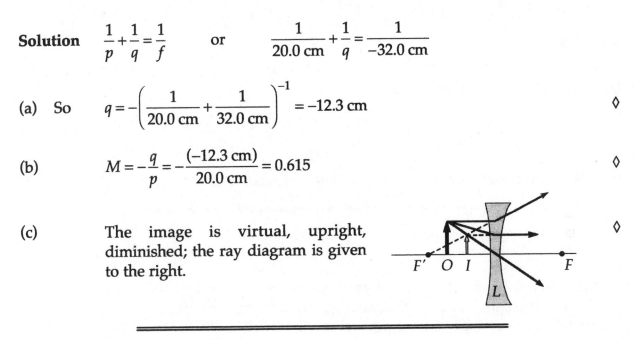

47. A nearsighted person cannot see objects clearly beyond 25.0 cm (her far point). If she has no astigmatism and contact lenses are prescribed for her, what power and type of lens are required to correct her vision?

Solution

The lens should take parallel light rays from a very distant object ($p = \infty$) and make them diverge from a virtual image at the woman's far point, which is 25.0 cm beyond the lens, at $q = -25.0$ cm.

Thus, $\dfrac{1}{f} = \dfrac{1}{p} + \dfrac{1}{q} = \dfrac{1}{\infty} + \dfrac{1}{-25.0 \text{ cm}}$

Hence, the power of the lens is $P = \dfrac{1}{f} = -\dfrac{1}{0.250 \text{ m}} = -4.00$ diopter

This is a diverging lens. ◊

53. The Yerkes refracting telescope has a 1.00-m diameter objective lens of focal length 20.0 m. Assume it is used with an eyepiece of focal length 2.50 cm. (a) Determine the magnification of the planet Mars as seen through this telescope. (b) Are the Martian polar caps right side up or upside down?

Solution

(a)   The angular magnification is

$$m = -\frac{f_o}{f_e} = \frac{-20.0 \text{ m}}{0.0250 \text{ m}} = -800$$ ◊

(b)   The minus sign means the image is inverted relative to the object.   ◊

65. A parallel beam of light enters a glass hemisphere perpendicular to the flat face, as shown in Figure P36.65. The magnitude of the radius is 6.00 cm, and the index of refraction is 1.560. Determine the point at which the beam is focused. (Assume paraxial rays.)

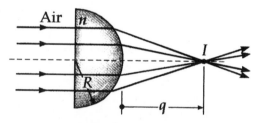

Figure P36.65

Solution

A hemisphere is too thick to be described as a thin lens. The light is undeviated on entry into the flat face. Because of this, we instead consider the light's exit from the second surface, for which $R = -6.00$ cm. The incident rays are parallel, so $p = \infty$.

Then $\dfrac{n_1}{p} + \dfrac{n_2}{q} = \dfrac{n_2 - n_1}{R}$ becomes $0 + \dfrac{1}{q} = \dfrac{1 - 1.56}{-6.00 \text{ cm}}$

and $q = 10.7$ cm   ◊

67. An object is placed 12.0 cm to the left of a diverging lens of focal length −6.00 cm. A converging lens of focal length 12.0 cm is placed a distance d to the right of the diverging lens. Find the distance d so that the final image is at infinity. Draw a ray diagram for this case.

Solution

From Equation 36.16, $q_1 = \dfrac{f_1 p_1}{p_1 - f_1} = \dfrac{(-6.00 \text{ cm})(12.0 \text{ cm})}{12.0 \text{ cm} - (-6.00 \text{ cm})} = -4.00 \text{ cm}$

When we require that $q_2 \rightarrow \infty$

Equation 36.16 becomes $p_2 = f_2 = 12.0 \text{ cm}$

Since the object for the converging lens must be 12.0 cm to its left, and since this is the image for the diverging lens which is 4.00 cm to **its** left, the two lenses must be separated by 8.00 cm.

Mathematically, $p_2 = d - (-4.00 \text{ cm})$

$d + 4.00 \text{ cm} = f_2 = 12.0 \text{ cm}$ and $d = 8.00 \text{ cm}$ ◊

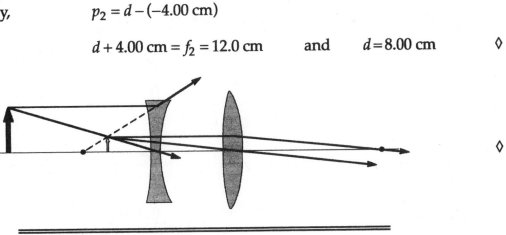

69. The disk of the Sun subtends an angle of 0.533° at the Earth. What are the position and diameter of the solar image formed by a concave spherical mirror with a radius of curvature of 3.00 m?

Solution For the mirror, $f = R/2 = +1.50$ m. In addition, because the distance to the Sun is so much larger than any other figures, we can take $p = \infty$.

The mirror equation, $\dfrac{1}{p} + \dfrac{1}{q} = \dfrac{1}{f}$ then gives $q = f = 1.50$ m. ◊

Now, in $M = -q/p = h'/h$, the magnification is nearly zero, but we can be more precise: the definition of radian measure means that h/p is the angular diameter of the object. Thus the image diameter is

$$h' = -\frac{hq}{p} = (-0.533°)\left(\frac{\pi}{180} \text{ rad}/\text{deg}\right)(1.50 \text{ m}) = -0.014\,0 \text{ m} = -1.40 \text{ cm} \qquad ◊$$

71. In a darkened room, a burning candle is placed 1.50 m from a white wall. A lens is placed between candle and wall at a location that causes a larger, inverted image to form on the wall. When the lens is moved 90.0 cm toward the wall, another image of the candle is formed. Find (a) the two object distances that produce the specified images and (b) the focal length of the lens. (c) Characterize the second image.

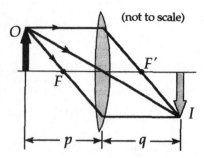

Solution

Originally,

$$p_1 + q_1 = 1.50 \text{ m}$$

In the final situation,

$$p_2 = p_1 + 0.900 \text{ m}$$

and

$$q_2 = q_1 - 0.900 \text{ m} = (1.50 \text{ m} - p_1) - 0.900 \text{ m} = 0.600 \text{ m} - p_1$$

Our lens equation is

$$\frac{1}{p_1} + \frac{1}{q_1} = \frac{1}{f} = \frac{1}{p_2} + \frac{1}{q_2}$$

Substituting, we have

$$\frac{1}{p_1} + \frac{1}{1.50 \text{ m} - p_1} = \frac{1}{p_1 + 0.900} + \frac{1}{0.600 - p_1}$$

Adding the fractions,

$$\frac{1.50 \text{ m} - p_1 + p_1}{p_1(1.50 \text{ m} - p_1)} = \frac{0.600 - p_1 + p_1 + 0.900}{(p_1 + 0.900)(0.600 - p_1)}$$

Simplified, this becomes

$$p_1(1.50 \text{ m} - p_1) = (p_1 + 0.900)(0.600 - p_1)$$

(a) Thus,

$$p_1 = \frac{0.540}{1.80} \text{ m} = 0.300 \text{ m} \qquad \diamond$$

$$p_2 = p_1 + 0.900 \text{ m} = 1.20 \text{ m} \qquad \diamond$$

(b)

$$\frac{1}{f} = \frac{1}{0.300 \text{ m}} + \frac{1}{1.50 \text{ m} - 0.300 \text{ m}}$$

and

$$f = 0.240 \text{ m} \qquad \diamond$$

(c) The second image is real, inverted, and diminished, with

$$M = \frac{-q_2}{p_2} = -0.250 \qquad \diamond$$

Chapter 37
INTERFERENCE OF LIGHT WAVES

EQUATIONS AND CONCEPTS

In **Young's double-slit experiment**, two slits, S_1 and S_2, separated by a distance d, serve as monochromatic coherent sources. The light intensity at any point on the screen is the resultant of light reaching the screen from both slits. *As illustrated in the figure, a point P on the screen can be identified by the angle θ or by the distance y from the center of the screen.*

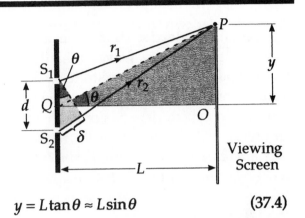

$$y = L\tan\theta \approx L\sin\theta \tag{37.4}$$

A **path difference** δ arises because waves from S_1 and S_2 travel unequal distances to reach a point on the screen (except the center point). *The value of δ determines whether the waves from the two slits arrive in phase or out of phase.*

$$\delta = r_2 - r_1 = d\sin\theta \tag{37.1}$$

Constructive interference (bright fringes) will appear at points on the screen for which the path difference is equal to an integer multiple of the wavelength. The positions of bright fringes can also be located by calculating their vertical distances from the center of the screen (y). *In each case, the number m is called the order number of the fringe. The central bright fringe ($\theta = 0$, $m = 0$) is called the zeroth-order maximum.*

$$\delta = d\sin\theta_{\text{bright}} = m\lambda \tag{37.2}$$
$$m = 0, \pm 1, \pm 2, \ldots$$

$$y_{\text{bright}} \approx \frac{\lambda L}{d}m \tag{37.5}$$
$$m = 0, \pm 1, \pm 2, \ldots$$

Destructive interference (dark fringes) will occur at points on the screen which correspond to path differences of an odd multiple of half wavelengths. *For these points, waves which leave the two slits in phase arrive at the screen 180° out of phase.*

$$\delta = d\sin\theta_{\text{dark}} = \left(m + \tfrac{1}{2}\right)\lambda \tag{37.3}$$
$$m = 0, \pm 1, \pm 2, \ldots$$

$$y_{\text{dark}} \approx \frac{\lambda L}{d}\left(m + \tfrac{1}{2}\right) \tag{37.6}$$
$$m = 0, \pm 1, \pm 2, \ldots$$

The **phase difference** ϕ between the two waves at any point on the screen depends on the path difference at that point.

$$\phi = \frac{2\pi}{\lambda}\delta = \frac{2\pi}{\lambda}d\sin\theta \qquad (37.8)$$

The **resultant electric field** (due to the superposition of two waves initially in phase and of equal amplitude, E_0) at some point on the screen has a magnitude which depends on the phase difference. *The frequency of the resultant wave has the same frequency as the that of the two coherent sources; the amplitude, $2E_0\cos(\phi/2)$, is multiplied by a factor that depends on the phase difference.*

$$E_p = 2E_0\cos\left(\frac{\phi}{2}\right)\sin\left(\omega t + \frac{\phi}{2}\right) \qquad (37.10)$$

The **average light intensity** I at any point P on the screen is proportional to the square of the amplitude of the resultant wave. The average intensity can be expressed

- as a function of phase difference ϕ;

$$I = I_{max}\cos^2\left(\frac{\phi}{2}\right) \qquad (37.11)$$

- as a function of $\sin\theta$, where θ is the angle that locates a point on the screen (see figure above showing the double-slit arrangement); or

$$I = I_{max}\cos^2\left(\frac{\pi d\sin\theta}{\lambda}\right) \qquad (37.12)$$

- as a function of the vertical distance y from the center of the screen.

$$I \approx I_{max}\cos^2\left(\frac{\pi d}{\lambda L}y\right) \qquad (37.13)$$

Increasing the number of equally spaced slits will increase the number of secondary maxima; the principal maxima will become narrower but remain fixed in position. See Figure 37.14 of the textbook.

The wavelength of light in a medium having an index of refraction n is smaller than the wavelength in vacuum.

$$\lambda_n = \frac{\lambda}{n} \qquad (37.14)$$

Interference in thin films depends on wavelength, film thickness, and the indices of refraction of the film and surrounding media. *Differences in phase may be due to path difference or phase change upon reflection.* There are two general cases (refer to diagram at right):

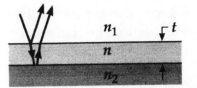

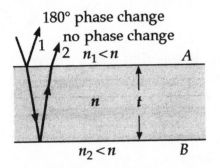

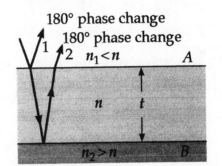

Case (1) **Phase change at only one film surface** ($n_1 < n$ and $n_2 < n$) or ($n_1 > n$ and $n_2 > n$)

Indices of refraction of media on both sides of the film are less than that of the film (figure above left) or both greater than that of the film.

Constructive interference,

$$2nt = \left(m + \tfrac{1}{2}\right)\lambda \qquad (37.16)$$

Destructive interference,

$$2nt = m\lambda \qquad (37.17)$$

$$(m = 0, 1, 2, \ldots)$$

Case (2) **Phase changes at both or neither surface** ($n_1 < n < n_2$ or $n_1 > n > n_2$)

Film is between two media either of which has an index of refraction greater than that of the film and the other a smaller index (figure above right).

Constructive interference,

$$2nt = m\lambda \qquad (37.17)$$

Destructive interference,

$$2nt = \left(m + \tfrac{1}{2}\right)\lambda \qquad (37.16)$$

$$(m = 0, 1, 2, \ldots)$$

Note that in Case (2) the roles of equations for Case (1) are reversed.

SUGGESTIONS, SKILLS, AND STRATEGIES

THIN-FILM INTERFERENCE PROBLEMS

• Identify the thin film from which interference effects are being observed.

• The type of interference that occurs in a specific problem is determined by the phase relationship between that portion of the wave reflected at the upper surface of the film and that portion reflected at the lower surface of the film.

• Phase differences between the two portions of the wave occur because of differences in the distances traveled by the two portions and by phase changes occurring upon reflection.

Phase Difference due to Path Difference

The wave reflected from the lower surface of the film has to travel a distance equal to twice the thickness of the film before it returns to the upper surface of the film where it interferes with that portion of the wave reflected at the upper surface.

Phase Change due to Reflection

When a wave traveling in a particular medium reflects off a surface having a higher index of refraction than the one it is in, a 180° phase shift occurs. This has the same effect as if the wave lost $\frac{1}{2}\lambda$. This effect must be considered in addition to the phase difference related to the greater distance traveled by one of the waves.

• When distance and phase changes upon reflection are both taken into account:

Constructive interference will occur when the phase difference is an integral multiple of λ — zero, λ, 2λ, 3λ . . .

Destructive interference will occur when the phase difference is an odd number of half wavelengths — $\frac{1}{2}\lambda$, $\frac{3}{2}\lambda$, $\frac{5}{2}\lambda$, . . .

PHASOR ADDITION

This technique offers a convenient alternative to the algebraic method for finding the resultant wave amplitude at some point on a screen. This is especially true when a large number of waves are to be combined. The method of phasor addition is outlined in the following steps and is illustrated in the figure for the case of three equal amplitude waves differing in phase by an angle of ϕ.

- Draw the phasors representing the waves end to end. The angle between successive phasors is equal to the phase angle between the waves from successive source slits. The length of each phasor, which must be drawn to scale, is proportional to the magnitude of the wave it represents.

- The resultant phasor is the vector sum (the vector from the tail of the first phasor to the head of the last one) of the individual phasors.

- The phase angle (α) of the resultant wave is the angle between the direction of the resultant phasor and the direction of the first phasor.

- The resultant electric field is $E_P = E_R \sin(\omega t + \alpha)$. The resultant phasor E_R is a rotating phasor of constant magnitude and angular frequency ω and is the vector sum of the three individual phasors. E_P is the projection of E_R on the vertical axis and represents the combination of the three phasors at any instant of time. E_P varies in magnitude from zero to $|E_R|$.

REVIEW CHECKLIST

You should be able to:

▷ Describe Young's double-slit experiment to demonstrate the wave nature of light. Account for the phase difference between light waves from the two sources as they arrive at a given point on the screen. State the conditions for constructive and destructive interference in terms of each of the following: path difference δ, phase difference ϕ, distance from the center of the screen y, and angle subtended by the observation point at the source mid-point θ. (Section 37.2)

▷ Calculate the ratio of average intensities at two points in a double-slit interference pattern. (Section 37.3)

▷ Construct and use a phasor diagram to determine the amplitude and phase of the wave which is the resultant of two or three coherent sources. (Section 37.4)

▷ State the conditions, and write the corresponding equations, for constructive and destructive interference in thin films considering both path difference and any expected phase changes due to reflection. Calculate the minimum film thickness to produce constructive/destructive interference in a film between media of known indices of refraction. (Section 37.6)

▷ Describe the Michelson interferometer and make calculations based on this instrument. (Section 37.7)

ANSWERS TO SELECTED QUESTIONS

1. What is the necessary condition on the path length difference between two waves that interfere (a) constructively and (b) destructively?

Answer (a) Two waves interfere constructively if their path difference is either zero or some integral multiple of the wavelength; that is, if the path difference equals $m\lambda$. (b) Two waves interfere destructively if their path difference is an odd multiple of one-half of a wavelength; that is, if the path difference equals $(m+1/2)\lambda$.

□ □ □ □

4. In Young's double-slit experiment, why do we use monochromatic light? If white light is used, how would the pattern change?

Answer Every color produces its own pattern, with a spacing between the maxima that is characteristic of the wavelength. With several colors, the patterns are superimposed, and it can become difficult to pick out a single maximum. Using monochromatic light can eliminate this problem.

With white light, the central maximum is white. The first side maximum is a full spectrum, with violet on the inside and red on the outside. The second side maximum is a full spectrum also, but red in the second maximum overlaps the violet in the third maximum. At larger angles, the light soon starts mixing to white again, though it is often so faint that you would call it gray.

□ □ □ □

SOLUTIONS TO SELECTED PROBLEMS

3. Two radio antennas separated by 300 m as shown in Figure P37.3 simultaneously broadcast identical signals at the same wavelength. A radio in a car traveling due north receives the signals. (a) If the car is at the position of the second maximum, what is the wavelength of the signals? (b) How much farther must the car travel to encounter the next minimum in reception? (**Note:** Do not use the small-angle approximation in this problem.)

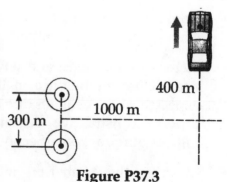

Figure P37.3

Solution   Note that with the conditions given, the small angle approximation **does not work well**. That is, $\sin\theta$, $\tan\theta$, and θ are significantly different. We use the Fraunhofer interference model, treating the waves from the two sources as moving along essentially parallel rays.

(a)   At the $m=2$ maximum,

$$\tan\theta = \frac{400 \text{ m}}{1\,000 \text{ m}} = 0.400$$

and $\theta = 21.8°$ so $\lambda = \frac{d\sin\theta}{m} = \frac{(300 \text{ m})(\sin 21.8°)}{2} = 55.7 \text{ m}$ ◊

(b)   The next minimum encountered is the $m=2$ minimum.

At that point $d\sin\theta = \left[m+\frac{1}{2}\right]\lambda$: $d\sin\theta = \frac{5}{2}\lambda$

or $\sin\theta = \frac{5\lambda}{2d} = \frac{5(55.7 \text{ m})}{2(300 \text{ m})} = 0.464$

and $\theta = 27.7°$ so $y = (1\,000 \text{ m})(\tan 27.7°) = 524 \text{ m}$

Therefore, the car must travel an additional 124 m.   ◊

If we considered Fresnel interference, we would more precisely find

(a) $\lambda = \frac{1}{2}\left(\sqrt{550^2 + 1\,000^2} - \sqrt{250^2 + 1\,000^2}\right) = 55.2 \text{ m}$

(b) $\Delta y = 123 \text{ m}$

═══════════════════════════════════

5.   Young's double-slit experiment is performed with 589-nm light and a distance of 2.00 m between the slits and the screen. The tenth interference minimum is observed 7.26 mm from the central maximum. Determine the spacing of the slits.

Solution

Conceptualize: For the situation described, the observed interference pattern is very narrow (the minima are less than 1 mm apart when the screen is 2 m away). In fact, the minima and maxima are so close together that it would probably be difficult to resolve adjacent maxima, so the pattern might look like a solid blur to the naked eye. Since the angular spacing of the pattern is inversely proportional to the slit width, we should expect that for this narrow pattern, the space between the slits will be larger than the typical fraction of a millimeter, and certainly much greater than the wavelength of the light ($d \gg \lambda = 589$ nm).

Categorize: Since we are given the location of the tenth minimum for this interference pattern, we should use the equation for **destructive interference** from a double slit. The figure for Problem 7 shows the critical variables for this problem.

Analyze: In the equation $d\sin\theta = \left[m+\frac{1}{2}\right]\lambda$, the first minimum is described by $m=0$ and the tenth by $m=9$.

So, $\sin\theta = \frac{\lambda}{d}\left[9+\frac{1}{2}\right]$

Also, $\tan\theta = y/L$, but for small θ, $\sin\theta \approx \tan\theta$. Thus, the distance between the slits is,

$$d = \frac{9.5\lambda}{\sin\theta} = \frac{9.5\lambda L}{y} = \frac{9.5(5890\times10^{-10}\text{ m})(2.00\text{ m})}{7.26\times10^{-3}\text{ m}} = 1.54\times10^{-3}\text{ m} = 1.54\text{ mm} \qquad \diamond$$

Finalize: The spacing between the slits is relatively large, as we expected (about 3000 times greater than the wavelength of the light). In order to more clearly distinguish between maxima and minima, the pattern could be expanded by increasing the distance to the screen. However, as L is increased, the overall pattern would be less bright as the light expands over a larger area, so that beyond some distance, the light would be too dim to see.

---

7. Two narrow, parallel slits separated by 0.250 mm are illuminated by green light ($\lambda=546.1$ nm). The interference pattern is observed on a screen 1.20 m away from the plane of the slits. Calculate the distance (a) from the central maximum to the first bright region on either side of the central maximum and (b) between the first and second dark bands.

Solution

Conceptualize: The spacing between adjacent maxima and minima should be fairly uniform across the pattern as long as the width of the pattern is much less than the distance to the screen (so that the small angle approximation is valid). The separation between fringes should be at least a millimeter if the pattern can be easily observed with a naked eye.

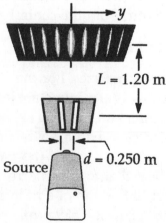

Categorize: The bright regions are areas of constructive interference and the dark bands are destructive interference, so the corresponding double-slit equations will be used to find the y distances.

It can be confusing to keep track of four different symbols for distances. Three are shown in the drawing to the right. Note that:

> y is the unknown distance from the bright central maximum ($m=0$) to another maximum or minimum on either side of the center of the interference pattern.

> λ is the wavelength of the light, determined by the source.

Analyze:

(a) For **very small** θ $\qquad$ $\sin\theta \approx \tan\theta$ $\quad$ and $\quad$ $\tan\theta = y/L$

and the equation for constructive interference

$$\sin\theta = m\lambda/d \qquad\qquad \textbf{[Eq. 37.2]}$$

becomes $\qquad\qquad y_{bright} \approx (\lambda L/d)m \qquad\qquad \textbf{[Eq. 37.5]}$

Substituting values, $\qquad y_{bright} = \dfrac{(546.1\times10^{-9}\text{ m})(1.20\text{ m})}{0.250\times10^{-3}\text{ m}}(1) = 2.62\text{ mm}$ $\qquad$ ◊

(b) If you have trouble remembering whether Equation 37.5 or Equation 37.6 applies to a given situation, you can instead remember that the first bright band is in the center, and dark bands are halfway between bright bands. Thus, Equation 37.5 describes them all, with $m = 0, 1, 2, \ldots$ for bright bands, and with $m = 0.5, 1.5, 2.5, \ldots$ for dark bands.

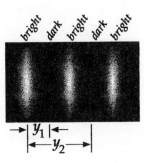

The dark band version of Eq. 37.5 is simply Eq. 37.6:

$$y_{dark} = \frac{\lambda L}{d}\left[m+\frac{1}{2}\right] \quad \text{or} \quad \Delta y_{dark} = \frac{\lambda L}{d}\left[1+\frac{1}{2}\right] - \frac{\lambda L}{d}\left[0+\frac{1}{2}\right] = \frac{\lambda L}{d} = 2.62\text{ mm} \qquad ◊$$

Finalize: This spacing is large enough for easy resolution of adjacent fringes. The distance between minima is the same as the distance between maxima. We expected this equality since the angles are small:

$$\theta = (2.62\text{ mm})/(1.20\text{ m}) = 0.00218\text{ rad} = 0.125°$$

When the angular spacing exceeds about 3°, then $\sin\theta$ differs from $\tan\theta$ when written to three significant figures.

13. In Figure 37.5 let $L = 1.20$ m and $d = 0.120$ mm and assume that the slit system is illuminated with monochromatic 500-nm light. Calculate the phase difference between the two wavefronts arriving at P when (a) $\theta = 0.500°$ and (b) $y = 5.00$ mm. (c) What is the value of θ for which the phase difference is 0.333 rad? (d) What is the value of θ for which the path difference is $\lambda/4$?

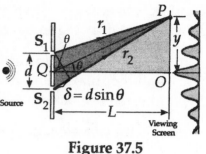

Figure 37.5

Solution

(a) The path difference is $\delta = d\sin\theta = (0.120\times10^{-3} \text{ m})(\sin 0.500°) = 1.05\times10^{-6}$ m

The phase difference is $\phi = \dfrac{2\pi\delta}{\lambda} = \dfrac{2\pi(1.05\times10^{-6} \text{ m})}{500\times10^{-9} \text{ m}} = 13.2$ rad   ◊

This is equivalent to $\phi = 0.593$ rad $= 34.0°$ ◊

(b) $\tan\theta = \dfrac{y}{L} = \dfrac{5.00\times10^{-3} \text{ m}}{1.20 \text{ m}} \approx \sin\theta$

$\phi = \dfrac{2\pi d\sin\theta}{\lambda} = \dfrac{2\pi(0.120\times10^{-3} \text{ m})(4.17\times10^{-3})}{500\times10^{-9} \text{ m}} = 2\pi$ rad $= 0$ ◊

(c) $\phi = \dfrac{2\pi d\sin\theta}{\lambda}$: $\theta = \sin^{-1}\left[\dfrac{\lambda\phi}{2\pi d}\right] = \sin^{-1}\left[\dfrac{(500\times10^{-9} \text{ m})(0.333)}{2\pi(1.20\times10^{-4} \text{ m})}\right] = 0.0127°$ ◊

(d) $\dfrac{\lambda}{4} = d\sin\theta$: $\theta = \sin^{-1}\left[\dfrac{\lambda}{4d}\right] = \sin^{-1}\left[\dfrac{500\times10^{-9} \text{ m}}{4(1.20\times10^{-4} \text{ m})}\right] = 0.0597°$ ◊

17. In Figure 37.5, let $L = 120$ cm and $d = 0.250$ cm. The slits are illuminated with coherent 600-nm light. Calculate the distance y above the central maximum for which the average intensity on the screen is 75.0% of the maximum.

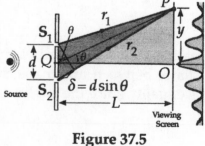

Figure 37.5

Solution   For small θ, $I = I_{max}\cos^2\left[\dfrac{\pi d\sin\theta}{\lambda}\right]$

From the drawing, $\sin\theta \approx \dfrac{y}{L}$: $y = \dfrac{\lambda L}{\pi d}\cos^{-1}\sqrt{\dfrac{I}{I_{max}}}$

In addition, since $I = 0.750I_{max}$, we can substitute a value for each variable:

$$y = \frac{(6.00 \times 10^{-7} \text{ m})(1.20 \text{ m})}{\pi(2.50 \times 10^{-3} \text{ m})} \cos^{-1}\sqrt{0.750} = 0.0480 \text{ mm} \quad \Diamond$$

19. Two narrow parallel slits separated by 0.850 mm are illuminated by 600-nm light, and the viewing screen is 2.80 m away from the slits. (a) What is the phase difference between the two interfering waves on a screen at a point 2.50 mm from the central bright fringe? (b) What is the ratio of the intensity at this point to the intensity at the center of a bright fringe?

Solution

Conceptualize: It is difficult to accurately predict the relative intensity at the point of interest without actually doing the calculation. The waves from each slit could meet in phase ($\phi = 0$) to produce a bright spot of **constructive interference**, out of phase ($\phi = 180°$) to produce a dark region of **destructive interference**, or most likely the phase difference will be somewhere between these extremes, $0 < \phi < 180°$, so that the relative intensity will be $0 < I/I_{max} < 1$.

Categorize: The phase angle depends on the path difference of the waves according to Equation 37.8. This phase difference is used to find the average intensity at the point of interest. Then the relative intensity is simply this intensity divided by the maximum intensity.

Analyze:

(a) Using the variables shown in the diagram for problem 7 we have,

$$\phi = \frac{2\pi d}{\lambda}\sin\theta = \frac{2\pi d}{\lambda}\left(\frac{y}{\sqrt{y^2 + L^2}}\right) \approx \frac{2\pi yd}{\lambda L}$$

$$\phi = \frac{2\pi(0.850 \times 10^{-3} \text{ m})(0.00250 \text{ m})}{(600 \times 10^{-9} \text{ m})(2.80 \text{ m})} = 7.95 \text{ rad} = 2\pi + 1.66 \text{ rad} = 95.4° \quad \Diamond$$

(b) $\dfrac{I}{I_{max}} = \dfrac{\cos^2\left(\dfrac{\pi d}{\lambda}\sin\theta\right)}{\cos^2\left(\dfrac{\pi d}{\lambda}\sin\theta_{max}\right)} = \dfrac{\cos^2\left(\dfrac{\phi}{2}\right)}{\cos^2(m\pi)} = \cos^2\left(\dfrac{\phi}{2}\right)$

$$\frac{I}{I_{max}} = \cos^2\left(\frac{95.4°}{2}\right) = 0.453 \quad \Diamond$$

Finalize: It appears that at this point, the waves show **partial interference** so that the combination is about half the brightness found at the central maximum. We should remember that the equations used in this solution do not account for the diffraction caused by the finite width of each slit. This diffraction effect creates an "envelope" that diminishes in intensity away from the central maximum as shown by the dotted line in Figures 37.14 and P37.61. Therefore, the relative intensity at $y = 2.50$ mm will actually be slightly less than 0.453.

---

23. Determine the resultant of the two waves given by

$$E_1 = 6.0 \sin(100 \pi t)$$

and $$E_2 = 8.0 \sin(100 \pi t + \pi/2).$$

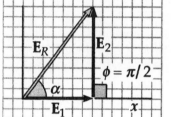

Solution

Let the x axis lie along $\mathbf{E}_1$ in phase space. Its component form is then $\mathbf{E}_1 = 6.00 \hat{\imath} + 0 \hat{\jmath}$.

The components of $\mathbf{E}_2$ are $$\mathbf{E}_2 = 8.00 \cos\left(\frac{\pi}{2}\right)\hat{\imath} + 8.00 \sin\left(\frac{\pi}{2}\right)\hat{\jmath} = 8.00\hat{\jmath}$$

The resultant is $$\mathbf{E}_R = \mathbf{E}_1 + \mathbf{E}_2 = 6.00\hat{\imath} + 8.00\hat{\jmath}$$

with amplitude $$\sqrt{6.00^2 + 8.00^2} = 10.0$$

and phase $$\alpha = \tan^{-1}\left(\frac{8.00}{6.00}\right) = 0.927 \text{ rad}$$

Thus, $$E_R = 10.0 \sin(100\pi t + 0.927) \qquad \lozenge$$

---

29. Consider N coherent sources described as follows: $E_1 = E_0 \sin(\omega t + \phi)$, $E_2 = E_0 \sin(\omega t + 2\phi)$, $E_3 = E_0 \sin(\omega t + 3\phi)$ ..., $E_N = E_0 \sin(\omega t + N\phi)$. Find the minimum value of ϕ for which $E_R = E_1 + E_2 + E_3 + \ldots + E_N$ is zero.

Solution   Take $\phi = 360°/N$ $\lozenge$

Then
$$E_R = \sum_{m=1}^{N} E_0 \sin(\omega t + m\phi) = 0$$

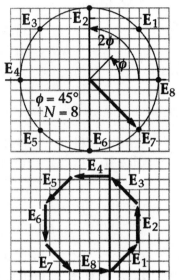

where N defines the number of coherent sources, assumed greater than 1.

In essence, the set of electric field components completes a full 360° circle, returning to zero.

In this situation, when the vectors are added, the sources are symmetric about $E_R = 0$, and thus sum to zero.

Alternatively, one can note that the vectors must create a regular polygon when summed, in order to return to the origin. (The figures to the right may help.)

If N has an integer divisor m, then the choice $\phi = 360°/m$ will also make all the phasors add to zero.

---

31. An oil film ($n = 1.45$) floating on water is illuminated by white light at normal incidence. The film is 280 nm thick. Find (a) the color of the light in the visible spectrum most strongly reflected and (b) the color of the light in the spectrum most strongly transmitted. Explain your reasoning.

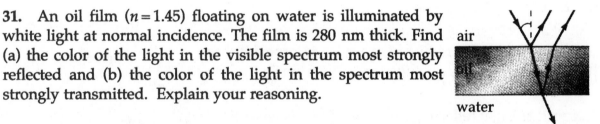

Solution   The light reflected from the top of the oil film undergoes phase reversal. Since $1.45 > 1.33$, the light reflected from the bottom undergoes no reversal. For constructive interference of reflected light, we then have

$$2nt = \left[m + \tfrac{1}{2}\right]\lambda \qquad \text{or} \qquad \lambda_m = \frac{2nt}{m + \frac{1}{2}} = \frac{2(1.45)(280\text{ nm})}{m + \frac{1}{2}}$$

(a) Substituting for m, we have

$$m = 0: \quad \lambda_0 = 1\,624\text{ nm} \qquad \text{(infrared)}$$
$$m = 1: \quad \lambda_1 = 541\text{ nm} \qquad \text{(green)}$$
$$m = 2: \quad \lambda_2 = 325\text{ nm} \qquad \text{(ultraviolet)}$$

Both infrared and ultraviolet light are invisible to the human eye, so the dominant color is green.   ◊

(b) Any light that is not reflected is transmitted. To find the color transmitted most strongly, we find the wavelengths reflected least strongly. According to the condition for destructive interference $2nt = m\lambda$.

Therefore,

$$\lambda = \frac{2nt}{m} = \frac{2(1.45)(280 \text{ nm})}{m}$$

For

$m = 1,$	$\lambda = 812 \text{ nm}$	(near infrared)
$m = 2,$	$\lambda = 406 \text{ nm}$	(violet) ◊
$m = 3,$	$\lambda = 271 \text{ nm}$	(ultraviolet)

Thus violet is the visible color least attenuated by reflection, and the dominant color in the transmitted light.

39. An air wedge is formed between two glass plates separated at one edge by a very fine wire, as shown in Figure P37.39. When the wedge is illuminated from above by 600-nm light and viewed from above, 30 dark fringes are observed. Calculate the radius of the wire.

Figure P37.39

Solution

Conceptualize: The radius of the wire is probably less than 0.1 mm since it is described as a "very fine wire."

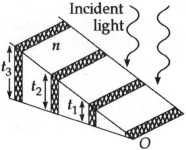

Categorize: Light reflecting from the bottom surface of the top plate undergoes no phase shift, while light reflecting from the top surface of the bottom plate is shifted by π, and also has to travel an extra distance $2t$, where t is the thickness of the air wedge.

Analyze:

For destructive interference, $2t = m\lambda$ ($m = 0, 1, 2, 3, \dots$)

The first dark fringe appears where $m = 0$ at the line of contact between the plates. The 30th dark fringe gives for the diameter of the wire $2t = 29\lambda$, and $t = 14.5\lambda$.

$$r = \frac{t}{2} = 7.25\lambda = 7.25(600 \times 10^{-9} \text{ m}) = 4.35 \ \mu\text{m} \qquad \diamond$$

Finalize: This wire is not only less than 0.1 mm; it is even thinner than a typical human hair (~50 μm).

41. Mirror M_1 in Figure 37.22 is displaced a distance ΔL. During this displacement, 250 fringe reversals (formation of successive dark or bright bands) are counted. The light being used has a wavelength of 632.8 nm. Calculate the displacement ΔL.

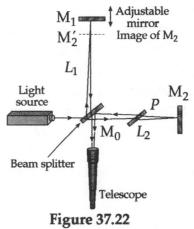

Figure 37.22

Solution

When the mirror on one arm is displaced by ΔL, the path difference changes by $2\Delta L$. A shift resulting in the reversal between dark and bright fringes requires a path length change of one-half wavelength. Therefore, since in this case $m = 250$,

$$2\Delta L = \frac{m\lambda}{2} \qquad \text{and} \qquad \Delta L = \frac{m\lambda}{4} = \frac{250(6.328\times10^{-7}\text{ m})}{4} = 3.955\times10^{-5}\text{ m} \qquad \Diamond$$

---

51. Astronomers observed a 60.0-MHz radio source both directly and by reflection from the sea. If the receiving dish is 20.0 m above sea level, what is the angle of the radio source above the horizon at first maximum?

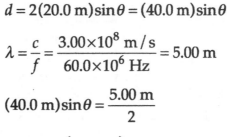

Solution One radio wave reaches the receiver R directly from the distant source at an angle θ above the horizontal. The other wave undergoes phase reversal as it reflects from the water at P.

Constructive interference first occurs for a path difference of

$$d = \lambda/2 \qquad\qquad [1]$$

It is equally far from P to R as from P to R', the mirror image of the telescope. The angles θ in the figure are equal because they each form part of a right triangle with a shared angle at R'.

So the path difference is

$$d = 2(20.0\text{ m})\sin\theta = (40.0\text{ m})\sin\theta$$

The wavelength is

$$\lambda = \frac{c}{f} = \frac{3.00\times10^8\text{ m/s}}{60.0\times10^6\text{ Hz}} = 5.00\text{ m}$$

Substituting for d and λ in Eq. [1],

$$(40.0\text{ m})\sin\theta = \frac{5.00\text{ m}}{2}$$

Solving for the angle θ,

$$\theta = \sin^{-1}\left(\frac{5.00\text{ m}}{80.0\text{ m}}\right) = 3.58° \qquad \Diamond$$

55. Measurements are made of the intensity distribution in a Young's interference pattern (see Fig. 37.7). At a particular value of y, it is found that $I/I_{max}=0.810$ when 600-nm light is used. What wavelength of light should be used to reduce the relative intensity at the same location to 64.0% of the maximum intensity?

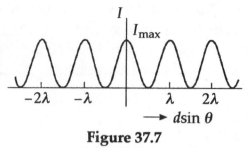

Figure 37.7

Solution   We assume the measurements are made in the central bright fringe of the interference pattern.

From Equation 37.13,

$$\frac{I_1}{I_{max}} = \cos^2\left(\frac{\pi y d}{\lambda_1 L}\right) = 0.810$$

Therefore,

$$\frac{\pi y d}{L} = \lambda_1 \cos^{-1}\sqrt{0.810} = (600 \text{ nm})\cos^{-1}(0.900) = 271 \text{ nm}$$

For the 2nd wavelength,

$$\frac{I_2}{I_{max}} = \cos^2\left(\frac{\pi y d}{\lambda_2 L}\right) = 0.640$$

and

$$\lambda_2 = \frac{\pi y d / L}{\cos^{-1}\sqrt{0.640}}$$

But this is simply

$$\lambda_2 = \frac{271 \text{ nm}}{\cos^{-1}\sqrt{0.640}} = 421 \text{ nm} \qquad \lozenge$$

---

57. The condition for constructive interference by reflection from a thin film in air as developed in Section 37.6 assumes nearly normal incidence. **What If?** Show that if the light is incident on the film at a nonzero angle ϕ_1 (relative to the normal), then the condition for constructive interference is $2nt\cos\theta_2 = \left(m+\frac{1}{2}\right)\lambda$, where θ_2 is the angle of refraction.

Solution

It may be helpful to draw a ray diagram of the interference. Because refraction out of the film is the opposite of refraction into the film, the refracted ray leaves the film parallel to the reflected ray. Then, since the angles marked α are equal,

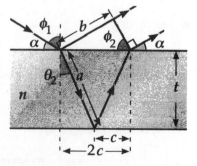

$$\phi_2 + \alpha + 90° = 180°$$

From the angle of incidence, $\qquad \phi_1 + \alpha = 90°$

so $\qquad\qquad\qquad\qquad\qquad \phi_2 = \phi_1$

Before they head off together along parallel paths, one beam travels a distance b and undergoes phase reversal on reflection; the other beam travels a distance $2a$ inside the film, where its wavelength is λ / n. The number of cycles it completes along this path is

$$\frac{2a}{\lambda / n} = \frac{2na}{\lambda}$$

The optical path length in the film is $2na$. Then the shift between the two outgoing rays is

$$\delta = 2na - b - \frac{\lambda}{2}$$

where a and b are as shown in the ray diagram, n is the index of refraction, and the term $\lambda/2$ is due to phase reversal at the top surface. For constructive interference, $\delta = m\lambda$ where m has integer values.

This condition becomes $\qquad 2na - b = \left[m + \frac{1}{2}\right]\lambda \qquad\qquad\qquad\qquad$ **[1]**

From the figure's geometry, $\qquad a = \dfrac{t}{\cos\theta_2}$

and $\qquad\qquad\qquad\qquad c = a\sin\theta_2 = \dfrac{t\sin\theta_2}{\cos\theta_2}$

Therefore, $\qquad\qquad\qquad b = 2c\sin\phi_1 = \dfrac{2t\sin\theta_2}{\cos\theta_2}\sin\phi_1$

Also, from Snell's law, $\qquad \sin\phi_1 = n\sin\theta_2$

so $\qquad\qquad\qquad\qquad b = \dfrac{2nt\sin^2\theta_2}{\cos\theta_2}$

With these results, the condition for constructive interference given in Equation [1] becomes:

$$2n\left(\frac{t}{\cos\theta_2}\right) - \frac{2nt\sin^2\theta_2}{\cos\theta_2} = \frac{2nt}{\cos\theta_2}(1 - \sin^2\theta_2) = \left[m + \frac{1}{2}\right]\lambda$$

or $\qquad\qquad\qquad 2nt\cos\theta_2 = \left[m + \frac{1}{2}\right]\lambda \qquad\qquad\qquad\qquad\qquad\lozenge$

61. Consider the double-slit arrangement shown in Figure P37.61, where the slit separation is d and the slit to screen distance is L. A sheet of transparent plastic having an index of refraction n and thickness t is placed over the upper slit. As a result, the central maximum of the interference pattern moves upward a distance y'. Find y'.

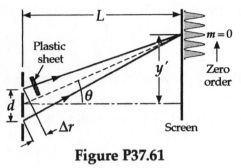

Figure P37.61

Solution

Conceptualize: Since the film shifts the pattern upward, we should expect y' to be proportional to n, t, and L.

Categorize: The film increases the optical path length of the light passing through the upper slit, so the physical distance of this path must be shorter for the waves to meet in phase ($\phi=0$) to produce the central maximum. Thus, the added distance Δr traveled by the light from the lower slit must introduce a phase difference equal to that introduced by the plastic film.

Analyze: First calculate the additional phase difference due to the plastic. Recall that the relation between phase difference and path difference is $\phi=2\pi\delta/\lambda$. The presence of plastic affects this by changing the wavelength of the light, so that the phase change of the light in air and plastic, as it travels over the thickness t is

$$\phi_{air} = \frac{2\pi t}{\lambda_{air}} \quad \text{and} \quad \phi_{plastic} = \frac{2\pi t}{\lambda_{air}/n}$$

Thus, plastic causes an additional phase change of $\quad \Delta\phi = \frac{2\pi t}{\lambda_{air}}(n-1)$

Next, in order to interfere constructively, we must calculate the additional distance that the light from the bottom slit must travel.

$$\Delta r = \frac{\Delta\phi\lambda_{air}}{2\pi} = t(n-1)$$

In the small angle approximation we can write $\quad \Delta r = y'd/L$,

so $\quad\quad y' = \dfrac{t(n-1)L}{d}$ ◊

Finalize: As expected, y' is proportional to t and L. It increases with increasing n, being proportional to $(n-1)$. It is also inversely proportional to the slit separation d, which makes sense since slits that are closer together make a wider interference pattern.

Chapter 38
DIFFRACTION PATTERNS AND POLARIZATION

EQUATIONS AND CONCEPTS

In **single-slit diffraction**, the total phase difference β between waves from the top and bottom portions of the slit will depend on the angle θ which determines the direction to a point on the screen.

$$\beta = \frac{2\pi}{\lambda} a \sin\theta \qquad (38.3)$$

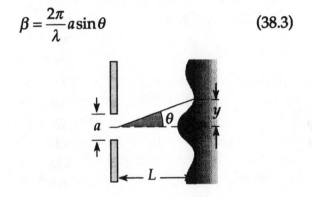

The **general condition for destructive interference** in single slit diffraction requires that β, the phase difference, equal an integer multiple of 2π. *Under this condition the resultant phasor will be zero.*

$$\sin\theta_{\text{dark}} = m\frac{\lambda}{a} \qquad \left(\begin{array}{c}\text{destructive} \\ \text{interference}\end{array}\right) \qquad (38.1)$$

$$(m = \pm 1, \pm 2, \pm 3, \ldots)$$

The **intensity** at any point on the screen is given in terms of the intensity I_{max} at $\theta = 0$ where β is given by Equation 38.3.

$$I = I_{\text{max}}\left[\frac{\sin(\beta/2)}{\beta/2}\right]^2 \qquad (38.4)$$

Rayleigh's criterion states the condition for the resolution of two images due to nearby sources.

For a **slit**, the angular separation between the sources must be greater than the ratio of the wavelength to slit width.

$$\theta_{min} = \frac{\lambda}{a} \qquad \text{(Slit)} \qquad (38.8)$$

For a **circular aperture**, the minimum angular separation depends on D, the diameter of the aperture (or lens).

$$\theta_{min} = 1.22\frac{\lambda}{D} \qquad \left(\begin{smallmatrix}\text{Circular}\\\text{aperture}\end{smallmatrix}\right) \qquad (38.9)$$

A **diffraction grating** (an array of a large number of parallel slits separated by a distance d) will produce an interference pattern in which there is a series of maxima for each wavelength. *Maxima due to wavelengths of different values comprise a spectral order denoted by order number m.*

$$d\sin\theta_{bright} = m\lambda \qquad (38.10)$$
$$(m = 0, \pm1, \pm2, \ldots)$$

The **resolving power** of a grating determines the minimum difference between two closely spaced wavelengths that can just be resolved as separate lines in a spectrum. In Equation 38.11, $\lambda = (\lambda_1 + \lambda_2)/2$ and $\Delta\lambda = \lambda_2 - \lambda_1$.

$$R = \frac{\lambda}{\lambda_2 - \lambda_1} = \frac{\lambda}{\Delta\lambda} \qquad (38.11)$$

The **resolving power is proportional to the order number** m and increases as the number of lines illuminated is increased. *A grating with a large value of R can distinguish small differences in wavelength.*

$$R = Nm \qquad (38.12)$$

Bragg's law gives the conditions for constructive interference of x-rays reflected from the parallel planes of a crystalline solid separated by a distance d. θ is the angle between the incident beam and the surface.

$$2d\sin\theta = m\lambda \qquad (38.13)$$
$$(m = 1, 2, 3, \ldots)$$

Malus's law states that the fraction of initially polarized light (from a polarizer) that will be transmitted by a second sheet of polarizing material (the analyzer) depends on the square of the cosine of the angle θ between the transmission axis of the polarizer and that of the analyzer.

$$I = I_{max} \cos^2 \theta \qquad (38.14)$$

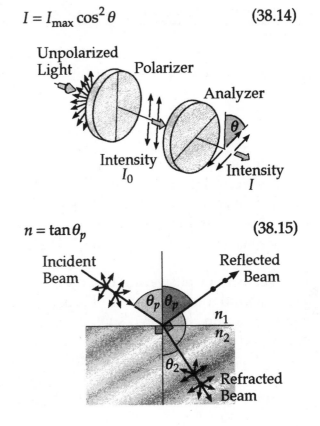

Brewster's law gives the angle of incidence (the polarizing angle, θ_p) for which the reflected beam will be completely polarized.

$$n = \tan \theta_p \qquad (38.15)$$

REVIEW CHECKLIST

You should be able to:

▷ Determine the positions of the maxima and minima in a single-slit diffraction pattern and calculate the intensities of the secondary maxima relative to the intensity of the central maximum. (Sections 38.1 and 38.2)

▷ Calculate the intensities of interference maxima due to a double slit, expressed as a fraction of the intensity at the center of the pattern. (Section 38.2)

▷ Determine whether or not two sources under a given set of conditions are resolvable as defined by Rayleigh's criterion. (Section 38.3)

▷ Determine the positions of the principle maxima in the diffraction pattern of a diffraction grating. Calculate the resolving power of a grating under specified conditions. (Section 38.4)

▷ Describe the technique of x-ray diffraction and make calculations of the lattice spacing using Bragg's law. (Section 38.5)

▷ Describe how the state of polarization of a light beam can be determined by use of a polarizer-analyzer combination. Describe qualitatively the polarization of light by selective absorption, reflection, scattering, and double refraction. Make appropriate calculations using Malus's law and Brewster's law. (Section 38.6)

ANSWERS TO SELECTED QUESTIONS

1. Why can you hear around corners, but not see around corners?

Answer

Audible sound has wavelengths on the order of meters or centimeters, while visible light has wavelengths on the order of half a micron. In this world of breadbox-size objects, λ is comparable to the object size a for sound, and sound diffracts around walls and through doorways. But λ/a is much smaller for visible light passing ordinary-size objects or apertures, so light diffracts only through very small angles.

Another way of answering this question would be as follows. We can see by a small angle around a small obstacle or around the edge of a small opening. The side fringes in Figure 38.1 and the Arago spot in the center of Figure 38.3 (in the textbook) show this diffraction. Conversely, we cannot always hear around corners. Out-of-doors, away from reflecting surfaces, have someone a few meters distant face away from you and whisper. The high-frequency, short-wavelength, information-carrying components of the sound do not diffract around his head enough for you to understand his words.

□   □   □   □

5. Describe the change in width of the central maximum of the single-slit diffraction pattern as the width of the slit is made narrower.

Answer

Equation 38.1 describes the angles at which you get destructive interference; from it, we can obtain an estimate of the width of the central maximum. For small angles, the equation can be rewritten as

$$\theta_m = \sin^{-1}(m\lambda/a) \approx m\lambda/a$$

Thus, as the width of the slit a decreases, the angle of the first destructive interference θ_1 grows, and the width of the central maximum grows as well.

SOLUTIONS TO SELECTED PROBLEMS

3. A screen is placed 50.0 cm from a single slit, which is illuminated with 690-nm light. If the distance between the first and third minima in the diffraction pattern is 3.00 mm, what is the width of the slit?

Solution In the equation for single-slit diffraction minima at small angles

$$\frac{y}{L} \approx \sin\theta_{dark} = \frac{m\lambda}{a}$$

take differences between the first and third minima,

to see that $\dfrac{\Delta y}{L} = \dfrac{\Delta m\lambda}{a}$ with $\Delta y = 3.00\times10^{-3}$ m and $\Delta m = 3-1 = 2$

The width of the slit is then

$$a = \frac{\lambda L \Delta m}{\Delta y} = \frac{(690\times10^{-9}\ \text{m})(0.500\ \text{m})(2)}{3.00\times10^{-3}\ \text{m}}$$

$$a = 2.30\times10^{-4}\ \text{m} \qquad \diamond$$

---

9. A diffraction pattern is formed on a screen 120 cm away from a 0.400-mm-wide slit. Monochromatic 546.1-nm light is used. Calculate the fractional intensity I/I_{max} at a point on the screen 4.10 mm from the center of the principal maximum.

Solution $\sin\theta \approx \dfrac{y}{L} = \dfrac{4.10\times10^{-3}\ \text{m}}{1.20\ \text{m}} = 3.417\times10^{-3}$

$$\frac{\beta}{2} = \frac{\pi a \sin\theta}{\lambda} = \frac{\pi(4.00\times10^{-4}\ \text{m})(3.417\times10^{-3})}{546.1\times10^{-9}\ \text{m}} = 7.862\ \text{rad}$$

$$\frac{I}{I_{max}} = \left[\frac{\sin(\beta/2)}{(\beta/2)}\right]^2 = \left[\frac{\sin(7.862\ \text{rad})}{7.862\ \text{rad}}\right]^2 = 1.62\times10^{-2} \qquad \diamond$$

---

13. A helium-neon laser emits light that has a wavelength of 632.8 nm. The circular aperture through which the beam emerges has a diameter of 0.500 cm. Estimate the diameter of the beam 10.0 km from the laser.

Solution

Conceptualize: A typical laser pointer makes a spot about 5 cm in diameter at 100 m, so the spot size at 10 km would be about 100 times bigger, or about 5 m across. Assuming that this HeNe laser is similar, we could expect a comparable beam diameter.

Categorize: We assume that the light is parallel and not diverging as it passes through and fills the circular aperture. However, as the light passes through the circular aperture, it will spread from diffraction according to Equation 38.9.

Analyze: The beam spreads into a cone of half-angle

$$\theta_{min} = 1.22\frac{\lambda}{D} = 1.22\left(\frac{632.8\times10^{-9}\text{ m}}{0.005\,00\text{ m}}\right) = 1.54\times10^{-4}\text{ rad}$$

The radius of the beam ten kilometers away is, from the definition of radian measure

$$r_{beam} = \theta_{min}(1.00\times10^4\text{ m}) = 1.54\text{ m}$$

and its diameter is $d_{beam} = 2r_{beam} = 3.09\text{ m}$ ◊

Finalize: The beam is several meters across as expected, and is about 600 times larger than the laser aperture. Since most HeNe lasers are low power units in the mW range, the beam at this range would be so spread out that it would be too dim to see on a screen.

17. The Impressionist painter Georges Seurat created paintings with an enormous number of dots of pure pigment, each of which was approximately 2.00 mm in diameter. The idea was to have colors such as red and green next to each other to form a scintillating canvas (Fig. P38.17). Outside what distance would one be unable to discern individual dots on the canvas? (Assume that $\lambda=500$ nm and that the pupil diameter is 4.00 mm.)

Solution

We will assume that the dots are just touching and do not overlap, so that the distance between their centers is 2.00 mm. By Rayleigh's criterion, two dots separated center to center by 2.00 mm would be seen to overlap when

$$\theta_{min} = \frac{d}{L} = 1.22\frac{\lambda}{D} \qquad \text{with} \quad d=2.00\text{ mm, } \lambda=500\text{ nm, and } D=4.00\text{ mm}$$

Thus, $$L = \frac{Dd}{1.22\lambda} = \frac{(4.00\times10^{-3}\text{ m})(2.00\times10^{-3}\text{ m})}{1.22(500\times10^{-9}\text{ m})} = 13.1\text{ m}$$ ◊

23. White light is spread out into its spectral components by a diffraction grating. If the grating has $2\,000$ grooves per centimeter, at what angle does red light of wavelength 640 nm appear in first order?

Solution The grating spacing is $\qquad d = \dfrac{1.00 \times 10^{-2} \text{ m}}{2\,000} = 5.00 \times 10^{-6} \text{ m}$

The light is deflected according to $\qquad \sin\theta = \dfrac{m\lambda}{d} = \dfrac{1(640 \times 10^{-9} \text{ m})}{5.00 \times 10^{-6} \text{ m}} = 0.128$

at an angle of $\qquad\qquad\qquad\qquad \theta = 7.35°$ ◊

---

25. The hydrogen spectrum has a red line at 656 nm and a blue line at 434 nm. What are the angular separations between these two spectral lines obtained with a diffraction grating that has $4\,500$ grooves/cm?

Solution

Conceptualize: Most diffraction gratings yield several spectral orders within the 180° viewing range, which means that the angle between red and blue lines is probably 10° to 30°.

Categorize: The angular separation is the difference between the angles corresponding to the red and blue wavelengths for each visible spectral order according to the diffraction grating equation, $d\sin\theta = m\lambda$.

Analyze: The grating spacing is

$$d = 1.00 \times 10^{-2} \text{ m} / 4\,500 \text{ lines} = 2.22 \times 10^{-6} \text{ m}$$

In the first-order spectrum $(m = 1)$, the angles of diffraction are given by $\sin\theta = \lambda/d$:

$$\sin\theta_{1r} = \frac{656 \times 10^{-9} \text{ m}}{2.22 \times 10^{-6} \text{ m}} = 0.295 \qquad \text{so} \qquad \theta_{1r} = 17.17°$$

$$\sin\theta_{1b} = \frac{434 \times 10^{-9} \text{ m}}{2.22 \times 10^{-6} \text{ m}} = 0.195 \qquad \text{so} \qquad \theta_{1b} = 11.26°$$

The angular separation is $\quad \Delta\theta_1 = \theta_{1r} - \theta_{1b} = 17.17° - 11.26° = 5.91°$ ◊

In the 2nd-order $(m = 2)$ $\quad \Delta\theta_2 = \sin^{-1}\left(\dfrac{2\lambda_r}{d}\right) - \sin^{-1}\left(\dfrac{2\lambda_b}{d}\right) = 13.2°$ ◊

In the third order ($m=3$), $\Delta\theta_3 = \sin^{-1}\left(\dfrac{3\lambda_r}{d}\right) - \sin^{-1}\left(\dfrac{3\lambda_b}{d}\right) = 26.5°$ ◊

Examining the fourth order, we find the red line is not visible:

$$\theta_{4r} = \sin^{-1}(4\lambda_r/d) = \sin^{-1}(1.18) \text{ does not exist}$$ ◊

Finalize: The full spectrum is visible in the first 3 orders with this diffraction grating, and the fourth is partially visible. We can also see that the pattern is dispersed more for higher spectral orders so that the angular separation between the red and blue lines increases as m increases. It is also worth noting that the spectral orders can overlap (as is the case for the second and third order spectra above), which makes the pattern look confusing if you do not know what you are looking for.

29. A diffraction grating of width 4.00 cm has been ruled with 3000 grooves/cm. (a) What is the resolving power of this grating in the first three orders? (b) If two monochromatic waves incident on this grating have a mean wavelength of 400 nm, what is their wavelength separation if they are just resolved in the third order?

Solution

From Equation 38.12, $R = mN$

where $N = (3000 \text{ lines/cm})(4.00 \text{ cm}) = 12000 \text{ lines}$

(a) In the 1ˢᵗ order, $R = (1)(12000 \text{ lines}) = 12000$ ◊

In the 2ⁿᵈ order, $R = (2)(12000 \text{ lines}) = 24000$ ◊

In the 3ʳᵈ order, $R = (3)(12000 \text{ lines}) = 36000$ ◊

(b) From Eq. 38.11, $R = \lambda/\Delta\lambda$

In the 3ʳᵈ order, $\Delta\lambda = \dfrac{\lambda}{R} = \dfrac{4.00\times10^{-7}\text{ m}}{3.60\times10^{4}} = 1.11\times10^{-11}\text{ m} = 0.0111\text{ nm}$ ◊

31. A source emits 531.62-nm and 531.81-nm light. (a) What minimum number of grooves is required for a grating that resolves the two wavelengths in the first-order spectrum? (b) Determine the slit spacing for a grating 1.32 cm wide that has the required minimum number of grooves.

Solution The resolving power of the diffraction grating is $Nm = \lambda/\Delta\lambda$.

(a) Resolving the first-order spectrum,
$$N(1) = \frac{531.7\,\text{nm}}{0.190\,\text{nm}} = 2800\,\text{lines} \qquad \diamond$$

(b) The slits are spaced at intervals of
$$\frac{1.32\times10^{-2}\,\text{m}}{2800} = 4.72\,\mu\text{m} \qquad \diamond$$

---

33. A grating with 250 grooves/mm is used with an incandescent light source. Assume the visible spectrum to range in wavelength from 400 to 700 nm. In how many orders can one see (a) the entire visible spectrum and (b) the short-wavelength region?

Solution

The grating spacing is
$$d = \frac{1.00\,\text{mm}}{250} = 4.00\times10^{-6}\,\text{m}$$

(a) In each order of interference m, red light diffracts at a larger angle than the other colors with shorter wavelengths. We find the largest integer m satisfying

$$d\sin\theta = m\lambda \qquad \text{with} \qquad \lambda = 700\,\text{nm}$$

With $\sin\theta$ having its largest value, $(4.00\times10^{-6}\,\text{m})(\sin 90°) = m(700\times10^{-9}\,\text{m})$

so that $\qquad\qquad m = 5.71$

Thus the red light cannot be seen in the 6th order, and the full visible spectrum appears in only five orders. $\qquad \diamond$

(b) Now consider light at the boundary between violet and ultraviolet.

$d\sin\theta = m\lambda \qquad$ becomes $\qquad (4.00\times10^{-6}\,\text{m})(\sin 90°) = m(400\times10^{-9}\,\text{m})$

and $\qquad\qquad m = 10 \qquad \diamond$

37. If the interplanar spacing of NaCl is 0.281 nm, what is the predicted angle at which 0.140-nm x-rays are diffracted in a first-order maximum?

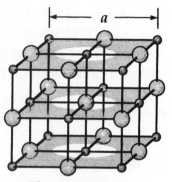

Solution

The atomic planes in this crystal are shown in Figure 38.26 of the text. The diffraction they produce is described by the Bragg condition,

Figure 38.26

that $$2d\sin\theta = m\lambda:$$

$$\sin\theta = \frac{m\lambda}{2d} = \frac{1(0.140\times10^{-9}\ \text{m})}{2(0.281\times10^{-9}\ \text{m})} = 0.249$$

$$\theta = 14.4°$$ ◊

41. Plane-polarized light is incident on a single polarizing disk with the direction of E_0 parallel to the direction of the transmission axis. Through what angle should the disk be rotated so that the intensity in the transmitted beam is reduced by a factor of (a) 3.00, (b) 5.00, (c) 10.0?

Solution

We define the initial angle, at which all the light is transmitted to be $\theta = 0$. Turning the disk to another angle will then reduce the transmitted light by an intensity factor of $I = I_0\cos^2\theta$.

(a) For $I = \dfrac{I_0}{3.00}$, $\quad\cos\theta = \dfrac{1}{\sqrt{3.00}}\quad$ and $\quad\theta = 54.7°$ ◊

(b) For $I = \dfrac{I_0}{5.00}$, $\quad\cos\theta = \dfrac{1}{\sqrt{5.00}}\quad$ and $\quad\theta = 63.4°$ ◊

(c) For $I = \dfrac{I_0}{10.0}$, $\quad\cos\theta = \dfrac{1}{\sqrt{10.0}}\quad$ and $\quad\theta = 71.6°$ ◊

45. The critical angle for total internal reflection for sapphire surrounded by air is 34.4°. Calculate the polarizing angle for sapphire.

Solution $n = \tan\theta_p$: $\theta_p = \tan^{-1} n$

and $\sin\theta_c = \dfrac{1}{n}$: $n = \dfrac{1}{\sin\theta_c}$

Therefore, $\theta_p = \tan^{-1}\left(\dfrac{1}{\sin\theta_c}\right) = \tan^{-1}\left(\dfrac{1}{\sin 34.4°}\right) = 60.5°$ ◊

57. Light of wavelength 500 nm is incident normally on a diffraction grating. If the third-order maximum of the diffraction pattern is observed at 32.0°, (a) what is the number of rulings per centimeter for the grating? (b) Determine the total number of primary maxima that can be observed in this situation.

Solution

Conceptualize: The diffraction pattern described in this problem seems to be similar to previous problems that have diffraction gratings with 2 000 to 5 000 lines/mm. With the third-order maximum at 32°, there are probably 5 or 6 maxima on each side of the central bright fringe, for a total of 11 or 13 primary maxima.

Categorize: The diffraction grating equation can be used to find the grating spacing and the angles of the other maxima that should be visible within the 180° viewing range.

Analyze:

(a)   Use Equation 38.10, $d\sin\theta = m\lambda$: $d = \dfrac{m\lambda}{\sin\theta} = \dfrac{3(5.00\times10^{-7}\text{ m})}{\sin 32.0°} = 2.83\times10^{-6}\text{ m}$

   Thus, the grating gauge is $\dfrac{1}{d} = 3.534\times10^{5}\text{ lines/m} = 3\,530\text{ lines/cm}$ ◊

(b) $\sin\theta = m\left(\dfrac{\lambda}{d}\right) = \dfrac{m(5.00\times10^{-7}\text{ m})}{2.83\times10^{-6}\text{ m}} = m(0.177)$

   For $\sin\theta \le 1$, we require that $m(0.177) \le 1$ or $m \le 5.66$. Because m must be an integer, its maximum value is really 5. Therefore, the total number of maxima is $2m + 1 = 11$.   ◊

Finalize: The results agree with our predictions, and apparently there are 5 maxima on either side of the central maximum. If more maxima were desired, a grating with **fewer** lines/cm would be required; however, this would reduce the ability to resolve the difference between lines that appear close together.

63. Suppose that the single slit in Figure 38.6 is 6.00 cm wide and in front of a microwave source operating at 7.50 GHz. (a) Calculate the angle subtended by the first minimum in the diffraction pattern. (b) What is the relative intensity I/I_{max} at $\theta = 15.0°$? (c) Assume that two such sources, separated laterally by 20.0 cm, are behind the slit. What must the maximum distance between the plane of the sources and the slit be if the diffraction patterns are to be resolved? (In this case, the approximation $\sin\theta \approx \tan\theta$ is not valid because of the relatively small value of a/λ.)

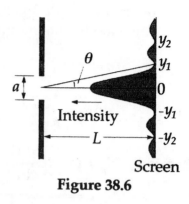

Figure 38.6

Solution

(a) From Eq. 38.1,

$$\theta = \sin^{-1}(m\lambda/a)$$

In this case, $m = 1$, and

$$\lambda = \frac{c}{f} = \frac{3.00\times10^8 \text{ m/s}}{7.50\times10^9 \text{ s}^{-1}} = 0.040\,0 \text{ m}$$

so

$$\theta = \sin^{-1}(0.666) = 41.8° \qquad \Diamond$$

(b) From Equation 38.4,

$$\frac{I}{I_{max}} = \left(\frac{\sin(\beta/2)}{(\beta/2)}\right)^2 \quad \text{where} \quad \beta = \frac{2\pi a\sin\theta}{\lambda}$$

When $\theta = 15.0°$,

$$\beta = 2.44 \text{ rad}$$

and

$$I/I_{max} = 0.593 \qquad \Diamond$$

(c) Let L' be the maximum distance between the plane of the two sources and the slit. The minimum angle subtended by the two sources at the slit is $\theta = 41.81°$, and the half-angle between the sources is $\alpha = \theta/2 = 20.91°$.

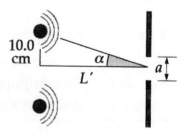

From the figure,

$$\tan\alpha = 10.0 \text{ cm}/L'$$

so

$$L' = \frac{10.0 \text{ cm}}{\tan 20.91°} = 26.2 \text{ cm} \qquad \Diamond$$

71. Another method to solve the equation $\phi = \sqrt{2}\sin\phi$ in Problem 70 is to guess a first value of ϕ, use a computer or calculator to see how nearly it fits, and continue to update your estimate until the equation balances. How many steps (iterations) does this take?

Solution

We can list each trial as we try to home in on the solution to $\phi = \sqrt{2}\sin\phi$ by narrowing the range in which it must lie:

ϕ	$\sqrt{2}\sin\phi$	
1	1.19	bigger than ϕ
2	1.29	smaller than ϕ
1.5	1.41	smaller
1.4	1.394	
1.39	1.391	bigger
1.395	1.392	
1.392	1.391 7	smaller
1.391 5	1.391 54	bigger
1.391 52	1.391 55	bigger
1.391 6	1.391 568	smaller
1.391 58	1.391 563	
1.391 57	1.391 561	
1.391 56	1.391 558	
1.391 559	1.391 557 8	
1.391 558	1.391 557 5	
1.391 557	1.391 557 3	
1.391 557 4	1.391 557 4	

We get the answer to seven digits after 17 steps. Clever guessing, like just using the value of $\sqrt{2}\sin\phi$ as the next guess for ϕ, could reduce this to around 13 steps. ◊

Chapter 39
RELATIVITY

EQUATIONS AND CONCEPTS

Galilean space-time transformation equations transform the location and time of an event in one inertial frame of reference, S(x,y,z,t) to a second frame of reference, S'(x',y',z',t') moving with constant velocity **v** relative to the first. *Equations 39.1 apply when S and S' have a common x axis. Observers in both frames measure the same time interval between two successive events.*

$$\left. \begin{array}{l} x' = x - vt \\ y' = y \\ z' = z \\ t' = t \end{array} \right\}$$

(39.1)

The **Galilean velocity transformation equation** relates the velocity of a particle measured as u' in the moving frame to the velocity of the particle measured as u in the rest frame.

$$u'_x = u_x - v$$

(39.2)

Basic postulates of the special theory of relativity:

- The laws of physics are the same in all inertial frames of reference.

- The speed of light has the same value in all inertial frames.

Important consequences of the theory of special relativity:

- **Simultaneity** — events observed as simultaneous in one frame of reference are not necessarily observed as simultaneous in a second frame moving relative to the first.

- **Time dilation** — A time interval Δt measured by an observer moving with respect to a clock is longer than the time interval Δt_p (the proper time) measured by an observer at rest with respect to the clock. *Moving clocks run slower than clocks at rest with respect to an observer.*

$$\Delta t = \frac{\Delta t_p}{\sqrt{1 - \dfrac{v^2}{c^2}}} = \gamma t_p \qquad (39.7)$$

$$\gamma = \frac{1}{\sqrt{1 - \dfrac{v^2}{c^2}}} \qquad (39.8)$$

- **Length contraction** — If an object has a length L_p (the proper length) when measured by an observer at rest with respect to the object, the length L measured by an observer as the object moves along a direction parallel to its length will be less than L_p. *Length contraction occurs only along the direction of motion.*

$$L = \frac{L_p}{\gamma} = L_p \sqrt{1 - \frac{v^2}{c^2}} \qquad (39.9)$$

- **Relativistic Doppler effect** — When a light source and observer approach each other, the observed frequency is greater than the frequency of the source. *When the light and observer recede from each other, negative values are used for v in the equation.*

$$f_{\text{obs}} = \frac{\sqrt{1 + v/c}}{\sqrt{1 - v/c}} f_{\text{source}} \qquad (39.10)$$

The **Lorentz transformation equations** transform space and time coordinates from the rest frame of reference (S) into the frame (S') moving with relative velocity v. *In order to transform coordinates in S' to coordinates in S, replace v by $-v$ and interchange primed and unprimed coordinates in Equation 39.11.*

$$\begin{aligned}
x' &= \gamma(x - vt) \\
y' &= y \\
z' &= z \\
t' &= \gamma\left(t - \frac{v}{c^2}x\right)
\end{aligned} \qquad (39.11)$$

Lorentz velocity transformation equations relate the observed velocity u' in the moving frame (S') to the measured velocity u in the rest frame (S). *The relative velocity of S' with respect to S is v.*

$$u'_x = \frac{u_x - v}{1 - \frac{u_x v}{c^2}}$$ (39.16)

$$u'_y = \frac{u_y}{\gamma\left(1 - \frac{u_x v}{c^2}\right)}$$ (39.17)

$$u'_z = \frac{u_z}{\gamma\left(1 - \frac{u_x v}{c^2}\right)}$$

The **relativistic linear momentum** of a particle with mass m and moving with speed u satisfies the following conditions:

$$\mathbf{p} = \frac{m\mathbf{u}}{\sqrt{1 - \frac{u^2}{c^2}}} = \gamma m \mathbf{u}$$ (39.19)

(i) Momentum is conserved in all collisions, and

(ii) The relativistic value of momentum approaches the classical value ($m\mathbf{u}$) as $\mathbf{u}$ approaches zero.

The **relativistic kinetic energy** of a particle of mass m moving with a speed u, includes the **rest energy** term mc^2.

$$K = \gamma mc^2 - mc^2 = (\gamma - 1)mc^2$$ (39.23)

The **rest energy** of a particle is independent of the speed of a particle. The mass m must have the same value in all inertial frames.

$$E_R = mc^2$$ (39.24)

The **total energy** E of a particle is the sum of the kinetic energy and the rest energy. *This expression shows that mass is a form of energy.*

$$E = \frac{mc^2}{\sqrt{1 - \frac{u^2}{c^2}}} = \gamma mc^2$$ (39.26)

The **total energy** of a relativistic particle is related to the relativistic momentum and the rest energy. *This expression for total energy is useful when the momentum or energy of a particle is known (rather than the speed).*

$$E^2 = p^2 c^2 + (mc^2)^2 \qquad (39.27)$$

Photons ($m = 0$) travel with the speed of light. *Equation 39.28 is an exact expression relating energy and momentum for particles which have zero mass.*

$$E = pc \qquad (39.28)$$

The **electron volt** (eV) is a convenient energy unit to use to express the energies of electrons and other subatomic particles.

$$1\,eV = 1.60 \times 10^{-19}\,J$$

REVIEW CHECKLIST

You should be able to:

▷ Explain the Michelson-Morley experiment, its objectives, results, and the significance of its outcome. (Section 39.2)

▷ State Einstein's two postulates of the special theory of relativity. (Section 39.3)

▷ Make calculations using the equations for time dilation, length contraction, and relativistic Doppler effect. (Section 39.4)

▷ Make calculations using the Lorentz transformation equations and the Lorentz velocity transformation equations. (Sections 39.5 and 39.6)

▷ Make calculations using relativistic expressions for momentum, kinetic energy, and total energy of a particle. (Sections 39.7, 39.8 and 39.9)

ANSWERS TO SELECTED QUESTIONS

5. Explain why it is necessary, when defining the length of a rod, to specify that the positions of the ends of the rod are to be measured simultaneously.

Answer Suppose a railroad train is moving past you. One way to measure its length is this: You mark the tracks at the front of the moving engine at 9:00:00 AM, while your assistant marks the tracks at the back of the caboose at the same time. Then you find the distance between the marks on the tracks with a tape measure. You and your assistant must make the marks simultaneously (in your reference frame), for otherwise the motion of the train would make its length different from the distance between marks.

□   □   □   □

7. List some ways our day-to-day lives would change if the speed of light were only 50 m/s.

Answer For a wonderful fictional exploration of this question, get a "Mr. Tompkins" book by George Gamow. All of the relativity effects would be obvious in our lives. Time dilation and length contraction would both occur. Driving home in a hurry, you would push on the gas pedal not to increase your speed very much, but to make the blocks shorter. Big Doppler shifts in wave frequencies would make red lights look green as you approached, and make car horns and radios useless. High-speed transportation would be both very expensive, requiring huge fuel purchases, as well as dangerous, since a speeding car could knock down a building. When you got home, hungry for lunch, you would find that you had missed dinner; there would be a five-day delay in transit when you watched the Olympics in Australia on live TV. Finally, we would not be able to see the Milky Way, since the fireball of the Big Bang would surround us at the distance of Rigel or Deneb.

□   □   □   □

11. Give a physical argument that shows that it is impossible to accelerate an object of mass *m* to the speed of light, even with a continuous force acting on it.

Answer As an object approaches the speed of light, its energy approaches infinity. Hence, it would take an infinite amount of work to accelerate the object to the speed of light under the action of a constant force, or it would take an infinitely large force.

□   □   □   □

264

SOLUTIONS TO SELECTED PROBLEMS

3.   In a laboratory frame of reference, an observer notes that Newton's second law is valid. Show that it is also valid for an observer moving at a constant speed, small compared with the speed of light, relative to the laboratory frame.

Solution   The first observer watches some object accelerate along the x direction under applied forces. Call the instantaneous velocity of the object u_x. The second observer moves along the x axis at constant speed v relative to the first, and measures the object to have velocity

$$u_x' = u_x - v$$

The acceleration is

$$\frac{du_x'}{dt} = \frac{du_x}{dt} - 0$$

This is the same as that measured by the first observer. In this nonrelativistic case, they measure also the same mass and forces; so the second observer also confirms that $\Sigma F = ma$.   ◊

---

9.   An atomic clock moves at 1 000 km/h for 1.00 h as measured by an identical clock on the Earth. How many nanoseconds slow will the moving clock be compared with the Earth clock, at the end of the 1.00-h interval?

Solution   This problem is slightly more difficult than most, for the simple reason that your calculator probably cannot hold enough decimal places to yield an accurate answer. However, we can bypass the difficulty by noting the approximation:

$$\sqrt{1 - \frac{v^2}{c^2}} \approx 1 - \frac{v^2}{2c^2}$$

Squaring both sides will show that when v/c is small, these two terms are equivalent.

Evaluating v/c,

$$\frac{v}{c} = \left(\frac{1\,000 \times 10^3 \text{ m / h}}{3.00 \times 10^8 \text{ m / s}}\right)\left(\frac{1\,\text{h}}{3\,600 \text{ s}}\right) = 9.26 \times 10^{-7}$$

From Equation 39.7,

$$\Delta t = \gamma \Delta t_p = \frac{\Delta t_p}{\sqrt{1 - v^2/c^2}}$$

Rearranging, our approximation yields

$$\Delta t_p = \left(\sqrt{1 - \frac{v^2}{c^2}}\right)\Delta t \approx \left(1 - \frac{v^2}{2c^2}\right)\Delta t$$

and
$$\Delta t - \Delta t_p = \frac{v^2}{2c^2}\Delta t$$

Substituting,
$$\Delta t - \Delta t_p = \frac{(9.26\times10^{-7})^2}{2}(3\,600\text{ s})$$

Thus, the time lag of the moving clock is $\Delta t - \Delta t_p = 1.54\times10^{-9}$ s $= 1.54$ ns   ◊

---

11. A spacecraft with a proper length of 300 m takes 0.750 μs to pass an Earth observer. Determine the speed of the spacecraft as measured by the Earth observer.

Solution

Conceptualize: We should first determine if the spaceship is traveling at a relativistic speed: classically, $v = (300\text{m})/(0.750\ \mu s) = 4.00\times10^8$ m/s, which is faster than the speed of light (impossible)! Quite clearly, the relativistic correction must be used to find the correct speed of the spaceship, which we can guess will be close to the speed of light.

Categorize: We can use the contracted length equation to find the speed of the spaceship in terms of the proper length and the time. The time of 0.750 μs is the **proper time** measured by the Earth observer, because it is the time interval between two events that she sees as happening at the same point in space. The two events are the passage of the front end of the spaceship over her stopwatch, and the passage of the back end of the ship.

Analyze: $L = L_p/\gamma$, with $L = v\Delta t$:
$$v\Delta t = L_p\left(1 - v^2/c^2\right)^{1/2}$$

Squaring both sides,
$$v^2\Delta t^2 = L_p^2\left(1 - v^2/c^2\right)$$

$$v^2c^2 = L_p^2c^2/\Delta t^2 - v^2L_p^2/\Delta t^2$$

Solving for the velocity, $v = \dfrac{cL_p/\Delta t}{\sqrt{c^2 + L_p^2/\Delta t^2}} = \dfrac{(3.00\times10^8)(300\text{ m})/(0.750\times10^{-6}\text{ s})}{\sqrt{(3.00\times10^8)^2 + (300\text{ m})^2/(0.750\times10^{-6}\text{ s})^2}}$

So
$$v = 2.40\times10^8 \text{ m/s}$$ ◊

Finalize: The spaceship is traveling at 0.8c. We can also verify that the general equation for the speed reduces to the classical relation $v = L_p/\Delta t$ when the time is relatively large.

---

27. Two jets of material from the center of a radio galaxy are ejected in opposite directions. Both jets move at 0.750c relative to the galaxy. Determine the speed of one jet relative to the other.

Solution Take the galaxy as the unmoving frame. Arbitrarily define the jet moving upwards to be the object, and the jet moving downwards to be the "moving" frame.

$$u_x = 0.750\ c \qquad\qquad v = -0.750\ c$$

$$u'_x = \frac{u_x - v}{1 - u_x v/c^2} = \frac{0.750c - (-0.750c)}{1 - (0.750c)(-0.750c)/c^2} = \frac{1.50c}{1 + 0.750^2} = 0.960c \qquad\qquad \Diamond$$

33. An unstable particle at rest breaks into two fragments of unequal mass. The mass of the first fragment is 2.50×10^{-28} kg, and that of the other is 1.67×10^{-27} kg. If the lighter fragment has a speed of 0.893c after the breakup, what is the speed of the heavier fragment?

Solution

Conceptualize: The heavier fragment should have a speed less than that of the lighter piece since the momentum of the system must be conserved. However, due to the relativistic factor, the ratio of the speeds will not equal the simple ratio of the particle masses, which would give a speed of 0.134c for the heavier particle.

Categorize: Relativistic momentum of the system must be conserved. For the total momentum to be zero after the fission, as it was before, $p_1 + p_2 = 0$, where we will refer to the lighter particle with the subscript '1', and to the heavier particle with the subscript '2.'

Analyze:

$\gamma_2 m_2 v_2 + \gamma_1 m_1 v_1 = 0:$
$\qquad \gamma_2 m_2 v_2 + \left(\dfrac{2.50 \times 10^{-28}\ \text{kg}}{\sqrt{1 - 0.893^2}} \right)(0.893c) = 0$

Rearranging,
$\qquad \left(\dfrac{1.67 \times 10^{-27}\ \text{kg}}{\sqrt{1 - v_2^2/c^2}} \right) \dfrac{v_2}{c} = -4.96 \times 10^{-28}\ \text{kg}$

Squaring both sides,
$\qquad \left(2.79 \times 10^{-54}\ \text{kg}^2 \right) \left(\dfrac{v_2}{c} \right)^2 = \left(2.46 \times 10^{-55}\ \text{kg}^2 \right) \left(1 - \dfrac{v_2^2}{c^2} \right)$

and
$\qquad v_2 = -0.285c$

We choose the negative sign only to mean that the two particles must move in opposite directions. The speed, then, is $|v_2| = 0.285c$ ◊

Finalize: The speed of the heavier particle is less than the lighter particle, as expected. We can also see that for this situation, the relativistic speed of the heavier particle is about twice as great as was predicted by a simple non-relativistic calculation.

35. A proton in a high-energy accelerator moves with a speed of $c/2$. Use the work-kinetic energy theorem to find the work required to increase its speed to (a) $0.750c$ and (b) $0.995c$.

Solution

Conceptualize: Since particle accelerators have typical maximum energies on the order of a GeV ($1\ eV = 1.60 \times 10^{-19}$ J), we could expect the work to be $\sim 10^{-10}$ J.

Categorize: The work-energy theorem is $W = \Delta K = K_f - K_i$ which for relativistic speeds ($v \sim c$) is:

$$W = \left(\frac{1}{\sqrt{1 - v_f^2/c^2}} - 1 \right) mc^2 - \left(\frac{1}{\sqrt{1 - v_i^2/c^2}} - 1 \right) mc^2$$

or (simplified), $W = \left(1/\sqrt{1 - v_f^2/c^2} - 1/\sqrt{1 - v_i^2/c^2} \right) mc^2$

Analyze:

(a) $W = \left(\dfrac{1}{\sqrt{1 - 0.750^2}} - \dfrac{1}{\sqrt{1 - 0.500^2}} \right) (1.67 \times 10^{-27}\ \text{kg})(3.00 \times 10^8\ \text{m/s})^2$

$W = (1.512 - 1.155)(1.50 \times 10^{-10}\ \text{J}) = 5.37 \times 10^{-11}\ \text{J}$ ◊

(b) $W = \left(\dfrac{1}{\sqrt{1 - 0.995^2}} - \dfrac{1}{\sqrt{1 - 0.500^2}} \right) (1.67 \times 10^{-27}\ \text{kg})(3.00 \times 10^8\ \text{m/s})^2$

$W = (10.01 - 1.155)(1.50 \times 10^{-10}\ \text{J}) = 1.33 \times 10^{-9}\ \text{J}$ ◊

Finalize: Even though these energies may seem like small numbers, we must remember that the proton has very small mass, so these input energies are comparable to the rest mass energy of the proton (1.50×10^{-10} J). To produce a speed higher by 33%, the answer to part (b) is 25 times larger than the answer to part (a). Even with arbitrarily large accelerating energies, the particle will never reach or exceed the speed of light.

37. Find the momentum of a proton in MeV/c units assuming its total energy is twice its rest energy.

Solution We solve this problem in several steps. First, we solve for the rest energy of the proton; then we use that in the equation for total energy, to find the speed of the proton. Last, we substitute the speed and rest energy into the relativistic momentum equation to obtain the momentum.

For a proton, $\qquad\qquad mc^2 = (1.67\times10^{-27}\text{ kg})(2.998\times10^8 \text{ m/s})^2 = 1.50\times10^{-10}\text{ J}$

In MeV, $\qquad\qquad mc^2 = \left(1.50\times10^{-10}\text{ J}\right)\left(\dfrac{1\text{ eV}}{1.60\times10^{-19}\text{ kg}\cdot\text{m}^2/\text{s}^2}\right) = 938\text{ MeV}$

Total energy is $\qquad\qquad \gamma mc^2 = 2mc^2 \quad$ so that $\quad \gamma = 2$

Therefore $\qquad\qquad \dfrac{1}{\sqrt{1-v^2/c^2}} = 2 \quad$ and $\qquad v/c = 0.866$

Momentum is $\qquad\qquad p = \gamma mv = \dfrac{(\gamma mc^2)}{c}(v/c)$

So in this case $\qquad\qquad p = \dfrac{2(938\text{ MeV})}{c}(0.866) = 1\,620\text{ MeV}/c \qquad\qquad\qquad \Diamond$

By using Eq. 39.27, we could calculate the momentum without first finding the speed.

39. A proton moves at $0.950c$. Calculate its (a) rest energy, (b) total energy, and (c) kinetic energy.

Solution At $v = 0.950c$, $\quad \gamma = \dfrac{1}{\sqrt{1-v^2/c^2}} = \dfrac{1}{\sqrt{1-0.950^2}} = 3.20$

(a) $\quad E_R = mc^2 = (1.67\times10^{-27}\text{ kg})(2.998\times10^8\text{ m/s})^2 = 1.50\times10^{-10}\text{ J} = 938\text{ MeV} \qquad \Diamond$

(b) $\quad E = \gamma mc^2 = \gamma E_R = (3.20)(938\text{ MeV}) = 3.00\text{ GeV} \qquad\qquad\qquad\qquad \Diamond$

(c) $\quad K = E - E_R = 3.00\text{ GeV} - 938\text{ MeV} = 2.07\text{ GeV} \qquad\qquad\qquad\qquad \Diamond$

43. Show that the energy-momentum relationship $E^2 = p^2c^2 + (mc^2)^2$ follows from the expressions $E = \gamma mc^2$ and $p = \gamma mu$.

Solution $\qquad\qquad\qquad\qquad\qquad E = \gamma mc^2 \qquad\qquad\qquad\qquad p = \gamma mu$

Squaring both equations, $\qquad\qquad\qquad E^2 = (\gamma mc^2)^2 \qquad\qquad\qquad p^2 = (\gamma mu)^2$

Multiplying the second equation by c^2, and subtracting it from the first,

$E^2 - p^2c^2 = (\gamma mc^2)^2 - (\gamma mu)^2 c^2:$ $\qquad E^2 - p^2c^2 = \gamma^2\big((mc^2)(mc^2) - (mc^2)(mu^2)\big)$

Extracting the (mc^2) factors, $\qquad E^2 - p^2c^2 = \gamma^2(mc^2)^2\left(1 - \dfrac{u^2}{c^2}\right)$

and applying the definition of γ, $\qquad E^2 - p^2c^2 = \left(1 - \dfrac{u^2}{c^2}\right)^{-1}(mc^2)^2\left(1 - \dfrac{u^2}{c^2}\right)$

The γ factors divide out, leaving $\qquad E^2 - p^2c^2 = (mc^2)^2$ ◊

---

47. A pion at rest ($m_\pi = 273m_e$) decays to a muon ($m_\mu = 207m_e$) and an antineutrino ($m_{\bar\nu} \approx 0$). The reaction is written $\pi^- \to \mu^- + \bar\nu$. Find the kinetic energy of the muon and the energy of the antineutrino in electron volts. (*Suggestion:* Conserve both energy and momentum.)

Solution   We use, together, both the energy version and the momentum version of the isolated system model.

By conservation of energy, $\qquad m_\pi c^2 = \gamma m_\mu c^2 + |p_\nu|c$

By conservation of momentum, $\qquad p_\nu = -p_\mu = -\gamma m_\mu v$

Substituting the second equation into the first, $\qquad m_\pi c^2 = \gamma m_\mu c^2 + \gamma m_\mu v c$

Simplified, this equation then reads $\qquad m_\pi = m_\mu(\gamma + \gamma v/c)$

Substituting for the masses, $\qquad 273m_e = (207m_e)(\gamma + \gamma v/c)$

where $\qquad m_e c^2 = 0.511 \text{ MeV}$

Numerically, $\qquad \dfrac{273m_e}{207m_e} = \dfrac{1 + v/c}{\sqrt{1 - (v/c)^2}} = \sqrt{\dfrac{1 + v/c}{1 - v/c}}$

Solving for the muon speed, $\qquad \dfrac{v}{c} = \dfrac{273^2 - 207^2}{273^2 + 207^2} = 0.270$

Therefore, $\qquad \gamma = \dfrac{1}{\sqrt{1 - v^2/c^2}} = 1.038\,5$

and $\qquad K_\mu = (0.0385)(207 \times 0.511 \text{ MeV}) = 4.08 \text{ MeV}$ ◊

$\qquad K_{\bar\nu} = (273 \times 0.511 \text{ MeV}) - (207 \times 0.511 \text{ MeV} + 4.08 \text{ MeV}) = 29.6 \text{ MeV}$ ◊

53. The power output of the Sun is 3.77×10^{26} W. How much mass is converted to energy in the Sun each second?

Solution

From $E_R = mc^2$, we have

$$m = \frac{E_R}{c^2} = \frac{3.77 \times 10^{26} \text{ J}}{(3.00 \times 10^8 \text{ m / s})^2} = 4.19 \times 10^9 \text{ kg} \qquad \lozenge$$

57. The cosmic rays of highest energy are protons that have kinetic energy on the order of 10^{13} MeV. (a) How long would it take a proton of this energy to travel across the Milky Way galaxy, having a diameter ~10^5 ly, as measured in the proton's frame? (b) From the point of view of the proton, how many kilometers across is the galaxy?

Solution

Conceptualize: We can guess that the energetic cosmic rays will be traveling close to the speed of light, so the time it takes a proton to traverse the Milky Way will be much less in the proton's frame than 10^5 years. The galaxy will also appear smaller to the high-speed protons than the galaxy's proper diameter of 10^5 light-years.

Categorize: The kinetic energy of the protons can be used to determine the relativistic γ-factor, which can then be applied to the time dilation and length contraction equations to find the time and distance in the proton's frame of reference.

Analyze: The relativistic kinetic energy of a proton is $K = (\gamma - 1)mc^2 = 10^{13}$ MeV

Its rest energy is

$$mc^2 = (1.67 \times 10^{-27} \text{ kg})(2.998 \times 10^8 \text{ m / s})^2 \left(\frac{1 \text{ eV}}{1.60 \times 10^{-19} \text{ kg} \cdot \text{m}^2 / \text{s}^2} \right) = 938 \text{ MeV}$$

So $\qquad 10^{13}$ MeV $= (\gamma - 1)(938 \text{ MeV})$, and therefore $\gamma = 1.07 \times 10^{10}$

The proton's speed in the galaxy's reference frame can be found from

$$\gamma = 1 / \sqrt{1 - v^2 / c^2}: \qquad\qquad 1 - v^2 / c^2 = 8.80 \times 10^{-21}$$

and $\qquad v = c\sqrt{1 - 8.80 \times 10^{-21}} = (1 - 4.40 \times 10^{-21})c \approx 3.00 \times 10^8 \text{ m / s}$

The proton's speed is nearly as large as the speed of light. In the galaxy frame, the traversal time is

$$\Delta t = x/v = 10^5 \text{ light-years}/c = 10^5 \text{ years}$$

(a) This is dilated from the proper time measured in the proton's frame. The proper time is found from $\Delta t = \gamma \Delta t_p$:

$$\Delta t_p = \Delta t/\gamma = 10^5 \text{ yr}/1.07 \times 10^{10} = 9.38 \times 10^{-6} \text{ years} = 296 \text{ s}$$

$$\Delta t_p \sim \text{ a few hundred seconds} \qquad \Diamond$$

(b) The proton sees the galaxy moving by at a speed nearly equal to c, passing in 296 s:

$$\Delta L_p = v \Delta t_p = (3.00 \times 10^8)(296 \text{ s}) = 8.88 \times 10^7 \text{ km} \sim 10^8 \text{ km} \qquad \Diamond$$

$$\Delta L_p = (8.88 \times 10^{10} \text{ m}) \left(\frac{1 \text{ ly}}{9.46 \times 10^{15} \text{ m}} \right) = 9.39 \times 10^{-6} \text{ ly} \sim 10^{-5} \text{ ly}$$

Finalize: The results agree with our predictions, although we may not have guessed that the protons would be traveling so close to the speed of light! The calculated results should be rounded to zero significant figures since we were given order of magnitude data. We should also note that the relative speed of motion v and the value of γ are the same in both the proton and galaxy reference frames.

===

61. The net nuclear fusion reaction inside the Sun can be written as $4\,^1\text{H} \rightarrow\ ^4\text{He} + \Delta E$. The rest energy of each hydrogen atom is 938.78 MeV and the rest energy of the helium-4 atom is 3728.4 MeV. Calculate the percentage of the starting mass that is transformed to other forms of energy.

Solution

The original rest energy is $\qquad\qquad E_R = 4(938.78 \text{ MeV}) = 3\,755.12 \text{ MeV}$

The energy given off is $\qquad\qquad \Delta E = (3\,755.12 - 3\,728.4) \text{ MeV} = 26.7 \text{ MeV}$

The fractional energy released is $\qquad\qquad \dfrac{\Delta E}{E_R} = \dfrac{26.7 \text{ MeV}}{3\,755 \text{ MeV}} \times 100\% = 0.712\% \qquad \Diamond$

65. Spacecraft I, containing students taking a physics exam, approaches the Earth with a speed of 0.600c (relative to the Earth), while spacecraft II, containing professors proctoring the exam, moves at 0.280c (relative to the Earth) directly toward the students. If the professors stop the exam after 50.0 min have passed on their clock, how long does the exam last as measured by (a) the students (b) an observer on the Earth?

Solution

Suppose that in the Earth frame, the students are moving to the right at $u_x = 0.600c$ and the professors are moving to the left, with velocity component $v = -0.280c$

In the professors' frame, the Earth moves to the right at $u_e' = 0.280c$ and the students move to the right at

$$u_x' = \frac{u_x - v}{1 - u_x v / c^2}: \qquad u_x' = \frac{0.600c - (-0.280c)}{1 - (0.600c)(-0.280c)/c^2} = 0.753c$$

The professors measure 50 minutes on a clock at rest in their frame: they measure proper time and everyone else sees longer, dilated time intervals.

(a) For the students, $\qquad \Delta t = \gamma \Delta t_p = \dfrac{50.0 \text{ min}}{\sqrt{1 - (0.753)^2}} = 76.0 \text{ min} \qquad \Diamond$

(b) On Earth, $\qquad \Delta t = \gamma \Delta t_p = \dfrac{50 \text{ min}}{\sqrt{1 - (0.280)^2}} = 52 \text{ min, 5 sec} \qquad \Diamond$

---

67. A supertrain (proper length 100 m) travels at a speed of 0.950c as it passes through a tunnel (proper length 50.0 m). As seen by a trackside observer, is the train ever completely within the tunnel? If so, with how much space to spare?

Solution

The observer sees the proper length of the tunnel, 50.0 m, but sees the train Lorentz-contracted to length

$$L = L_p \sqrt{1 - v^2/c^2} = (100 \text{ m})\sqrt{1 - (0.950c)^2} = 31.2 \text{ m}$$

This is shorter than the tunnel by 18.8 m, so it is completely within the tunnel. $\qquad \Diamond$

---

69. A particle with electric charge q moves along a straight line in a uniform electric field **E** with a speed of u. The electric force exerted on the particle is qE. The motion and the electric field are both in the x direction. (a) Show that the acceleration of the particle in the x direction is given by

$$a = \frac{du}{dt} = \frac{qE}{m}\left(1 - \frac{u^2}{c^2}\right)^{3/2}$$

(b) Discuss the significance of the dependence of the acceleration on the speed. (c) **What If?** If the particle starts from rest at $x=0$ at $t=0$, how would you proceed to find the speed of the particle and its position at time t?

Solution   The force on a charge in an electric field is given by $F = qE$. Further, at any speed, the **momentum** of the particle is given by

$$p = \gamma m u = \frac{mu}{\sqrt{1 - u^2/c^2}}$$

(a)   We use the momentum version of the nonisolated system model.

Since $F = qE = \dfrac{dp}{dt}$, $\qquad qE = \dfrac{d}{dt}\left(mu\left(1 - u^2/c^2\right)^{-1/2}\right)$

$$qE = m\left(1 - \frac{u^2}{c^2}\right)^{-1/2}\frac{du}{dt} + \frac{1}{2}mu\left(1 - \frac{u^2}{c^2}\right)^{-3/2}\left(\frac{2u}{c^2}\right)\frac{du}{dt}$$

Simplifying, we find that $\quad \dfrac{qE}{m} = \dfrac{du}{dt}\left(1 - \dfrac{u^2}{c^2}\right)^{-3/2} \quad$ and $\quad a = \dfrac{du}{dt} = \dfrac{qE}{m}\left(1 - \dfrac{u^2}{c^2}\right)^{3/2} \quad \lozenge$

(b)   As $u \to c$, we see that $a \to 0$. The particle thus never attains the speed of light. If $u \ll c$, $a \approx qE/m$, in agreement with the nonrelativistic account.

(c)   Taking the acceleration equation, isolating the velocity terms and integrating,

$$\int_0^u \left(1 - \frac{u^2}{c^2}\right)^{-3/2} du = \int_0^t \frac{qE}{m}\,dt: \quad u = \frac{qEct}{\sqrt{m^2c^2 + q^2E^2t^2}} = \frac{dx}{dt} \qquad \lozenge$$

$$x = \int_0^x dx = qEc\int_0^t \frac{t\,dt}{\sqrt{m^2c^2 + q^2E^2t^2}} = \frac{c}{qE}\left(\sqrt{m^2c^2 + q^2E^2t^2} - mc\right) \qquad \lozenge$$

Chapter 40

INTRODUCTION TO QUANTUM PHYSICS

EQUATIONS AND CONCEPTS

Stefan's law states that the total power emitted as thermal radiation depends on the fourth power of the temperature in kelvins. The parameter σ is the Stefan-Boltzmann constant and e is the emissivity of the surface of area A.

$$\mathcal{P} = \sigma A e T^4 \tag{40.1}$$

$$\sigma = 5.669\,6 \times 10^{-8} \ \text{W/m}^2 \cdot \text{K}^4$$

According to the **Wien displacement law**, as the temperature of a blackbody increases the radiation intensity increases and the peak of the distribution shifts to shorter wavelengths. *The total radiation emitted is the area under the curve and increases with temperature.*

$$\lambda_{max} T = 2.898 \times 10^{-3} \ \text{m} \cdot \text{K} \tag{40.2}$$

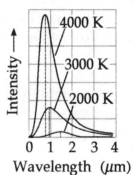

Discrete energy values of an atomic oscillator are determined by a quantum number, n. Each discrete energy value corresponds to a quantum state.

$$E_n = nhf \tag{40.4}$$

$$(n = 1, 2, 3 \ldots)$$

The **energy of a quantum** or photon corresponds to the energy difference between initial and final quantum states. *An oscillator emits or absorbs energy only when there is a transition between quantum states.*

$$E = hf \tag{40.5}$$

Planck's constant is a fundamental constant of nature.

$$h = 6.626 \times 10^{-34} \ \text{J} \cdot \text{s} \tag{40.7}$$

275

The **maximum kinetic energy of an ejected photoelectron** depends on the work function of the metal, ϕ, which is typically a few eV.

$$K_{max} = hf - \phi \qquad (40.9)$$

The **cutoff wavelength** (and corresponding frequency) depends on the value of the work function of a specific surface; for wavelengths greater than λ_c no photoelectric effect will be observed.

$$\lambda_c = \frac{hc}{\phi} \qquad (40.10)$$

$$hc = 1\,240\,\text{eV·nm}$$

The **Compton shift** is the change in wavelength of an x-ray when scattered from an electron. *The scattered x-ray makes an angle θ with the direction of the incident x-ray. The Compton wavelength of the electron is $h/m_e c = 0.002\,43$ nm.*

$$\lambda' - \lambda_0 = \frac{h}{m_e c}(1 - \cos\theta) \qquad (40.11)$$

The **de Broglie wavelength** of a particle is inversely proportional to the momentum of the particle.

$$\lambda = \frac{h}{mv} \qquad (40.15)$$

Phase speed refers to the speed of a single wave crest in a wave packet.

$$v_{phase} = \frac{\omega}{k} \qquad (40.18)$$

Group speed refers to the speed of the packet (envelope) or group of waves.

$$v_g = \frac{d\omega}{dk} \qquad (40.19)$$

The **Heisenberg uncertainty principle** can be stated in two forms:

- Simultaneous measurements of **position and momentum** with respective uncertainties Δx and Δp_x.

$$\Delta x \Delta p_x \geq \frac{\hbar}{2} \qquad (40.23)$$

- Simultaneous measurements of **energy and lifetime** with uncertainties ΔE and Δt.

$$\Delta E \Delta t \geq \frac{\hbar}{2} \qquad (40.24)$$

REVIEW CHECKLIST

You should be able to:

▷ Describe the formula for blackbody radiation proposed by Planck, the assumption made in deriving this formula, and the related experimental results. (Section 40.1)

▷ Describe the Einstein model for the photoelectric effect, including the important experimental results. Make calculations using the photoelectric equation. (Section 40.2)

▷ Describe the Compton effect (the scattering of x-rays by electrons) and make calculations using the equation for the Compton shift. (Section 40.3)

▷ Make calculations using the de Broglie wavelength equation. (Section 40.5)

▷ Calculate the phase speed and group speed associated with a quantum particle. (Section 40.6)

▷ Make calculations using both forms of the Heisenberg uncertainty principle. (Section 40.8)

ANSWERS TO SELECTED QUESTIONS

5. If the photoelectric effect is observed for one metal, can you conclude that the effect will also be observed for another metal under the same conditions? Explain.

Answer No. Suppose that the incident light frequency at which you first observed the photoelectric effect is above the cutoff frequency of the first metal, but less than the cutoff frequency of the second metal. In that case, the photoelectric effect would not be observed at all in the second metal.

□   □   □   □

7. Why does the existence of a cutoff frequency in the photoelectric effect favor a particle theory for light over a wave theory?

Answer Wave theory predicts that the photoelectric effect should occur at any frequency, provided that the light intensity is high enough. As is implied by the question, this is in contradiction to experimental results.

□   □   □   □

15. An x-ray photon is scattered by an electron. What happens to the frequency of the scattered photon relative to that of the incident photon?

Answer The x-ray photon transfers some of its energy to the electron. Thus, its energy, and therefore its frequency, must be decreased.

☐   ☐   ☐   ☐

21. If matter has a wave nature, why is this wave-like characteristic not observable in our daily experiences?

Answer For any object that we can perceive directly, the de Broglie wavelength $\lambda = h/mv$ is too small to be measured by any means; therefore, no wavelike characteristics can be observed. The object will not diffract noticeably when it goes through an aperture. It will not show resolvable interference maxima and minima when it goes through two openings. It will not show resolvable nodes and antinodes if it is in resonance.

☐   ☐   ☐   ☐

SOLUTIONS TO SELECTED PROBLEMS

1. The human eye is most sensitive to 560-nm light. What is the temperature of a black body that would radiate most intensely at this wavelength?

Solution We use Wien's law:
$$\lambda_{max} T = 2.898 \times 10^{-3} \text{ m} \cdot \text{K}$$

$$T = \frac{2.90 \text{ mm} \cdot \text{K}}{560 \times 10^{-6} \text{ mm}} = 5180 \text{ K} \qquad \diamond$$

Related Information: This is close to the temperature of the surface of the Sun (which acts as a pretty good black body). Living things on Earth evolved to be sensitive to electromagnetic waves near this wavelength because there is such a lot of it bouncing around, carrying information.

Chapter 40

7. Calculate the energy, in electron volts, of a photon whose frequency is (a) 620 THz, (b) 3.10 GHz, (c) 46.0 MHz. (d) Determine the corresponding wavelengths for these photons and state the classification of each on the electromagnetic spectrum.

Solution $E = hf$

(a) $E = (6.63\times10^{-34}\text{ J·s})(6.20\times10^{14}\text{ Hz}) = 4.11\times10^{-19}\text{ J} = 2.57\text{ eV}$ ◊

(b) $E = (6.63\times10^{-34}\text{ J·s})(3.10\times10^{9}\text{ Hz}) = 2.06\times10^{-24}\text{ J} = 12.8\ \mu eV$ ◊

(c) $E = (6.63\times10^{-34}\text{ J·s})(46.0\times10^{6}\text{ Hz}) = 3.05\times10^{-26}\text{ J} = 1.91\times10^{-7}\text{ eV}$ ◊

(d) $\lambda_a = \dfrac{c}{f} = \dfrac{3.00\times10^{8}\text{ m/s}}{6.20\times10^{14}\text{ s}^{-1}} = 4.84\times10^{-7}\text{ m} = 484\text{ nm}$ ◊

$\lambda_b = \dfrac{3.00\times10^{8}\text{ m/s}}{3.10\times10^{9}\text{ s}^{-1}} = 0.096\,8\text{ m} = 9.68\text{ cm}$ ◊

$\lambda_c = \dfrac{3.00\times10^{8}\text{ m/s}}{46.0\times10^{6}\text{ s}^{-1}} = 6.52\text{ m}$ ◊

These wavelengths correspond, respectively, to blue light, microwave radiation, and radio waves in the public LO band.   ◊

9. An FM radio transmitter has a power output of 150 kW and operates at a frequency of 99.7 MHz. How many photons per second does the transmitter emit?

Solution   Each photon has an energy

$E = hf = (6.63\times10^{-34}\text{ J·s})(99.7\times10^{6}\text{ s}^{-1}) = 6.61\times10^{-26}\text{ J}$

The number of photons per second is the power divided by the energy per photon:

$R = \dfrac{\mathcal{P}}{E} = \dfrac{150\times10^{3}\text{ J/s}}{6.61\times10^{-26}\text{ J}} = 2.27\times10^{30}\text{ photons/s}$ ◊

17. Two light sources are used in a photoelectric experiment to determine the work function for a particular metal surface. When green light from a mercury lamp ($\lambda = 546.1$ nm) is used, a stopping potential of 0.376 V reduces the photocurrent to zero. (a) Based on this measurement, what is the work function for this metal? (b) What stopping potential would be observed when using the yellow light from a helium discharge tube ($\lambda = 587.5$ nm)?

279

Solution

Conceptualize: According to Table 40.1, the work function for most metals is on the order of a few eV, so this metal is probably similar. We can expect the stopping potential for the yellow light to be slightly lower than 0.376 V since the yellow light has a longer wavelength (lower frequency) and therefore less energy than the green light.

Categorize: In this photoelectric experiment, the green light has sufficient energy hf to overcome the work function of the metal ϕ so that the ejected electrons have a maximum kinetic energy of 0.376 eV. With this information, we can use the photoelectric effect equation to find the work function, which can then be used to find the stopping potential for the less energetic yellow light.

Analyze:

(a) Einstein's photoelectric effect equation is $K_{max} = hf - \phi$ and the energy required to raise an electron through a 1-V potential is 1 eV, so that

$$K_{max} = e\Delta V_s = 0.376 \text{ eV}$$

The energy of a photon from the mercury lamp is:

$$hf = \frac{hc}{\lambda} = \frac{(4.14\times10^{-15} \text{ eV·s})(3.00\times10^8 \text{ m/s})}{546.1\times10^{-9} \text{ m}} = 2.27 \text{ eV}$$

Therefore, the work function for this metal is:

$$\phi = hf - K_{max} = 2.27 \text{ eV} - 0.376 \text{ eV} = 1.90 \text{ eV} \qquad \lozenge$$

(b) For the yellow light, $\lambda = 587.5$ nm

and

$$hf = \frac{hc}{\lambda} = \frac{(4.14\times10^{-15} \text{ eV·s})(3.00\times10^8 \text{ m/s})}{587.5\times10^{-9} \text{ m}} = 2.11 \text{ eV}$$

Therefore, $K_{max} = hf - \phi = 2.11 \text{ eV} - 1.90 \text{ eV} = 0.216 \text{ eV}$

so $\Delta V_s = 0.216 \text{ V}$ $\qquad \lozenge$

Finalize: The work function for this metal is lower than we expected, and does not correspond with any of the values in Table 40.1. Further examination in the **CRC Handbook of Chemistry and Physics** reveals that all of the metal elements have work functions between 2 and 6 eV. However, a single metal's work function may vary by about 1 eV depending on impurities in the metal, so it is just barely possible that a metal might have a work function of 1.90 eV.

The stopping potential for the yellow light is indeed lower than for the green light as we expected. An interesting calculation is to find the wavelength for the lowest energy light that will eject electrons from this metal. That threshold wavelength giving $K_{max} = 0$ is 654 nm, which is red light in the visible portion of the electromagnetic spectrum.

23. A 0.001 60-nm photon scatters from a free electron. For what (photon) scattering angle does the recoiling electron have kinetic energy equal to the energy of the scattered photon?

Solution   The energy of the incoming photon is

$$E_0 = \frac{hc}{\lambda_0} = \frac{(6.63 \times 10^{-34} \text{ J} \cdot \text{s})(3.00 \times 10^8 \text{ m/s})}{0.001\,60 \times 10^{-9} \text{ m}} = 1.24 \times 10^{-13} \text{ J}$$

The outgoing photon and the electron share equally in this energy. The kinetic energy of the electron and the energy of the scattered photon are each one half of E_0.

$$E' = 6.22 \times 10^{-14} \text{ J} \qquad \text{and} \qquad \lambda' = \frac{hc}{E'} = 3.20 \times 10^{-12} \text{ m}$$

The shift in wavelength is $\Delta\lambda = \lambda' - \lambda_0 = 1.60 \times 10^{-12}$ m

But by Equation 40.11, $\Delta\lambda = \lambda_C(1 - \cos\theta)$ where λ_C is the Compton wavelength.

Then $$\cos\theta = 1 - \frac{\Delta\lambda}{\lambda_c} = 1 - \frac{1.60 \times 10^{-12} \text{ m}}{0.002\,43 \times 10^{-9} \text{ m}} = 0.342$$

and $$\theta = 70.0° \qquad \qquad \diamond$$

---

37. The nucleus of an atom is on the order of 10^{-14} m in diameter. For an electron to be confined to a nucleus, its de Broglie wavelength would have to be on this order of magnitude or smaller. (a) What would be the kinetic energy of an electron confined to this region? (b) Given that typical binding energies of electrons in atoms are measured to be on the order of a few eV, would you expect to find an electron in a nucleus? Explain.

Solution

Conceptualize: The de Broglie wavelength of a normal ground-state orbiting electron is on the order of 10^{-10} m (the diameter of a hydrogen atom) so with a shorter wavelength, the electron would have more kinetic energy if confined inside the nucleus. If the kinetic energy is much greater than the potential energy characterizing its attraction with the positive nucleus, then the electron will escape from its electrostatic potential well.

Categorize: If we try to calculate the velocity of the electron from the de Broglie wavelength,

we find that $$v = \frac{h}{m_e\lambda} = \frac{6.63 \times 10^{-34} \text{ J} \cdot \text{s}}{(9.11 \times 10^{-31} \text{ kg})(10^{-14} \text{ m})} = 7.27 \times 10^{10} \text{ m/s}$$

which is not possible since it exceeds the speed of light. Therefore, we must use the relativistic energy expression to find the kinetic energy of this fast-moving electron.

Analyze:

(a) We find the momentum of the particle:

$$p = \frac{h}{\lambda} = \frac{6.63 \times 10^{-34} \text{ J·s}}{10^{-14} \text{ m}} = 6.63 \times 10^{-20} \text{ N·s}$$

We find the particle's relativistic total energy from $E^2 = (pc)^2 + (mc^2)^2$:

$$E = \sqrt{(1.99 \times 10^{-11} \text{ J})^2 + (8.19 \times 10^{-14} \text{ J})^2} = 1.99 \times 10^{-11} \text{ J}$$

Its relativistic kinetic energy is $K = E - mc^2$:

$$K = \frac{1.99 \times 10^{-11} \text{ J} - 8.19 \times 10^{-14} \text{ J}}{1.60 \times 10^{-19} \text{ J/eV}} = 124 \text{ MeV} \sim 100 \text{ MeV} \qquad \lozenge$$

(b) The electric potential energy of an electron-proton system with a separation of 10^{-14} m is:

$$U = -\frac{k_e e^2}{r} = -\frac{(8.99 \times 10^9 \text{ N·m}^2/\text{C}^2)(1.60 \times 10^{-19} \text{ C})^2}{10^{-14} \text{ m}} = -2.30 \times 10^{-14} \text{ J} \sim -0.1 \text{ MeV}$$

Since the kinetic energy is nearly 1 000 times greater than the potential energy, the electron would immediately escape the proton's attraction and would not be confined to the nucleus. $\qquad \lozenge$

Finalize: It is also interesting to notice in the above calculations that the rest energy of the electron is negligible compared to the momentum contribution to the total energy.

---

41. The resolving power of a microscope depends on the wavelength used. If one wished to "see" an atom, a resolution of approximately 1.00×10^{-11} m would be required. (a) If electrons are used (in an electron microscope), what minimum kinetic energy is required for the electrons? (b) **What If?** If photons are used, what minimum photon energy is needed to obtain the required resolution?

Solution

(a) Since the de Broglie wavelength is $\lambda = h/p$,

$$p_e = \frac{h}{\lambda} = \frac{6.63 \times 10^{-34} \text{ J·s}}{1.00 \times 10^{-11} \text{ m}} = 6.63 \times 10^{-23} \text{ kg·m/s}$$

$$K_e = \frac{p_e{}^2}{2m_e} = \frac{(6.63\times10^{-23}\ \text{kg}\cdot\text{m}/\text{s})^2}{2(9.11\times10^{-31}\ \text{kg})} = 2.41\times10^{-15}\ \text{J} = 15.1\ \text{keV} \qquad \lozenge$$

For better accuracy, you can use the relativistic equation

$$(m_e c^2 + K)^2 = p_e{}^2 c^2 + m_e{}^2 c^4 \qquad \text{to find that} \qquad K = 14.9\ \text{keV} \qquad \lozenge$$

(b)   For photons:

$$E = hf = \frac{hc}{\lambda} = \frac{(6.63\times10^{-34}\ \text{J}\cdot\text{s})(3.00\times10^8\ \text{m}/\text{s})}{1.00\times10^{-11}\ \text{m}} = 1.99\times10^{-14}\ \text{J} = 124\ \text{keV} \qquad \lozenge$$

For the photon, this wavelength $\lambda = 10\ \text{pm}$ is in the x-ray range of the electromagnetic spectrum.

---

45.   Neutrons traveling at 0.400 m/s are directed through a pair of slits having a 1.00-mm separation. An array of detectors is placed 10.0 m from the slits. (a) What is the de Broglie wavelength of the neutrons?   (b) How far off axis is the first zero-intensity point on the detector array? (c) When a neutron reaches a detector, can we say which slit the neutron passed through? Explain.

Solution

We use the waves in interference model.

(a) $$\lambda = \frac{h}{mv} = \frac{6.63\times10^{-34}\ \text{J}\cdot\text{s}}{(1.67\times10^{-27}\ \text{kg})(0.400\ \text{m}/\text{s})} = 9.93\times10^{-7}\ \text{m} \qquad \lozenge$$

(b)   The condition for destructive interference in a multiple-slit experiment is

$$d\sin\theta = \left(m + \tfrac{1}{2}\right)\lambda \text{ with } m = 0 \text{ for the first minimum.}$$

Then $$\theta = \sin^{-1}\left(\frac{\lambda}{2d}\right) = 0.028\ 4°$$

$$\frac{y}{L} = \tan\theta : \quad y = L\tan\theta = (10.0\ \text{m})\tan(0.028\ 4°) = 4.96\ \text{mm} \qquad \lozenge$$

(c)   We cannot say the neutron passed through one slit. We can only say it passed through the pair of slits, as a water wave does to produce an interference pattern. $\qquad \lozenge$

---

49. An electron ($m_e = 9.11 \times 10^{-31}$ kg) and a bullet ($m = 0.0200$ kg) each have a velocity of magnitude of 500 m/s, accurate to within 0.0100 %. Within what limits could we determine the position of the objects along the direction of the velocity?

Solution

Conceptualize: It seems reasonable that a tiny particle like an electron could be located within a more narrow region than a bigger object like a bullet, but we often find that the realm of the very small does not obey common sense.

Categorize: Heisenberg's uncertainty principle can be used to find the uncertainty in position from the uncertainty in the momentum.

Analyze:

The uncertainty principle states: $\qquad \Delta x \Delta p_x \geq \hbar / 2$

where $\qquad \Delta p_x = m \Delta v \quad \text{and} \quad \hbar \equiv h / 2\pi$

Both the electron and bullet have a velocity uncertainty,

$$\Delta v = (0.000\,100)(500 \text{ m/s}) = 0.050\,0 \text{ m/s}$$

For the electron, the minimum uncertainty in position is

$$\Delta x = \frac{h}{4\pi m \Delta v} = \frac{6.63 \times 10^{-34} \text{ J} \cdot \text{s}}{4\pi (9.11 \times 10^{-31} \text{ kg})(0.050\,0 \text{ m/s})} = 1.16 \text{ mm} \qquad \Diamond$$

For the bullet $\quad \Delta x = \dfrac{h}{4\pi m \Delta v} = \dfrac{6.63 \times 10^{-34} \text{ J} \cdot \text{s}}{4\pi (0.020\,0 \text{ kg})(0.050\,0 \text{ m/s})} = 5.28 \times 10^{-32} \text{ m} \qquad \Diamond$

Finalize: Our intuition did not serve us well here, since the position of the center of the larger bullet can be determined much more precisely than the electron. Quantum mechanics describes all objects, but the quantum fuzziness in position is too small to observe for the bullet, yet large for the small-mass electron.

55. The following table shows data obtained in a photoelectric experiment. (a) Using these data, make a graph similar to Figure 40.11 that plots as a straight line. From the graph, determine (b) an experimental value for Planck's constant (in joule-seconds) and (c) the work function (in electron volts) for the surface. (Two significant figures for each answer are sufficient.)

Wavelength (nm)	Maximum Kinetic Energy of Photoelectrons (eV)
588	0.67
505	0.98
445	1.35
399	1.63

Solution Convert each wavelength to a frequency using the relation $\lambda f = c$, where c is the speed of light:

$$\lambda_1 = 588 \times 10^{-9} \text{ m} \qquad f_1 = 5.10 \times 10^{14} \text{ Hz}$$

$$\lambda_2 = 505 \times 10^{-9} \text{ m} \qquad f_2 = 5.94 \times 10^{14} \text{ Hz}$$

$$\lambda_3 = 445 \times 10^{-9} \text{ m} \qquad f_3 = 6.74 \times 10^{14} \text{ Hz}$$

$$\lambda_4 = 399 \times 10^{-9} \text{ m} \qquad f_4 = 7.52 \times 10^{14} \text{ Hz}$$

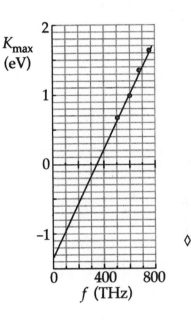

(a) Plot each point on an energy vs. frequency graph, as shown at the right. Extend a line through the set of 4 points, as far as the –y intercept.

(b) Our basic equation is $K_{max} = hf - \phi$. Therefore, Planck's constant should be equal to the slope of the linear K-f graph, which can be found from a least-squares fit or from reading the graph as:

$$h_{exp} = \frac{\text{Rise}}{\text{Run}} = \frac{1.25 \text{ eV} - 0.25 \text{ eV}}{6.5 \times 10^{14} \text{ Hz} - 4.0 \times 10^{14} \text{ Hz}} = 4.0 \times 10^{-15} \text{ eV} \cdot \text{s} = 6.4 \times 10^{-34} \text{ J} \cdot \text{s} \qquad \lozenge$$

From the scatter of the data points on the graph, we estimate the uncertainty of the slope to be about 3%. Thus we choose to show two significant figures in writing the experimental value of Planck's constant.

(c) From the linear equation $K_{max} = hf - \phi$, the work function for the metal surface is the negative of the y-intercept of the graph, so $\phi_{exp} = -(-1.4 \text{ eV}) = 1.4 \text{ eV}$. $\qquad \lozenge$

Based on the range of slopes that appear to fit the data, the estimated uncertainty of the work function is 5%.

61. The total power per unit area radiated by a black body at a temperature T is the area under the $I(\lambda,T)$-versus-λ curve, as shown in Figure 40.3. (a) Show that this power per unit area is

$$\int_0^\infty I(\lambda, T)\,d\lambda = \sigma T^4$$

where $I(\lambda,T)$ is given by Planck's radiation law and σ is a constant independent of T. This result is known as Stefan's law (see Section 20.7). To carry out the integration, you should make the change of variable $x = hc/\lambda kT$ and use the fact that

$$\int_0^\infty \frac{x^3 dx}{e^x - 1} = \frac{\pi^4}{15}$$

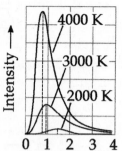

Intensity

4000 K
3000 K
2000 K

0 1 2 3 4
Wavelength (μm)

Figure 40.3

(b) Show that the Stefan-Boltzmann constant σ has the value

$$\sigma = \frac{2\pi^5 k_B{}^4}{15c^2 h^3} = 5.67\times10^{-8}\ \text{W}/\text{m}^2\cdot\text{K}^4$$

Solution   In order to make the suggested substitution, we find λ and $d\lambda$:

$$x = \frac{hc}{\lambda k_B T} \qquad\qquad \lambda = \frac{hc}{x k_B T} \qquad\qquad d\lambda = -\frac{hc\,dx}{x^2 k_B T}$$

We also note that the limits of integration change from $\lambda = (0, \infty)$ to $x = (\infty, 0)$. Substituting these variables into the integral, the intensity of the blackbody radiation is:

$$\int_0^\infty I(\lambda,T)\,d\lambda = \int_0^\infty \frac{2\pi hc^2}{\lambda^5(e^{hc/\lambda k_B T}-1)}\,d\lambda = \int_\infty^0 -\left(\frac{2\pi hc^2}{e^x - 1}\right)\left(\frac{x^5 k_B{}^5 T^5}{h^5 c^5}\right)\left(\frac{hc\,dx}{x^2 k_B T}\right)$$

$$\int_0^\infty I(\lambda,T)\,d\lambda = \frac{2\pi k_B{}^4 T^4}{h^3 c^2}\int_\infty^0 -\frac{x^3}{e^x - 1}\,dx = \frac{2\pi k_B{}^4 T^4}{h^3 c^2}\int_0^\infty \frac{x^3}{e^x - 1}\,dx$$

But the integral is $\pi^4/15$, so

$$\int_0^\infty I(\lambda,T)\,d\lambda = \frac{2\pi^5 k_B{}^4 T^4}{15 h^3 c^2} = \sigma T^4 \qquad\qquad\qquad \Diamond$$

with $$\sigma = \frac{2\pi^5 k_B{}^4}{15c^2 h^3} = \frac{2\pi^5 (1.3807\times10^{-23}\ \text{J}/\text{K})^4}{15(2.998\times10^8\ \text{m}/\text{s})^2(6.626\times10^{-34}\ \text{J}\cdot\text{s})^3} = 5.67\times10^{-8}\ \text{W}/\text{m}^2\cdot\text{K}^4 \quad \Diamond$$

65. Show that the ratio of the Compton wavelength λ_C to the de Broglie wavelength $\lambda = h/p$ for a relativistic electron is

$$\frac{\lambda_C}{\lambda} = \left[\left(\frac{E}{m_e c^2} \right)^2 - 1 \right]^{1/2}$$

where E is the total energy of the electron and m_e is its mass.

Solution

From the definition of the Compton wavelength, $\lambda_C = h/m_e c$. Taking the ratio of the Compton wavelength to the de Broglie wavelength,

$$\frac{\lambda_C^2}{\lambda^2} = \frac{p^2}{(m_e c)^2}$$

From Eq. 39.27, the relativistic momentum is

$$p^2 = \frac{E^2 - m_e^2 c^4}{c^2}$$

Substituting and simplifying,

$$\frac{\lambda_C^2}{\lambda^2} = \frac{(E^2 - m_e^2 c^4)}{(m_e c^2)^2}$$

Finally,

$$\frac{\lambda_C^2}{\lambda^2} = \left(\frac{E}{m_e c^2} \right)^2 - 1$$

and

$$\frac{\lambda_C}{\lambda} = \sqrt{\left(\frac{E}{m_e c^2} \right)^2 - 1}$$ ◊

Chapter 41

QUANTUM MECHANICS

EQUATIONS AND CONCEPTS

The **wave function of a free particle** moving along the x-axis is a sinusoidal wave. *The wave representing the particle has a constant amplitude A and angular wave number k.*

$$\psi(x) = Ae^{ikx} \tag{41.3}$$

The **probability of finding the particle within an arbitrary interval** equals the area under the curve of $|\psi|^2$ vs x, between the end points of the interval. *The probability density, $|\psi|^2$, is the relative probability of finding the particle at a given point along the interval.*

$$P_{ab} = \int_a^b |\psi|^2 dx \tag{41.5}$$

The **expectation value for the position** is the expected average value of the coordinate. *This is the average of values of position if calculated from the known wave function. To find the expectation value of any function f(x), replace x in Equation 41.7 with f(x).*

$$\langle x \rangle \equiv \int_{-\infty}^{\infty} \psi^* x \psi \, dx \tag{41.7}$$

For a **particle-in-a-box of width L** in one dimensional motion:

- The **wave function** can be represented as a sinusoidal function.

$$\psi(x) = A\sin\left(\frac{n\pi x}{L}\right) \tag{41.11}$$

$$(n = 1, 2, 3, \ldots)$$

- The **energy of the particle** is quantized. *The ground state corresponds to n = 1 and is the lowest energy state.*

$$E_n = \left(\frac{h^2}{8mL^2}\right)n^2 \tag{41.12}$$

$$(n = 1, 2, 3, \ldots)$$

288

The **time-independent Schrödinger equation for a bound system** (with energy E) allows, in principle, the determination of the wave functions and energies of the allowed states if the potential energy function U is known.

$$-\frac{\hbar^2}{2m}\frac{d^2\psi}{dx^2}+U\psi=E\psi \qquad (41.13)$$

The textbook describes the **application of the Schrödinger equation** to systems with three different potential energy functions: infinite square well, potential energy well of finite height, and harmonic oscillator.

For an **infinite square well,** shown to the right, U is zero inside the well and infinite outside; the Schrödinger equation can be stated as in Equation 41.14. *The wave function and quantized energy values are as found for a particle-in-a-box (see Equations 41.11 and 41.12).*

$$\frac{d^2\psi}{dx^2}=-\frac{2mE}{\hbar^2}\psi=-k^2\psi \qquad (41.14)$$

where $k=\frac{\sqrt{2mE}}{\hbar}$

For a **potential-energy well of finite height,** shown to the right, the potential U is zero inside the well and has a value greater than E outside.

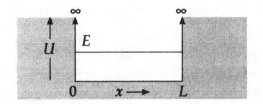

In **Regions I and III** (outside the potential well) the Schrödinger equation can be written as in Equation 41.18 or 41.19.

$$\frac{d^2\psi}{dx^2}=\left(\frac{2m(U-E)}{\hbar^2}\right)\psi \qquad (41.18)$$

$$\frac{d^2\psi}{dx^2}=C^2\psi \qquad (41.19)$$

where $C^2=\frac{2m(U-E)}{\hbar^2}$

The **wave functions for Regions I and III** decay exponentially with distance.

$$\psi_I = Ae^{Cx} \quad (x < 0)$$

$$\psi_{III} = Be^{-Cx} \quad (x > L)$$

In Region II (inside the potential well) the wave functions are sinusoidal. *The values of the constants A, B, F, and G can be determined from the boundary conditions and normalization.*

$$\psi_{II} = F \sin kx + G \cos kx$$

The **transmission coefficient** T represents the probability that a particle will tunnel or penetrate a barrier of width L where the potential energy is greater than the energy of the particle.

$$T \approx e^{-2CL} \qquad (41.20)$$

$$C = \frac{\sqrt{2m(U - E)}}{\hbar} \qquad (41.21)$$

For a **harmonic oscillator** of total energy E, angular frequency ω, and potential energy function as shown to the right, the Schrödinger equation can be written as in Equation 41.22.

The **ground state wave function for the harmonic oscillator** contains a constant B which must be determined from the normalization condition. *Wave functions for all excited states include an exponential factor.*

$$-\frac{\hbar^2}{2m}\frac{d^2\psi}{dx^2} + \frac{1}{2}m\omega^2 x^2 \psi = E\psi \qquad (41.22)$$

$$\psi = Be^{-(m\omega/2\hbar)x^2} \qquad (41.24)$$

The **energy levels of the harmonic oscillator** are quantized. *The ground state corresponds to $n = 0$.*

$$E_n = \left(n + \frac{1}{2}\right)\hbar\omega \qquad (41.25)$$

$$(n = 0, 1, 2, \dots)$$

REVIEW CHECKLIST

You should be able to:

▷ Describe the concept of wave function for the representation of matter waves and state in equation form the normalization condition and expectation value of the coordinate. (Section 41.1)

▷ Given the wave function for a particle in one-dimensional motion, calculate the wave length, momentum, energy and probability of finding the particle within a specified distance interval. (Section 41.1)

▷ State the sinusoidal wave function and calculate wavelength, energy levels, and quantum numbers for a particle in a box. (Section 41.2)

▷ Demonstrate that a given wave function is a solution to the Schrödinger equation (Eq. 41.13). Determine the expectation value for the coordinate of a particle in an infinitely deep square well; and calculate the probability of finding the particle within a finite range of coordinate values. Find and sketch the potential energy as a function of position given the wave function for a particle in an infinite square well potential. (Sections 41.3 and 41.4)

▷ State the boundary conditions on the wave functions for a particle in a finite height potential barrier. Sketch the wave function and probability density for the particle. Calculate the transmission coefficient for tunneling or penetration by an electron through a rectangular barrier. (Sections 41.5 and 41.6)

▷ Determine the amplitude constant B in the wave function of a harmonic oscillator by normalization. Find the total energy of an oscillator given the wave function. (Section 41.8)

ANSWERS TO SELECTED QUESTIONS

3. Discuss the relationship between ground-state energy and the uncertainty principle.

Answer   Consider a particle bound to a restricted region of space. If its minimum energy were zero, then the particle could have zero momentum and zero uncertainty in its momentum. At the same time, the uncertainty in its position would not be infinite, but equal to the width of the region. In such a case, the uncertainty product $\Delta x \Delta p_x$ would be zero, violating the uncertainty principle. This contradiction proves that the minimum energy of the particle is not zero.

□   □   □   □

SOLUTIONS TO SELECTED PROBLEMS

---

1. A free electron has a wave function $\psi(x) = Ae^{i(5.00\times10^{10}x)}$ where x is in meters. Find (a) its de Broglie wavelength, (b) its momentum, and (c) its kinetic energy in electron volts.

Solution

(a) The wave function, $\psi(x) = Ae^{i(5\times10^{10}x)} = A\cos(5\times10^{10}x) + iA\sin(5\times10^{10}x)$, will go through one full cycle between $x_1 = 0$ and $(5.00\times10^{10})x_2 = 2\pi$.

The wavelength is then $\qquad \lambda = x_2 - x_1 = \dfrac{2\pi}{5.00\times10^{10}\ \text{m}^{-1}} = 1.26\times10^{-10}\ \text{m}$ ◊

To say the same thing, we can inspect $Ae^{i(5\times10^{10}x)}$ to see that the wave number is $k = 5.00\times10^{10}\ \text{m}^{-1}$.

(b) Since $\lambda = h/p$, the momentum is $p = \dfrac{h}{\lambda} = \dfrac{6.63\times10^{-34}\ \text{J}\cdot\text{s}}{1.26\times10^{-10}\ \text{m}} = 5.28\times10^{-24}\ \text{kg}\cdot\text{m}/\text{s}$ ◊

(c) The electron's kinetic energy is

$$K = \tfrac{1}{2}mv^2 = \dfrac{p^2}{2m}:\qquad\qquad K = \dfrac{(5.28\times10^{-24}\ \text{kg}\cdot\text{m}/\text{s})^2}{2(9.11\times10^{-31}\ \text{kg})}\left(\dfrac{1\ \text{eV}}{1.60\times10^{-19}\ \text{J}}\right) = 95.5\ \text{eV}$$ ◊

Its relativistic total energy is $\qquad 511\ \text{keV} + 95.5\ \text{eV}$

---

5. An electron is contained in a one-dimensional box of length 0.100 nm. (a) Draw an energy-level diagram for the electron for levels up to $n = 4$. (b) Find the wavelengths of all photons that can be emitted by the electron in making downward transitions that could eventually carry it from the $n = 4$ to the $n = 1$ state.

Solution

(a) We can draw a diagram that parallels our treatment of mechanical waves under boundary conditions. In each standing-wave state, we measure the distance d from one node to another (N to N), and base our solution upon that:

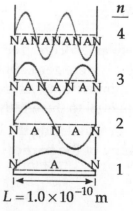

Since $\qquad d_{\text{N to N}} = \dfrac{\lambda}{2}$ and $\lambda = \dfrac{h}{p}$, $\qquad p = \dfrac{h}{\lambda} = \dfrac{h}{2d}$

Next, $\qquad K = \dfrac{p^2}{2m_e} = \dfrac{h^2}{8m_e d^2} = \dfrac{1}{d^2}\dfrac{(6.63\times10^{-34}\ \text{J}\cdot\text{s})^2}{8(9.11\times10^{-31}\ \text{kg})}$

$L = 1.0\times10^{-10}\ \text{m}$

Evaluating, $K = \dfrac{6.03 \times 10^{-38} \text{ J·m}^2}{d^2} = \dfrac{3.77 \times 10^{-19} \text{ eV·m}^2}{d^2}$

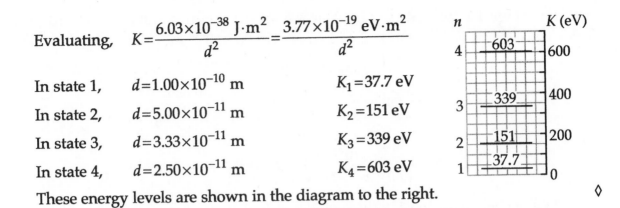

In state 1, $d = 1.00 \times 10^{-10}$ m $\qquad K_1 = 37.7$ eV

In state 2, $d = 5.00 \times 10^{-11}$ m $\qquad K_2 = 151$ eV

In state 3, $d = 3.33 \times 10^{-11}$ m $\qquad K_3 = 339$ eV

In state 4, $d = 2.50 \times 10^{-11}$ m $\qquad K_4 = 603$ eV

These energy levels are shown in the diagram to the right. ◊

(b) When the charged, massive electron inside the box makes a downward transition from one energy level to another, a chargeless, massless photon comes out of the box, carrying the difference in energy, ΔE. Its wavelength is

$$\lambda = \frac{c}{f} = \frac{hc}{\Delta E} = \frac{(6.63 \times 10^{-34} \text{ J·s})(3.00 \times 10^8 \text{ m/s})}{\Delta E (1.602 \times 10^{-19} \text{ J/eV})} = \frac{1.24 \times 10^{-6} \text{ eV·m}}{\Delta E}$$

Transition	$4 \to 3$	$4 \to 2$	$4 \to 1$	$3 \to 2$	$3 \to 1$	$2 \to 1$
ΔE (eV)	264	452	565	188	302	113
Wavelength (nm)	4.71	2.75	2.20	6.60	4.12	11.0

The wavelengths of light released for each transition are given in the table above. ◊

9. The nuclear potential energy that binds protons and neutrons in a nucleus is often approximated by a square well. Imagine a proton confined in an infinitely high square well of length 10.0 fm, a typical nuclear diameter. Calculate the wavelength and energy associated with the photon emitted when the proton moves from the $n = 2$ state to the ground state. In what region of the electromagnetic spectrum does this wavelength belong?

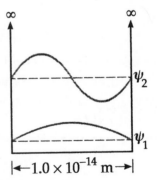

Solution

Conceptualize: Nuclear radiation from nucleon transitions is usually in the form of high energy gamma rays with short wavelengths.

Categorize: The energy of the particle can be obtained from the wavelengths of the standing waves corresponding to each level. The transition between energy levels will result in the emission of a photon with this energy difference.

Analyze: At level 1, the node-to-node distance of the standing wave is 1.00×10^{-14} m, so the wavelength is twice this distance: $h/p = 2.00 \times 10^{-14}$ m. The proton's kinetic energy is

$$K = \frac{1}{2}mv^2 = \frac{p^2}{2m} = \frac{h^2}{2m\lambda^2} = \frac{(6.63 \times 10^{-34} \text{ J·s})^2}{2(1.67 \times 10^{-27} \text{ kg})(2.00 \times 10^{-14} \text{ m})^2}$$

$$K = \frac{3.29 \times 10^{-13} \text{ J}}{1.60 \times 10^{-19} \text{ J/eV}} = 2.06 \times 10^6 \text{ eV} = 2.06 \text{ MeV}$$

In the first excited state, level 2, the node-to-node distance is two times smaller than in state 1. The momentum is two times larger and the energy is four times larger:

$$K = 8.23 \text{ MeV}$$

The proton has mass, has charge, moves slowly compared to light in a standing-wave state, and stays inside the nucleus. When it falls from level 2 to level 1, its energy change is $2.06 \text{ MeV} - 8.23 \text{ MeV} = -6.17 \text{ MeV}$. Therefore, we know that a photon (a traveling wave with no mass and no charge) is emitted at the speed of light, and that it has an energy of $+6.17$ MeV ◊

Its frequency is $\quad f = \dfrac{E}{h} = \dfrac{(6.17 \times 10^6 \text{ eV})(1.60 \times 10^{-19} \text{ J/eV})}{6.63 \times 10^{-34} \text{ J·s}} = 1.49 \times 10^{21} \text{ Hz}$

and its wavelength is $\quad \lambda = \dfrac{c}{f} = \dfrac{3.00 \times 10^8 \text{ m/s}}{1.49 \times 10^{21} \text{ s}^{-1}} = 2.02 \times 10^{-13} \text{ m}$ ◊

This is a gamma ray, according to the electromagnetic spectrum chart shown in Figure 34.12 of the text and in Problem 34.41 in this student manual. ◊

Finalize: The radiated photons are energetic gamma rays as we expected for a nuclear transition. In the above calculations, we assumed that the proton was not relativistic $(v < 0.1c)$, but we should check this assumption for the highest energy state we examined $(n = 2)$:

$$v = \sqrt{\frac{2K}{m}} = \sqrt{\frac{2(8.23 \times 10^6 \text{ eV})(1.60 \times 10^{-19} \text{ J/eV})}{1.67 \times 10^{-27} \text{ kg}}} = 3.97 \times 10^7 \text{ m/s} = 0.132c$$

This appears to be a borderline case where we should probably use relativistic equations, but our classical treatment should give reasonable results, within about $(0.132)^2 = 1\%$ accuracy.

11. Use the particle-in-a-box model to calculate the first three energy levels of a neutron trapped in a nucleus of diameter 20.0 fm. Do the energy-level differences have a realistic order of magnitude?

Solution

$$E_n = \frac{h^2 n^2}{8mL^2}: \quad E_1 = \frac{(6.626 \times 10^{-34} \text{ J·s})^2 (1)^2}{8(1.67 \times 10^{-27} \text{ kg})(2.00 \times 10^{-14} \text{ m})^2} = 8.22 \times 10^{-14} \text{ J} = 0.513 \text{ MeV} \qquad \Diamond$$

$$E_2 = 4E_1 = 2.05 \text{ MeV} \qquad \text{and} \qquad E_3 = 9E_1 = 4.62 \text{ MeV} \qquad \Diamond$$

Yes, the energy differences are of the order of 1 MeV, which is a typical energy for a γ-ray photon, as emitted by a nucleus in an excited state. $\qquad \Diamond$

---

13. Show that the wave function $\psi = Ae^{i(kx-\omega t)}$ is a solution to the Schrödinger equation (Eq. 41.13) where $k = 2\pi/\lambda$ and $U = 0$.

Solution From $\qquad \psi = Ae^{i(kx-\omega t)}$ $\qquad\qquad\qquad$ [1]

we evaluate $\qquad \dfrac{d\psi}{dx} = ikAe^{i(kx-\omega t)}$

and $\qquad \dfrac{d^2\psi}{dx^2} = -k^2 Ae^{i(kx-\omega t)}$ $\qquad\qquad\qquad$ [2]

We substitute Equations [1] and [2] into the Schrödinger equation, so that Eq. 41.13,

$$\frac{d^2\psi}{dx^2} = -\frac{2m}{\hbar^2}(E-U)\psi$$

becomes $\qquad -k^2 Ae^{i(kx-\omega t)} = \left(-\dfrac{2m}{\hbar^2}K\right)Ae^{i(kx-\omega t)}$ $\qquad\qquad$ [3]

where K is the kinetic energy. The wave function $\psi = Ae^{i(kx-\omega t)}$ is a solution to the Schrödinger equation if Eq. [3] is true. Both sides depend on A, x, and t in the same way, so we can divide out several factors and determine that we have a solution if

$$k^2 = \frac{2m}{\hbar^2}K$$

But this is true for a nonrelativistic particle with mass,

since $\qquad \dfrac{2m}{\hbar^2}K = \dfrac{2m}{(h/2\pi)^2}\left(\tfrac{1}{2}mv^2\right) = \dfrac{4\pi^2 m^2 v^2}{h^2} = \left(\dfrac{2\pi p}{h}\right)^2 = \left(\dfrac{2\pi}{\lambda}\right)^2 = k^2$

Therefore, the given wave function does satisfy Equation 41.13. $\qquad \Diamond$

25. Suppose a particle is trapped in its ground state in a box that has infinitely high walls (Fig. 41.4). Now suppose the left-hand wall is suddenly lowered to a finite height and width. (a) Qualitatively sketch the wave function for the particle a short time later. (b) If the box has a length L, what is the wavelength of the wave that penetrates the left-hand wall?

Solution

Before the wall is lowered, the wave function looks like the wave in the first figure to the right.

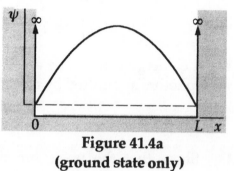

Figure 41.4a
(ground state only)

We assume that the wall is lowered to have a height in energy terms that is greater than the energy of the particle. We assume the potential energy outside the wall is equal to that inside the box. We assume the wall is so thick that the particle has a small tunneling probability.

(a)   The wave function will look much the same in region I. The wave function will show exponential decay in region II. This means that in region I the wave function must change slightly, to be slightly greater than zero at the left wall. A small-amplitude traveling wave in region III represents the tunneled probability leaving the scene. This is illustrated in the figure below.   ◊

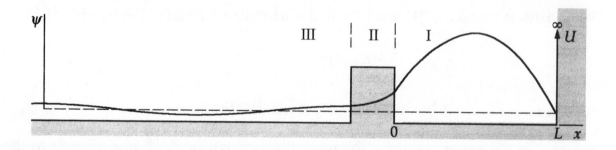

(b)   Since the wave function in region I has not changed much, it still has about the same energy. If the particle tunnels through the wall, its energy remains constant.

With the same potential on both sides of the wall, the particle will have the same kinetic energy after tunneling. Thus the momentum will remain the same and the wavelength in region III will be the same as before, namely

$$2d_{NN} = 2L.$$ ◊

27. An electron with kinetic energy $E = 5.00$ eV is incident on a barrier with thickness $L = 0.200$ nm and height $U = 10.0$ eV (Fig. P41.27). What is the probability that the electron (a) will tunnel through the barrier? (b) will be reflected?

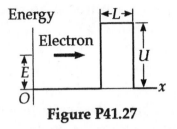

Figure P41.27

Solution

Conceptualize: Since the barrier energy is higher than the kinetic energy of the electron, transmission is not likely, but should be possible since the barrier is not infinitely high or thick.

Categorize: The probability of transmission is found from Equations 41.20 and 41.21.

Analyze: The decay constant for the wave function inside the barrier is:

$$C = \frac{\sqrt{2m(U-E)}}{\hbar} = \frac{\sqrt{2(9.11\times10^{-31}\ \text{kg})(10.0\ \text{eV} - 5.00\ \text{eV})(1.60\times10^{-19}\ \text{J/eV})}}{6.63\times10^{-34}\ \text{J}\cdot\text{s}/2\pi}$$

$$C = 1.14\times10^{10}\ \text{m}^{-1}$$

(a) The probability of transmission is

$$T \approx e^{-2CL} = e^{-2(1.14\times10^{10}\ \text{m}^{-1})(2.00\times10^{-10}\ \text{m})} = e^{-4.58} = 0.010\,3 \qquad \Diamond$$

(b) If the electron does not tunnel, it is reflected, with probability $1 - 0.010\,3 = 0.990$ $\Diamond$

Finalize: Our expectation was correct; there is only a 1% chance that the electron will penetrate the barrier. This tunneling probability would be greater if the barrier were thinner, shorter, or if the kinetic energy of the electron were greater.

Related Comment: A typical scanning-tunneling electron microscope (STM) can be built for less than $5000, and can be tied directly to a home computer.

The STM essentially consists of a needle (1) that is mounted on a few piezo-electric crystals (2). When a voltage is applied across the piezo-electric crystals, the crystals change their shape, and the needle moves.

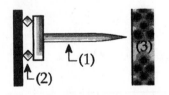

The STM charges the needle, and moves the tip of the needle to within a few tenths of a nanometer of the test sample (3); the remaining gap provides the energy barrier that is required for tunneling. When the electron cloud of one of the needle's atoms is close to the electron cloud of one of the sample's atoms, a relatively large number of electrons tunnels from one to the other, constituting a current. The current is then measured by a computer. As the needle moves across the surface of the sample, an image of the atoms is generated.

41. An electron is represented by the time-independent wave function

$$\psi(x) = \begin{cases} Ae^{-\alpha x} & \text{for } x > 0 \\ Ae^{+\alpha x} & \text{for } x < 0 \end{cases}$$

(a) Sketch the wave function as a function of x. (b) Sketch the probability density representing the likelihood that the electron is found between x and $x + dx$. (c) Argue that this can be a physically reasonable wave function. (d) Normalize the wave function. (e) Determine the probability of finding the electron somewhere in the range

$$x_1 = -1/2\alpha \quad \text{to} \quad x_2 = 1/2\alpha$$

Solution

(a)

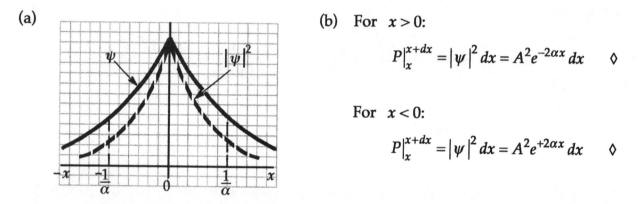

(b) For $x > 0$:

$$P\Big|_x^{x+dx} = |\psi|^2 \, dx = A^2 e^{-2\alpha x} \, dx \quad \lozenge$$

For $x < 0$:

$$P\Big|_x^{x+dx} = |\psi|^2 \, dx = A^2 e^{+2\alpha x} \, dx \quad \lozenge$$

(c) This might be reasonable since (1) ψ is continuous, (2) $\psi \to 0$ as $x \to \pm\infty$, and (3) the wave function can represent an electron bound by an infinitely deep, infinitely narrow potential well at $x = 0$. The wave function's derivative has no single value at the origin, but the wave function need not be differentiable at a point where the potential is infinite. (4) The wave function is integrable and can be normalized, as we show in part (d). $\lozenge$

(d) As ψ is symmetric, $\int_{-\infty}^{\infty} |\psi|^2 \, dx = 2 \int_0^{\infty} |\psi|^2 \, dx = 1$ or $2A^2 \int_0^{\infty} e^{-2\alpha x} \, dx = 1$

Integrating, $\left[\dfrac{2A^2}{-2\alpha}\right]\left[e^{-\infty} - e^0\right] = 1$ gives $A = \sqrt{\alpha}$ $\lozenge$

(e) $P\Big|_{-1/2\alpha}^{1/2\alpha} = 2\int_0^{1/2\alpha} (\sqrt{\alpha})^2 e^{-2\alpha x} \, dx = \left[\dfrac{2\alpha}{-2\alpha}\right]\left[e^{-2\alpha/2\alpha} - 1\right] = \left[1 - e^{-1}\right] = 0.632$ $\lozenge$

43. Particles incident from the left are confronted with a step in potential energy shown in Figure P41.42. The step has a height U, and the particles have energy $E = 2U$. Classically, all the particles would pass into the region of higher potential energy at the right. However, according to quantum mechanics, a fraction of the particles are reflected at the barrier. Use the result of Problem 42 to determine the fraction of the incident particles that are reflected. (This situation is analogous to the partial reflection and transmission of light striking an interface between two different media.)

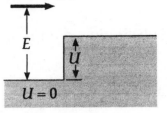

Figure P41.42

Solution   The reflection coefficient R, found from Problem 42, is

$$R = \frac{(k_1 - k_2)^2}{(k_1 + k_2)^2} \quad \text{or} \quad R = \frac{(1 - k_2/k_1)^2}{(1 + k_2/k_1)^2}$$

where $k_1 = 2\pi/\lambda_1$ and $k_2 = 2\pi/\lambda_2$ are the angular wave numbers for the incident and transmitted particles, respectively.   Taking K to be the kinetic energy of the particle,

$$k = \frac{p}{\hbar} = \frac{\sqrt{2mK}}{\hbar}$$

For the incident particles, $K = E = 2U$: $\quad k_1\hbar = \sqrt{2mE} = \sqrt{4mU}$

For the transmitted particles, $K = E - U = U$: $\quad k_2\hbar = \sqrt{2mU}$

Therefore, the ratio of the wave numbers is $\quad \dfrac{k_2}{k_1} = \dfrac{k_2\hbar}{k_1\hbar} = \dfrac{\sqrt{2mU}}{\sqrt{4mU}} = \dfrac{1}{\sqrt{2}}$

Solving for R, $\quad R = \dfrac{(1 - 1/\sqrt{2})^2}{(1 + 1/\sqrt{2})^2} = \left[\dfrac{(\sqrt{2} - 1)}{(\sqrt{2} + 1)}\right]^2 = 0.029\,4$ $\quad\Diamond$

---

47. For a particle described by a wave function $\psi(x)$, the expectation value of a physical quantity $f(x)$ associated with the particle is defined by

$$\langle f(x) \rangle \equiv \int_{-\infty}^{\infty} \psi^* f(x)\psi\, dx$$

For a particle in a one-dimensional box extending from $x = 0$ to $x = L$, show that

$$\langle x^2 \rangle = \frac{L^2}{3} - \frac{L^2}{2n^2\pi^2}$$

Solution We note that $\psi_n(x) = A\sin\left(\dfrac{n\pi x}{L}\right)$ where $A = \sqrt{2/L}$

Substituting x^2 for $f(x)$, $\langle x^2 \rangle = \displaystyle\int_{-\infty}^{\infty} x^2|\psi|^2\,dx$

Integrating, $\langle x^2 \rangle = \left(\dfrac{2}{L}\right)\displaystyle\int_0^L x^2\sin^2\left(\dfrac{n\pi x}{L}\right)dx = \dfrac{L^2}{3} - \dfrac{L^2}{2n^2\pi^2}$ ◇

---

49. A particle has a wave function

$$\psi(x) = \begin{cases} \sqrt{2/a}\,e^{-x/a} & \text{for } x > 0 \\ 0 & \text{for } x < 0 \end{cases}$$

(a) Find and sketch the probability density. (b) Find the probability that the particle will be at any point where $x < 0$. (c) Show that ψ is normalized, and then find the probability that the particle will be found between $x = 0$ and $x = a$.

Solution

(a)

$$|\psi|^2 = \begin{cases} \dfrac{2}{a}e^{-2x/a} & \text{for } x > 0 \\ 0 & \text{for } x < 0 \end{cases}$$

◇

(b) The particle has zero probability of being at any point where $x < 0$. ◇

(c) For normalization, $\displaystyle\int_{\text{all } x} |\psi|^2\,dx = 1$

$$0 + \int_0^{\infty}\frac{2}{a}e^{-2x/a}\,dx = -\int_0^{\infty}e^{-2x/a}(-2dx/a) = -e^{-2x/a}\Big|_0^{\infty} = -[0-1] = 1 \qquad ◇$$

Thus, $\sqrt{2/a}$ was the right normalization coefficient for ψ in the first place. Probability of finding the particle in $0 < x < a$ is

$$\int_0^a \frac{2}{a}e^{-2x/a}\,dx = -e^{-2x/a}\Big|_0^a = -[e^{-2}-1] = 0.865 \qquad ◇$$

Chapter 42
ATOMIC PHYSICS

EQUATIONS AND CONCEPTS

Balmer series wavelengths for the four visible emission lines in the spectrum of hydrogen can be calculated using the empirical Equation 42.1. R_H is called the Rydberg constant. The figure below shows the lines in the visible region of the spectrum and the short wavelength limit of the Balmer series.

$$\frac{1}{\lambda} = R_H\left(\frac{1}{2^2} - \frac{1}{n^2}\right) \tag{42.1}$$
$$(n = 3, 4, 5 \ldots)$$

$$R_H = \frac{k_e e^2}{2a_0 hc} = 1.097\,373\,2 \times 10^7 \text{ m}^{-1}$$

Basic **postulates of Bohr's model** for the hydrogen atom:

- The **electron moves in circular orbits** around the proton.

- The **electron in stationary (stable) orbits** does not emit energy as radiation.

- **A photon is emitted, carrying away energy,** when the atomic electron undergoes a transition from a higher-energy orbit to a lower energy orbit. *The frequency of the emitted photon is proportional to the difference between energies of the initial and final states.*

- The **angular momentum** of the electron about the nucleus is quantized in units of $n\hbar$. *This condition determines the radii of the allowed electron orbits.*

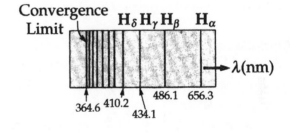

$$E_i - E_f = hf \tag{42.5}$$

$$m_e vr = n\hbar \qquad (n = 1, 2, 3 \ldots) \tag{42.6}$$

The **total energy of the hydrogen atom** (electron-proton bound system) is negative.

$$E = -\frac{k_e e^2}{2r} \tag{42.9}$$

Radii of the allowed orbits have discrete (quantized) values which can be expressed in terms of the Bohr radius, a_0. The **Bohr radius** corresponds to $n = 1$ (see the figure below).

$$r_n = \frac{n^2 \hbar^2}{m_e k_e e^2} \quad (n = 1, 2, 3, \ldots) \quad (42.10)$$

$$a_0 = \frac{\hbar^2}{m_e k_e e^2} = 0.052\,9 \text{ nm} \quad (42.11)$$

$$r_n = n^2 a_0 = n^2 (0.052\,9 \text{ nm}) \quad (42.12)$$

The **first three circular orbits** predicted by the Bohr model of the hydrogen atom are shown in the figure (not to scale).

Quantized energy level values can be expressed in units of electron volts (eV). *The lowest allowed energy state or ground state corresponds to the principal quantum number $n = 1$. The absolute value of the ground state energy is equal to the ionization energy of the atom. The energy level approaches $E = 0$ as r approaches infinity.*

$$E_n = -\frac{13.606}{n^2} \text{ eV} \quad (n = 1, 2, 3 \ldots) \quad (42.14)$$

A **photon of frequency** f (and wavelength λ) is emitted when an electron undergoes a transition from an outer orbit (higher energy level) to an inner orbit (lower energy level).

$$f = \frac{E_i - E_f}{h} \quad (42.15)$$

Photon wavelengths can be calculated using Equation 42.17. *Wavelength, rather than frequency, is the experimentally measured quantity.*

$$\frac{1}{\lambda} = R_H \left(\frac{1}{n_f^2} - \frac{1}{n_i^2} \right) \quad (42.17)$$

The **lines observed in the hydrogen spectrum** can be arranged into several series. Each series of lines corresponds to an assigned value of the quantum number of the final energy level, n_f. Individual lines within a given series result from transitions from successively higher energy levels to the n_f level. Within each series, the difference in energy between adjacent emission lines becomes smaller as n_i increases. Transitions in the Lyman, Balmer, and Paschen series are shown in the figure. The minimum wavelength in each series corresponds to $n_i \rightarrow \infty$. The **correspondence principle** states that for large quantum numbers, n, quantum physics agrees with classical physics.

Substituting into Equation 42.17:

Lyman series: $n_f = 1$; $n_i = 2, 3, 4, \ldots$
Balmer series: $n_f = 2$; $n_i = 3, 4, 5, \ldots$
Paschen series: $n_f = 3$; $n_i = 4, 5, 6, \ldots$
Brackett series: $n_f = 4$; $n_i = 5, 6, 7, \ldots$

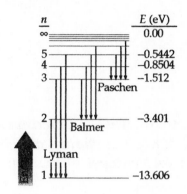

The **potential energy function** for the hydrogen atom depends on the radius of the allowed orbit of the electron. *There is no dependence on angular coordinates.*

$$U(r) = -k_e \frac{e^2}{r} \qquad (42.20)$$

The **energy level values for hydrogen** obtained from the quantum model are the same as the results of the Bohr model (see Eq. 42.14). The parameter a_0 is the Bohr radius.

$$E_n = -\left(\frac{k_e e^2}{2a_0}\right)\frac{1}{n^2} = -\frac{13.606 \text{ eV}}{n^2} \qquad (42.21)$$

$$(n = 1, 2, 3, \ldots)$$

The **1s state wave function for hydrogen** depends only on the radial distance r. *All s states have spherical symmetry.*

$$\psi_{1s}(r) = \frac{1}{\sqrt{\pi a_0{}^3}} e^{-r/a_0} \qquad (42.22)$$

The **radial probability density** for the 1s state of hydrogen is defined as the probability of finding the electron in a spherical shell of radius r and thickness dr. As shown in the figure, the largest value of $P_{1s}(r)$ corresponds to $r = a_0$.

$$P_{1s}(r) = \left(\frac{4r^2}{a_0^3}\right)e^{-2r/a_0} \qquad (42.25)$$

$a_0 = 0.052\ 9$ nm

Quantum numbers determine the allowed wave functions and energy levels for the hydrogen atom.

The **principal quantum number** n is associated with allowed orbits and determines the energy states of the atom (see Equation 42.21). *All states having the same principal quantum number form a shell.*

The **orbital quantum number** ℓ determines discrete values of the magnitude of the angular momentum, L. *There are n different possible values of ℓ. All states having the same values of n and ℓ form a sub-shell.*

$$L = \sqrt{\ell(\ell+1)}\,\hbar \qquad (42.27)$$

$$(\ell = 0, 1, 2, \ldots, n-1)$$

The **orbital magnetic quantum number** m_ℓ specifies the allowed values of the z component of the orbital momentum vector **L**. *When an atom is placed in an external magnetic field, the vector **L** lies on the surface of a cone and precesses about the z axis.*

$$L_z = m_\ell \hbar \qquad (42.28)$$

$$(m_\ell = -\ell, -\ell+1, -\ell+2, \ldots, 0, 1, 2, \ldots \ell)$$

The **Zeeman effect** ("splitting" of spectral lines) is due to the quantization of the angle θ that **L** makes with the direction of an external magnetic field. For example, when $n = 2$ and $\ell = 1$, a single line corresponding to the first excited state in hydrogen is split into three lines with slightly different energy levels in the presence of an external magnetic field (see figure to the right). There are $(2\ell + 1)$ possible values of m_ℓ.

$$\cos\theta = \frac{L_z}{|\mathbf{L}|} = \frac{m_\ell}{\sqrt{\ell(\ell+1)}} \qquad (42.29)$$

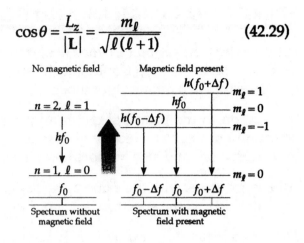

The **spin quantum number** s has only one value. This quantum number is not derived from the Schrödinger equation; it originates in the relativistic properties of the electron and it determines the magnitude of the spin angular momentum of an electron, **S**.

$$S = \sqrt{s(s+1)}\,\hbar = \frac{\sqrt{3}}{2}\hbar \qquad (42.30)$$

$$s = \frac{1}{2}$$

The **spin magnetic quantum number** m_s specifies the space orientation, relative to a z axis, of the spin angular momentum vector.

$$S_z = m_s\hbar = \pm\frac{1}{2}\hbar \qquad (42.31)$$

$$m_s = \pm\frac{1}{2}$$

The **exclusion principle** dictates that no two electrons in the same atom can have the exact same set of quantum numbers.

Selection rules govern the allowed changes in the orbital quantum number and the orbital magnetic quantum number when an electron undergoes a transition between stationary states.

$$\Delta\ell = \pm 1 \quad \text{and} \quad \Delta m_\ell = 0, \pm 1 \qquad (42.34)$$

Energy levels of multielectron atoms are calculated by taking into account the shielding effect of the nuclear charge by inner-core electrons. *The atomic number is replaced by an effective atomic number Z_{eff}, which depends on the values of n and ℓ.*

$$E_n = -\frac{13.6Z_{eff}^2}{n^2}\,\text{eV} \qquad (42.36)$$

SUGGESTIONS, SKILLS, AND STRATEGIES

After reading the chapter in your text, review the significance of each quantum number used to describe the electronic state of an electron in an atom and recall the allowed set of values for each quantum number.

In addition to the principal quantum number n (which can range from 1 to ∞), other quantum numbers are necessary to specify completely the possible energy levels in the hydrogen atom and also in more complex atoms.

All energy states with the same principal quantum number, n, form a shell. These shells are identified by the spectroscopic notation K, L, M corresponding to $n = 1, 2, 3, \ldots$

The orbital quantum number ℓ [which can range from 0 to $(n-1)$], determines the allowed value of orbital angular momentum. All energy states having the same values of n and ℓ form a subshell. The letter designations $s, p, d, f \ldots$ correspond to values of $\ell = 0, 1, 2, \ldots$

The orbital magnetic quantum number m_ℓ (which can range from $-\ell$ to ℓ) determines the possible orientations of the electron's orbital angular momentum vector in the presence of an external magnetic field.

The spin magnetic quantum number m_s, can have only two values, $m_s = -\frac{1}{2}$ and $m_s = \frac{1}{2}$, which in turn correspond to the two possible directions of the electron's intrinsic spin.

REVIEW CHECKLIST

You should be able to:

▷   Calculate the wavelength and principal quantum number for lines in the hydrogen spectrum. (Section 42.1)

▷   Calculate, based on the Bohr model, the orbital radius, linear momentum, angular momentum, and kinetic energy of an electron for a given value of n. Calculate the potential energy of a hydrogen atom for a given value of n. Calculate the ionization energy and construct an energy level diagram for hydrogen. (Section 42.3)

▷   Plot the wave function and probability density function for the $1s$ state of hydrogen. Find the expectation value for the radius of an electron in hydrogen given the wave function for a specific state. Show that a given wave function is a solution to the Schrödinger equation. (Section 42.5)

▷   Describe qualitatively what each of the quantum numbers n, ℓ, m_ℓ, s and m_s implies concerning atomic structure. State the allowed values which may be assigned to each. Associate the customary shell and subshell notations with allowed combinations of quantum numbers n and ℓ. (Section 42.6)

▷ Find possible sets of quantum numbers for given values of n. Calculate the possible values of the orbital angular momentum, L, corresponding to a given value of the principal quantum number. Find the allowed values for L_z (the component of the angular momentum vector along the direction of an external magnetic field) for a given value of L. (Section 42.6)

▷ State the Pauli exclusion principal and describe its relevance to the periodic table of the elements. Show how the exclusion principal leads to the known electronic ground state configuration of the light elements. (Section 42.7)

▷ Determine the accelerating potential difference for electron bombardment required to produce x-rays of a given wavelength. (Section 42.8)

ANSWERS TO SELECTED QUESTIONS

12. Could the Stern-Gerlach experiment be performed with ions rather than neutral atoms? Explain.

Answer Practically speaking, the answer would be no. Since ions have a net charge, the magnetic force $q\mathbf{v} \times \mathbf{B}$ would deflect the beam, making it very difficult to separate ions with different magnetic-moment orientations.

□   □   □   □

14. Discuss some of the consequences of the exclusion principle.

Answer If the Pauli exclusion principle were not valid, the elements and their chemical behavior would be grossly different because every electron would end up in the lowest energy level of the atom. All matter would therefore be nearly alike in its chemistry and composition, since the shell structures of each element would be identical. Most materials would have a much higher density, and the spectra of atoms and molecules would be very simple, resulting in the existence of less color in the world.

□   □   □   □

16. Why do lithium, potassium, and sodium exhibit similar chemical properties?

Answer The three elements have similar electronic configurations, with filled inner shells plus a single electron in an s orbital. Since atoms typically interact through their unfilled outer shells, and the outer shell of each of these atoms is similar, the chemical interactions of the three atoms is also similar.

□   □   □   □

SOLUTIONS TO SELECTED PROBLEMS

1. (a) What value of n_i is associated with the 94.96-nm spectral line in the Lyman series of hydrogen? (b) **What If?** Could this wavelength be associated with the Paschen or Balmer series?

Solution Our equation is
$$\frac{1}{\lambda} = R_H \left(\frac{1}{n_f^2} - \frac{1}{n_i^2} \right) \quad \text{where} \quad R_H = 1.097 \times 10^7 \text{ m}^{-1}$$

and for the Lyman series, $n_f = 1$, and $n_i = 2, 3, 4 \ldots$

Substituting the given values,
$$\frac{1}{94.96 \times 10^{-9} \text{ m}} = (1.097 \times 10^7 \text{ m}^{-1}) \left(1 - \frac{1}{n_i^2} \right)$$

(a) Solving for n_i, $n_i = 5$ ◊

(b) By Figure 42.9, spectral lines in the Balmer and Paschen series all have much longer wavelengths, since much smaller energy losses put the atom into energy levels 2 or 3.   ◊

3. According to classical physics, a charge e moving with an acceleration a radiates at a rate

$$\frac{dE}{dt} = -\frac{1}{6\pi\epsilon_0} \frac{e^2 a^2}{c^3}$$

(a) Show that an electron in a classical hydrogen atom (see Figure 42.6) spirals into the nucleus at a rate $dr/dt = -e^4/12\pi^2\epsilon_0^2 r^2 m_e^2 c^3$. (b) Find the time interval over which the electron will reach $r = 0$, starting from $r_0 = 2.00 \times 10^{-10}$ m.

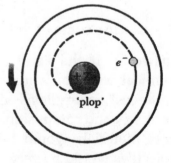

Figure 42.6

Solution

According to a classical model, the electron moving as a particle in uniform circular motion about the proton in the hydrogen atom experiences a force $k_e e^2/r^2$; and from Newton's second law, $F = ma$, its acceleration is $k_e e^2/m_e r^2$.

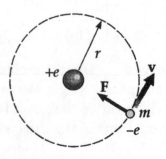

(a) Using the fact that the Coulomb constant

$$k_e = \frac{1}{4\pi\epsilon_0}, \qquad\qquad a = \frac{v^2}{r} = \frac{k_e e^2}{m_e r^2} = \frac{e^2}{4\pi\epsilon_0 m_e r^2} \qquad \textbf{[1]}$$

From the Bohr structural model of the atom, we can write the total energy of the atom as

$$E = -\frac{k_e e^2}{2r} = -\frac{e^2}{8\pi\epsilon_0 r} \qquad \text{so} \qquad \frac{dE}{dt} = \frac{e^2}{8\pi\epsilon_0 r^2}\frac{dr}{dt} = -\frac{1}{6\pi\epsilon_0}\frac{e^2 a^2}{c^3} \qquad \textbf{[2]}$$

Substituting [1] into [2] for a, solving for dr/dt, and simplifying gives

$$\frac{dr}{dt} = -\frac{4r^2}{3c^3}\left(\frac{e^2}{4\pi\epsilon_0 m_e r^2}\right)^2 = -\frac{e^4}{12\pi^2\epsilon_0^2 r^2 m_e^2 c^3} \qquad\qquad \lozenge$$

(b) We can express dr/dt in the simpler form:

$$\frac{dr}{dt} = -\frac{A}{r^2} = -\frac{3.15\times10^{-21}}{r^2} \qquad \text{thus} \qquad -\int_{2.00\times10^{-10}\text{ m}}^{0} r^2\,dr = 3.15\times10^{-21}\int_0^T dt$$

and $\quad T = \left(3.17\times10^{20}\right)\dfrac{r^3}{3}\Big|_0^{2.00\times10^{-10}\text{ m}} = 8.46\times10^{-10}\text{ s} = 0.846\text{ ns} \qquad\qquad \lozenge$

We know that atoms last much longer than 0.8 ns; thus, classical physics does not hold (fortunately) for atomic systems.

————————

7.   A hydrogen atom is in its first excited state ($n = 2$). Using the Bohr theory of the atom, calculate (a) the radius of the orbit, (b) the linear momentum of the electron, (c) the angular momentum of the electron, (d) the kinetic energy of the electron, (e) the potential energy of the system, and (f) the total energy of the system.

Solution   We note, during our calculations, that the nominal velocity of the electron is less than 1% of the speed of light; therefore, we do not need to use relativistic equations.

(a) By Bohr's model, $\qquad\qquad r_n = n^2(a_0) = 2^2(0.052\,9\text{ nm}) = 2.12\times10^{-10}\text{ m} \qquad\qquad \lozenge$

(b) Since $m_e vr = n\hbar$, $\qquad\qquad p = m_e v = \dfrac{n\hbar}{r} = \dfrac{2(1.0546\times10^{-34}\text{ J}\cdot\text{s})}{2.12\times10^{-10}\text{ m}}$

$$p = 9.97\times10^{-25}\text{ kg}\cdot\text{m/s} \qquad\qquad \lozenge$$

(c) $\mathbf{L} = \mathbf{r} \times \mathbf{p}$ becomes $L = rp = n\hbar = 2.11 \times 10^{-34}$ J·s $\diamond$

(d) Next, $v = p/m_e$: $\quad v = \dfrac{p}{m_e} = \dfrac{9.97 \times 10^{-25} \text{ kg·m/s}}{9.11 \times 10^{-31} \text{ kg}} = 1.09 \times 10^{6}$ m/s

So $\quad K = \frac{1}{2} m_e v^2$: $\quad K = \dfrac{(9.11 \times 10^{-31} \text{ kg})(1.09 \times 10^6 \text{ m/s})^2}{2} = \dfrac{5.45 \times 10^{-19} \text{ J}}{1.602 \times 10^{-19} \text{ J/eV}} = 3.40 \text{ eV}$ $\diamond$

(e) From Chapter 25, the electric potential energy is $U = qV$:

$$U = -\frac{k_e e^2}{r} = -\frac{(8.99 \times 10^9 \text{ N·m}^2/\text{C}^2)(1.602 \times 10^{-19} \text{ C})^2}{2.12 \times 10^{-10} \text{ m}}$$

and $\qquad U = -1.09 \times 10^{-18}$ J $= -6.80$ eV $\diamond$

(f) Thus, $\qquad E = K + U = -5.45 \times 10^{-19}$ J $= -3.40$ eV $\diamond$

The total energy ought to be equal to minus the kinetic energy: our answer confirms this, but we needed to use more than 3 significant figures for some constants. For an accurate answer from data with fewer significant figures, we could calculate the total energy first:

$$E = \frac{-k_e e^2}{2r} = \frac{-(8.99 \times 10^9 \text{ N·m}^2/\text{C}^2)(1.602 \times 10^{-19} \text{ C})^2}{2(2.12 \times 10^{-10} \text{ m})} = -5.45 \times 10^{-19} \text{ J}$$ $\diamond$

$K = E - U = 5.45 \times 10^{-19}$ J $= 3.40$ eV $\diamond$

---

9. A photon is emitted as a hydrogen atom undergoes a transition from the $n = 6$ state to the $n = 2$ state. Calculate (a) the energy, (b) the wavelength, and (c) the frequency of the emitted photon.

Solution By conservation of energy, the energy of the photon is equal to the energy lost in transition. From Bohr's theory,

(a) $E_6 - E_2 = (-13.6 \text{ eV})\left(\dfrac{1}{6^2} - \dfrac{1}{2^2}\right) = 3.02$ eV $\diamond$

(b) $\dfrac{1}{\lambda} = R_H\left(\dfrac{1}{n_f^2} - \dfrac{1}{n_i^2}\right) = (1.097 \, 4 \times 10^7 \text{ m}^{-1})\left(\dfrac{1}{4} - \dfrac{1}{36}\right); \qquad \lambda = 410$ nm $\diamond$

(This is the deep violet spectral line farthest to the left in the hydrogen spectrum of the text's Figure 42.1.)

(c) $f = \dfrac{c}{\lambda} = \dfrac{3.00 \times 10^8 \text{ m/s}}{410 \times 10^{-9} \text{ m}} = 7.32 \times 10^{14}$ Hz $\diamond$

---

13. (a) Construct an energy-level diagram for the He$^+$ ion, for which $Z = 2$. (b) What is the ionization energy for He$^+$?

Solution

From Equation 42.19,

$$E_n = -\frac{k_e e^2 Z^2}{2a_0 n^2} = \frac{(-13.6 \text{ eV})(2)^2}{n^2}$$

For the lowest energy level, $n = 1$ and $E_1 = -54.4$ eV

For the first excited state, $n = 2$ and $E_2 = -13.6$ eV

(a) The diagram is shown to the right. ◊

(b) The ionization energy is the energy required to lift the electron from the ground state to the point of breaking free from the nucleus, or from $n = 1$ to $n = \infty$, from energy -54.4 eV to 0 eV. The answer then, is 54.4 eV. ◊

16. A general expression for the energy levels of one-electron atoms and ions is

$$E_n = -\frac{\mu k_e^2 q_1^2 q_2^2}{2\hbar^2 n^2}$$

where k_e is the Coulomb constant, q_1 and q_2 are the charges of the electron and the nucleus, and μ is the reduced mass, given by $\mu = m_1 m_2 / (m_1 + m_2)$. The wavelength for the $n = 3$ to $n = 2$ transition of the hydrogen atom is 656.3 nm (visible red light). **What If?** What are the wavelengths for this same transition in (a) positronium, which consists of an electron and a positron, and (b) singly ionized helium? (*Note*: A positron is a positively charged electron.)

Solution

Conceptualize: The reduced mass of positronium is **less** than hydrogen, so the photon energy will be **less** for positronium than for hydrogen. This means that the wavelength of the emitted photon will be **longer** than 656.3 nm. On the other hand, helium has about the same reduced mass but more charge than hydrogen, so its transition energy will be **larger**, corresponding to a wavelength **shorter** than 656.3 nm.

Categorize: All the factors in the above equation are constant for this problem except for the reduced mass and the nuclear charge. Therefore, the wavelength corresponding to the energy difference for the transition can be found simply from the ratio of mass and charge variables.

Analyze: For hydrogen,

$$\mu = \frac{m_p m_e}{m_p + m_e} \approx m_e$$

The photon energy is

$$\Delta E = E_3 - E_2$$

Its wavelength is $\lambda = 656.3$ nm, where

$$\lambda = \frac{c}{f} = \frac{hc}{\Delta E}$$

(a) For positronium,

$$\mu = \frac{m_e m_e}{m_e + m_e} = \frac{m_e}{2}$$

so the energy of each level is one half as large as in hydrogen, which we could call "protonium." The photon energy is inversely proportional to its wavelength, so for positronium,

$$\lambda_{32} = 2(656.3 \text{ nm}) = 1.31 \ \mu\text{m} \quad \text{(in the infrared region)} \qquad \lozenge$$

(b) For He$^+$, $\mu \approx m_e$, $q_1 = e$, and $q_2 = 2e$, so the transition energy is $2^2 = 4$ times larger than hydrogen.

Then,

$$\lambda_{32} = \left(\frac{656}{4}\right) \text{nm} = 164 \text{ nm} \qquad \text{(in the ultraviolet region)} \qquad \lozenge$$

Finalize: As expected, the wavelengths for positronium and helium are respectively larger and smaller than for hydrogen. Other energy transitions should have wavelength shifts consistent with this pattern. It is important to remember that the reduced mass is not the total mass, but is generally close in magnitude to the smaller mass of the system (hence the name **reduced** mass).

─────────────────────────────────

17. An electron of momentum p is at a distance r from a stationary proton. The electron has kinetic energy $K = p^2/2m_e$. The atom has potential energy $U = -k_e e^2/r$, and total energy $E = K + U$. If the electron is bound to the proton to form a hydrogen atom, its average position is at the proton, but the uncertainty in its position is approximately equal to the radius r of its orbit. The electron's average vector momentum is zero, but its average squared momentum is approximately equal to the squared uncertainty in its momentum, as given by the uncertainty principle. Treating the atom as a one-dimensional system, (a) estimate the uncertainty in the electron's momentum in terms of r. (b) Estimate the electron's kinetic, potential, and total energies in terms of r. (c) The actual value of r is the one that *minimizes the total energy*, resulting in a stable atom. Find that value of r and the resulting total energy. Compare your answer with the predictions of the Bohr theory.

Solution

(a) $\Delta x \Delta p \geq \dfrac{\hbar}{2}$; so if $\Delta x = r$, $\Delta p \geq \dfrac{\hbar}{2r}$ ◊

(b)   Suppose that $\Delta p = \dfrac{\hbar}{r}$

Then, $K = \dfrac{p^2}{2m_e} = \dfrac{\hbar^2}{2m_e r^2}$; $U = -\dfrac{k_e e^2}{r}$; $E = \dfrac{\hbar^2}{2m_e r^2} - \dfrac{k_e e^2}{r}$ ◊

(c)   To minimize E, $\dfrac{dE}{dr} = -\dfrac{\hbar^2}{m_e r^3} + \dfrac{k_e e^2}{r^2} = 0$ or $r = \dfrac{\hbar^2}{m_e k_e e^2}$ = Bohr radius

Then, $E = \dfrac{\hbar^2}{2m_e}\left(\dfrac{m_e k_e e^2}{\hbar^2}\right)^2 - k_e e^2\left(\dfrac{m_e k_e e^2}{\hbar^2}\right) = -\dfrac{m_e k_e^2 e^4}{2\hbar^2} = -13.6$ eV ◊

This is the same value as that predicted by the Bohr theory for the ground state of hydrogen.   ◊

---

21.   For a spherically symmetric state of a hydrogen atom, the Schrödinger equation in spherical coordinates is

$$-\frac{\hbar^2}{2m}\left(\frac{d^2\psi}{dr^2} + \frac{2}{r}\frac{d\psi}{dr}\right) - \frac{k_e e^2}{r}\psi = E\psi$$

Show that the 1s wave function for an electron in hydrogen,

$$\psi(r) = \frac{1}{\sqrt{\pi a_0^3}}e^{-r/a_0}$$

satisfies the Schrödinger equation.

Solution   We use the quantum particle under boundary conditions model. We substitute the wave function and its derivatives into the Schrödinger equation, and then start simplifying the resulting equation. If we find that the resulting equation is true, then we know that the Schrödinger equation is satisfied.

$$\frac{d\psi}{dr} = \frac{1}{\sqrt{\pi a_0^3}}\frac{d}{dr}\left(e^{-r/a_0}\right) = -\frac{1}{\sqrt{\pi a_0^3}}\left(\frac{1}{a_0}\right)e^{-r/a_0} = -\frac{\psi}{a_0} \qquad [1]$$

Likewise, $$\frac{d^2\psi}{dr^2} = -\frac{1}{\sqrt{\pi a_0^5}}\frac{d}{dr}e^{-r/a_0} = \frac{1}{\sqrt{\pi a_0^7}}e^{-r/a_0} = \frac{1}{a_0^2}\psi \qquad [2]$$

Substituting [1] and [2] into the Schrödinger equation, and noting that m is the electron's mass m_e,

$$-\frac{\hbar^2}{2m_e}\left(\frac{1}{a_0^2}-\frac{2}{a_0 r}\right)\psi - \frac{k_e e^2}{r}\psi = E\psi \qquad [3]$$

Substituting

$$\hbar^2 = m_e k_e e^2 a_0$$

and canceling m_e and ψ,

$$-\frac{k_e e^2 a_0}{2}\left(\frac{1}{a_0^2}-\frac{2}{a_0 r}\right)-\frac{k_e e^2}{r}= E$$

We find that the second and third terms add to zero, leaving the equation that was given for the ground state energy of hydrogen:

$$E = -\frac{k_e e^2}{2a_0}$$

This is true, so the Schrödinger equation is satisfied.   ◊

---

27. How many sets of quantum numbers are possible for an electron for which (a) $n=1$, (b) $n=2$, (c) $n=3$, (d) $n=4$, and (e) $n=5$? Check your results to show that they agree with the general rule that the number of sets of quantum numbers for a shell is equal to $2n^2$.

Solution

(a)   For $n=1$,
we have $\ell=0$, $m_\ell=0$, $m_s=\pm\frac{1}{2}$.

n	ℓ	m_ℓ	m_s
1	0	0	$-1/2$
1	0	0	$+1/2$

This yields $2n^2 = 2(1)^2 = 2$ sets   ◊

(b)   For $n=2$, we have

n	ℓ	m_ℓ	m_s
2	0	0	$\pm1/2$
2	1	-1	$\pm1/2$
2	1	0	$\pm1/2$
2	1	$+1$	$\pm1/2$

This yields $2n^2 = 2(2)^2 = 8$ sets   ◊

Note that the number is twice the number of m_ℓ values. Also, for each ℓ there are $(2\ell + 1)$ different m_ℓ values. Finally, ℓ can take on values ranging from 0 to $n - 1$.

So the general expression is $\qquad\qquad\qquad$ number $= \sum_0^{n-1} 2(2\ell + 1)$

The series is the sum of an arithmetic progression like $\qquad 2 + 6 + 10 + 14$.

The sum is $\qquad \sum_0^{n-1} 4\ell + \sum_0^{n-1} 2 = 4\left[\dfrac{n^2 - n}{2}\right] + 2n = 2n^2$

(c) $n = 3$, $\qquad 2(1) + 2(3) + 2(5) = 2 + 6 + 10 = 18$: $\qquad 2n^2 = 2(3)^2 = 18 \qquad \lozenge$

(d) $n = 4$, $\qquad 2(1) + 2(3) + 2(5) + 2(7) = 32$: $\qquad 2n^2 = 2(4)^2 = 32 \qquad \lozenge$

(e) $n = 5$, $\qquad 32 + 2(9) = 32 + 18 = 50$: $\qquad 2n^2 = 2(5)^2 = 50 \qquad \lozenge$

31. The ρ-meson has a charge of $-e$, a spin quantum number of 1, and a mass $1\,507$ times that of the electron. **What If?** Imagine that the electrons in atoms were replaced by ρ-mesons. List the possible sets of quantum numbers for ρ-mesons in the $3d$ subshell.

Solution The $3d$ subshell has $\ell = 2$, and $n = 3$. Also, we have $s = 1$. Therefore, we can have $n = 3$, $\ell = 2$, $m_\ell = -2, -1, 0, 1, 2$, $s = 1$, and $m_s = -1, 0, 1$, leading to the following table:

n	3	3	3	3	3	3	3	3	3	3	3	3	3	3	3	
ℓ	2	2	2	2	2	2	2	2	2	2	2	2	2	2	2	
m_ℓ	-2	-2	-2	-1	-1	-1	0	0	0	1	1	1	2	2	2	
s	1	1	1	1	1	1	1	1	1	1	1	1	1	1	1	
m_s	-1	0	1	-1	0	1	-1	0	1	-1	0	1	-1	0	1	$\lozenge$

37. (a) Scanning through Table 42.4 in order of increasing atomic number, note that the electrons fill the subshells in such a way that those subshells with the lowest values of $n + \ell$ are filled first. If two subshells have the same value of $n + \ell$, the one with the lower value of n is filled first. Using these two rules, write the order in which the subshells are filled through $n + \ell = 7$. (b) Predict the chemical valence for the elements that have atomic numbers 15, 47, and 86 and compare your predictions with the actual valences (which may be found in a chemistry text).

Solution

(a)

$n + \ell$	1	2	3	4	5	6	7	
subshell	$1s$	$2s$	$2p, 3s$	$3p, 4s$	$3d, 4p, 5s$	$4d, 5p, 6s$	$4f, 5d, 6p, 7s$	◊

(b) The valence of an atom, also called its oxidation number, is the number of electrons it "loses" when the atom is in a molecule. A negative number indicates electron "gain." In stable molecules, atoms tend to be in states with full subshells. Full shells are even more preferred, but this is balanced by a preference for the net charge to be not too great.

$Z = 15$: Inner subshells: $1s, 2s, 2p$ (10 electrons)
Outer subshells: 2 electrons in the $3s$ subshell, and
3 electrons in the $3p$ subshell
Prediction: Valence +5 or –3
Element is phosphorus: Valence –3, +3, or +5 (Prediction works) ◊

$Z = 47$: Filled subshells: $1s, 2s, 2p, 3s, 3p, 4s, 3d, 4p, 5s$ (38 electrons)
Outer subshell: 9 electrons in the $4d$ subshell
Prediction: Valence –1
Element is silver: Valence +1 (Prediction fails) ◊

Note that the ground-state electron configuration of silver disobeys the second rule stated in the problem, by having 10 electrons in the $4d$ subshell and only 1 in the $5s$ subshell.

$Z = 86$: Filled subshells: $1s, 2s, 2p, 3s, 3p, 4s, 3d,$
$4p, 5s, 4d, 5p, 6s, 4f, 5d, 6p$

Outer subshell: Full
Prediction: Inert gas with valence 0
Element is radon: Inert gas (prediction works) ◊

43. Use the method illustrated in Example 42.9 to calculate the wavelength of the x-ray emitted from a molybdenum target ($Z = 42$) when an electron moves from the L shell ($n = 2$) to the K shell ($n = 1$).

Solution

Following Example 42.9, suppose the electron is originally in the L shell with just one other electron in the K shell between it and the nucleus, so it moves in a field of effective charge $(42 - 1)e$.

Its energy is then $E_L = -(42 - 1)^2 (13.6 \text{ eV} / 4)$

In its final state we estimate the screened charge holding it in orbit as again $(42-1)e$, so its energy is

$$E_K = -(42-1)^2(13.6 \text{ eV})$$

The photon energy emitted is the difference.

$$E_\gamma = \frac{3}{4}(42-1)^2(13.6 \text{ eV}) = 1.71 \times 10^4 \text{ eV} = 2.74 \times 10^{-15} \text{ J}$$

Then
$$f = \frac{E}{h} = 4.14 \times 10^{18} \text{ Hz} \quad \text{and} \quad \lambda = \frac{c}{f} = 72.5 \text{ pm} \qquad \diamond$$

49. A ruby laser delivers a 10.0-ns pulse of 1.00 MW average power. If the photons have a wavelength of 694.3 nm, how many are contained in the pulse?

Solution

Conceptualize: Lasers generally produce concentrated beams that are bright (except for IR or UV lasers that produce invisible beams). Since our eyes can detect light levels as low as a few photons, there are probably at least 1 000 photons in each pulse.

Categorize: From the pulse width and average power, we can find the energy delivered by each pulse. The number of photons can then be found by dividing the pulse energy by the energy of each photon, which is determined from the photon wavelength.

Analyze: The energy in each pulse is

$$E = \mathcal{P}\Delta t = (1.00 \times 10^6 \text{ W})(1.00 \times 10^{-8} \text{ s}) = 1.00 \times 10^{-2} \text{ J}$$

The energy of each photon is

$$E_\gamma = hf = \frac{hc}{\lambda} = \frac{(6.626 \times 10^{-34} \text{ J·s})(3.00 \times 10^8 \text{ m / s})}{694.3 \times 10^{-9} \text{ m}} = 2.86 \times 10^{-19} \text{ J}$$

So
$$N = \frac{E}{E_\gamma} = \frac{1.00 \times 10^{-2} \text{ J}}{2.86 \times 10^{-19} \text{ J / photon}} = 3.49 \times 10^{16} \text{ photons} \qquad \diamond$$

Finalize: With 10^{16} photons/pulse, this laser beam should produce a bright red spot when the light reflects from a surface, even though the time between pulses is generally much longer than the width of each pulse. For comparison, this laser produces more photons in a single ten-nanosecond pulse than a typical 5 mW helium-neon laser produces over a full second (about 1.6×10^{16} photons/second). Human eyes require many more than 1 000 photons for something to appear bright.

55. The positron is the antiparticle to the electron. It has the same mass and a positive electric charge of the same magnitude as that of the electron. Positronium is a hydrogen-like atom consisting of a positron and an electron revolving around each other. Using the Bohr model, find the allowed distances between the two particles and the allowed energies of the system.

Solution

Conceptualize: Since we are told that positronium is like hydrogen, we might expect the allowed radii and energy levels to be about the same as for hydrogen: $r = a_0 n^2 = (5.29 \times 10^{-11} \text{ m})n^2$ and $E_n = (-13.6 \text{ eV})/n^2$.

Categorize: Similar to the textbook calculations for hydrogen, we can use the quantization of angular momentum of positronium to find the allowed radii and energy levels.

Analyze: Let r represent the distance between the electron and the positron. The two move in a circle of radius $r/2$ around their center of mass with opposite velocities. The total angular momentum is quantized according to

$$L_n = \frac{mvr}{2} + \frac{mvr}{2} = n\hbar \quad \text{where} \quad n = 1, 2, 3 \ldots$$

For each particle, $\Sigma F = ma$ expands to $\quad \dfrac{k_e e^2}{r^2} = \dfrac{mv^2}{r/2}$

We can eliminate $v = \dfrac{n\hbar}{mr}$ to find $\quad \dfrac{k_e e^2}{r} = \dfrac{2mn^2\hbar^2}{m^2 r^2}$

So the separation distances are $\quad r = \dfrac{2n^2\hbar^2}{mk_e e^2} = 2a_0 n^2 = (1.06 \times 10^{-10})n^2 \qquad \Diamond$

The orbital radii are $r/2 = a_0 n^2$, the same as for the electron in hydrogen.

The energy can be calculated from $\quad E = K + U = \frac{1}{2}mv^2 + \frac{1}{2}mv^2 - \dfrac{k_e e^2}{r}$

Since $mv^2 = \dfrac{k_e e^2}{2r}$, $\qquad E = \dfrac{k_e e^2}{2r} - \dfrac{k_e e^2}{r} = -\dfrac{k_e e^2}{2r} = \dfrac{-k_e e^2}{4a_0 n^2} = -\dfrac{6.80 \text{ eV}}{n^2} \qquad \Diamond$

Finalize: It appears that the allowed separations for positronium are twice as large as for hydrogen, while the energy levels are half as big. One way to explain this is that in a hydrogen atom, the proton is much more massive than the electron, so the proton remains nearly stationary with essentially no kinetic energy. However, in positronium, the positron and electron have the same mass and therefore both have kinetic energy that separates them from each other and reduces their total energy compared with hydrogen.

57. *An example of the correspondence principle.* Use Bohr's model of the hydrogen atom to show that when the electron moves from the state n to the state $n - 1$, the frequency of the emitted light is

$$f = \left(\frac{2\pi^2 m_e k_e^2 e^4}{h^3 n^2}\right)\frac{2n-1}{(n-1)^2}$$

Show that as $n \to \infty$, this expression varies as $1/n^3$ and reduces to the classical frequency one expects the atom to emit. (*Suggestion:* To calculate the classical frequency, note that the frequency of revolution is $v/2\pi r$, where r is given by Eq. 42.10.)

Solution

Combining Eq. 42.11 and 42.13,
$$E_n = -\frac{m_e k_e^2 e^4}{2n^2 \hbar^2}$$

Because $\hbar = h/2\pi$,
$$hf = \Delta E = \frac{4\pi^2 m_e k_e^2 e^4}{2h^2}\left(\frac{1}{(n-1)^2} - \frac{1}{n^2}\right)$$

which reduces to
$$f = \frac{2\pi^2 m_e k_e^2 e^4}{h^3}\left(\frac{2n-1}{(n-1)^2 n^2}\right) \qquad \diamond$$

As $n \to \infty$, the '-1' terms lose importance and drop out, leaving the right-hand factor equal to

$$\lim_{n\to\infty}\left(\frac{2n-1}{(n-1)^2 n^2}\right) = \frac{2n}{n^4} = \frac{2}{n^3} \qquad \diamond$$

Thus the emission frequency becomes
$$f = \frac{4\pi^2 m_e k_e^2 e^4}{h^3}\left(\frac{1}{n^3}\right)$$

But in the Bohr model,
$$v^2 = \frac{k_e e^2}{m_e r} \quad \text{and} \quad r = \frac{n^2 h^2}{4\pi^2 m_e k_e e^2}$$

and the classical orbital frequency is
$$f = \frac{v}{2\pi r} = \frac{1}{2\pi r^{3/2}}\sqrt{\frac{k_e e^2}{m_e}}$$

or
$$f = \frac{1}{2\pi}\left(\sqrt{\frac{k_e e^2}{m_e}}\right)\left(\frac{4\pi^2 m_e k_e e^2}{n^2 h^2}\right)^{3/2}$$

But this is identical to
$$f = \frac{4\pi^2 m_e k_e^2 e^4}{h^3}\left(\frac{1}{n^3}\right) \qquad \diamond$$

59. Suppose a hydrogen atom is in the 2s state, with its wave function given by Equation 42.26. Taking $r = a_0$, calculate values for (a) $\psi_{2s}(a_0)$, (b) $|\psi_{2s}(a_0)|^2$, and (c) $P_{2s}(a_0)$.

Solution The wave function for the 2s state is given by

$$\psi_{2s}(r) = \frac{1}{4\sqrt{2\pi}} \left(\frac{1}{a_0}\right)^{3/2} \left(2 - \frac{r}{a_0}\right) e^{-r/2a_0}$$

(a) Taking $r = a_0 = 0.529 \times 10^{-10}$ m,

we find $\psi_{2s}(a_0) = \frac{1}{4\sqrt{2\pi}} \left(\frac{1}{0.529 \times 10^{-10} \text{ m}}\right)^{3/2} (2-1)e^{-1/2} = 1.57 \times 10^{14} \text{ m}^{-3/2}$ ◊

(b) $\quad |\psi_{2s}(a_0)| = \left(1.57 \times 10^{14} \text{ m}^{-3/2}\right)^2 = 2.47 \times 10^{28} \text{ m}^{-3}$ ◊

(c) Using Equation 42.24 and the results to (b) gives

$$P_{2s}(a_0) = 4\pi a_0^2 |\psi_{2s}(a_0)|^2 = 8.69 \times 10^8 \text{ m}^{-1}$$ ◊

---

65. (a) Show that the most probable radial position for an electron in the 2s state of hydrogen is $r = 5.236a_0$. (b) Show that the wave function given by Equation 42.26 is normalized.

Solution We are using the quantum particle model. We use Equation 42.26:

$$\psi_{2s}(r) = \frac{1}{4\sqrt{2\pi}} \left(\frac{1}{a_0}\right)^{3/2} \left[2 - \frac{r}{a_0}\right] e^{-r/2a_0}$$

By Equation 42.24, the radial probability distribution function is

$$P(r) = 4\pi r^2 \psi^2 = \frac{1}{8}\left(\frac{r^2}{a_0^3}\right)\left(2 - \frac{r}{a_0}\right)^2 e^{-r/a_0}$$

(a) Its extremes are given by

$$\frac{dP(r)}{dr} = \frac{1}{8}\left[\frac{2r}{a_0^3}\left(2 - \frac{r}{a_0}\right)^2 - \frac{2r^2}{a_0^3}\left(\frac{1}{a_0}\right)\left(2 - \frac{r}{a_0}\right) - \frac{r^2}{a_0^3}\left(2 - \frac{r}{a_0}\right)^2\left(\frac{1}{a_0}\right)\right]e^{-r/a_0} = 0$$

We factor in the following manner:

$$\frac{1}{8}\left(\frac{r}{a_0^3}\right)\left(2 - \frac{r}{a_0}\right)\left[2\left(2 - \frac{r}{a_0}\right) - \frac{2r}{a_0} - \frac{r}{a_0}\left(2 - \frac{r}{a_0}\right)\right]e^{-r/a_0} = 0$$

The roots of $dP/dr = 0$ at $r = 0$, $r = 2a_0$, and $r = \infty$ are minima, because $P(r) = 0$. Therefore, we focus on the roots given by

$$2\left(2 - \frac{r}{a_0}\right) - 2\frac{r}{a_0} - \left(\frac{r}{a_0}\right)\left(2 - \frac{r}{a_0}\right) = 4 - \frac{6r}{a_0} + \left(\frac{r}{a_0}\right)^2 = 0$$

which has solutions $r = (3 \pm \sqrt{5})a_0$. We substitute two roots into $P(r)$:

When $r = (3 - \sqrt{5})a_0 = 0.764\,a_0$ then $P(r) = \dfrac{0.0519}{a_0}$

When $r = (3 + \sqrt{5})a_0 = 5.24\,a_0$ then $P(r) = \dfrac{0.191}{a_0}$

Therefore, the most probable value of r is $(3 + \sqrt{5})a_0 = 5.24\,a_0$ ◊

(b) The total probability of finding the electron at any distance from the nucleus is

$$\int_0^\infty P(r)\,dr = \int_0^\infty \frac{1}{8}\left(\frac{r^2}{a_0^3}\right)\left(2 - \frac{r}{a_0}\right)^2 e^{-r/a_0}\,dr$$

Let $u = r/a_0$ and $dr = a_0\,du$

so that $\displaystyle\int_0^\infty P(r)\,dr = \int_0^\infty \frac{1}{8}u^2(4 - 4u + u^2)e^{-u}\,du = \int_0^\infty \frac{1}{8}\left(u^4 - 4u^3 + 4u^2\right)e^{-u}\,du$

Using a table of integrals, or integrating by parts repeatedly,

$$\int_0^\infty P(r)\,dr = -\frac{1}{8}\left(u^4 + 4u^2 + 8u + 8\right)e^{-u}\Big|_0^\infty = 1 \quad \text{as desired} \qquad ◊$$

---

67. An electron in chromium moves from the $n = 2$ state to the $n = 1$ state without emitting a photon. Instead, the excess energy is transferred to an outer electron (one in the $n = 4$ state), which is then ejected by the atom. (This is called an Auger [pronounced 'ohjay'] process, and the ejected electron is referred to as an Auger electron.) Use the Bohr theory to find the kinetic energy of the Auger electron.

Solution The chromium atom with nuclear charge $Z = 24$ starts with one vacancy in the $n = 1$ shell, perhaps produced by the absorption of an x-ray, which ionized the atom. An electron from the $n = 2$ shell tumbles down to fill the vacancy. We suppose that the electron is shielded from the electric field of the full nuclear charge by the one K-shell electron originally below it.

Its change in energy is $\qquad \Delta E = -(Z - 1)^2(13.6 \text{ eV})\left(\dfrac{1}{1^2} - \dfrac{1}{2^2}\right) = -5.40 \text{ keV}$

Then +5.40 keV can be transferred to the single 4s electron. Suppose that it is shielded by the 22 electrons in the K, L, and M shells. To break the outermost electron out of the atom, producing a Cr^{2+} ion, requires an energy investment of

$$E_{ionize} = \frac{(Z-22)^2(13.6 \text{ eV})}{4^2} = \frac{2^2(13.6 \text{ eV})}{16} = 3.40 \text{ eV}$$

As evidence that this relatively tiny amount of energy can still be the right order of magnitude, note that the (first) ionization energy for neutral chromium is tabulated as 6.76 eV. Then the remaining energy that can appear as kinetic energy is

$$K = \Delta E - E_{ionize} = 5\,395.8 \text{ eV} - 3.4 \text{ eV} = 5.39 \text{ keV} \qquad \Diamond$$

Because of conservation of momentum for the ion-electron system and the tiny mass of the electron compared to that of the Cr^{2+} ion, almost all of this kinetic energy will belong to the electron.

---

69. For hydrogen in the 1s state, what is the probability of finding the electron farther than $2.50a_0$ from the nucleus?

Solution

Conceptualize: From the graph shown in Figure 42.13, it appears that the probability of finding the electron beyond $2.5a_0$ is about 20%.

Categorize: The precise probability can be found by integrating the 1s radial probability distribution function from $r = 2.50a_0$ to ∞.

Analyze: The general radial probability distribution function is $P(r) = 4\pi r^2 |\psi|^2$

with $\psi_{1s} = (\pi a_0^3)^{-1/2} e^{-r/a_0}$, it is $\qquad P(r) = 4r^2 a_0^{-3} e^{-2r/a_0}$

The required probability is then $\qquad P = \int_{2.50a_0}^{\infty} P(r)dr = \int_{2.50a_0}^{\infty} \frac{4r^2}{a_0^3} e^{-2r/a_0} dr$

Let $z = 2r/a_0$ and $dz = 2dr/a_0$: $\qquad P = \frac{1}{2} \int_{5.00}^{\infty} z^2 e^{-z} dz$

Performing this integration by parts, $\qquad P = -\frac{1}{2}(z^2 + 2z + 2)e^{-z} \Big]_{5.00}^{\infty}$

$$P = -\frac{1}{2}(0) + \frac{1}{2}(25.0 + 10.0 + 2.00)e^{-5.00} = \left(\frac{37}{2}\right)(0.00674) = 0.125 \qquad \Diamond$$

Finalize: The probability of 12.5% is less than the 20% we estimated, but close enough to be a reasonable result. In comparing the 1s probability density function with others as in Figure 42.13, it appears that the ground state is the most narrow, indicating that a 1s electron will probably be found in the narrow range of 0 to 4 Bohr radii, and most likely at $r = a_0$.

71. In the technique known as electron spin resonance (ESR), a sample containing unpaired electrons is placed in a magnetic field. Consider the simplest situation, in which only one electron is present and therefore only two energy states are possible, corresponding to $m_s = \pm\frac{1}{2}$. In ESR, the absorption of a photon causes the electron's spin magnetic moment to flip from the lower energy state to the higher energy state. According to Section 29.3, the change in energy is $2\mu_B B$. (The lower energy state corresponds to the case where the z component of the magnetic moment μ_{spin} is aligned with the magnetic field, and the higher energy state is the case where the z component of μ_{spin} is aligned opposite to the field.) What is the photon frequency required to excite an ESR transition in a 0.350-T magnetic field?

Solution

As in Section 29.3, the magnetic moment feels torque $\boldsymbol{\tau} = \boldsymbol{\mu} \times \mathbf{B}$ in an external field. The electron spin vector cannot be geometrically parallel or antiparallel to the field; its z component must be either parallel or antiparallel to any applied field. To evaluate the energy difference, we imagine the z component m of the electron's magnetic moment as continuously variable. In turning it from alignment with the field to the opposite direction, the field does work according to Equation 10.22,

$$W = \int dW = \int_0^{180°} \tau \, d\theta = \int_0^\pi \mu B \sin\theta \, d\theta = -\mu B \cos\theta \Big|_0^\pi = 2\mu B$$

To make the electron flip, the photon must carry energy $\Delta E = 2\mu_B B = hf$

Therefore, $f = \dfrac{2\mu_B B}{h} = \dfrac{2(9.27 \times 10^{-24} \text{ J/T})(0.350 \text{ T})}{6.63 \times 10^{-34} \text{ J} \cdot \text{s}} = 9.79 \times 10^9 \text{ Hz}$ ◊

Chapter 43
MOLECULES AND SOLIDS

EQUATIONS AND CONCEPTS

A **potential-energy function to model a molecule** must account for a force of repulsion between the atoms of the molecule at small distances of separation and attractive forces at larger separations. In Equation 43.1, the parameters A and B are associated with attractive and repulsive forces respectively.

$$U(r) = -\frac{A}{r^n} + \frac{B}{r^m} \qquad (43.1)$$

The **total energy of a diatomic molecule** is due to the combined effects of electronic, translational, rotational and vibrational energy. This chapter of the textbook considers energy levels related to rotational and vibrational motion; these are important in the interpretation of molecular spectra.

$$E = E_{el} + E_{trans} + E_{rot} + E_{vib}$$

A **diatomic molecule orientated along the x axis** as shown in the figure has two degrees of rotational freedom corresponding to rotation about the y and z axes.

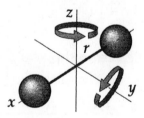

The **moment of inertia I for a diatomic molecule** (about an axis perpendicular to the axis of the molecule), can be written in terms of the reduced mass, μ and the atomic separation distance r.

$$I = \mu r^2 \qquad (43.3)$$

$$\mu = \frac{m_1 m_2}{m_1 + m_2} \qquad (43.4)$$

The **rotational quantum number** J determines the quantized rotational frequencies and angular momentum values of a molecule.

$$L = \sqrt{J(J+1)}\,\hbar \qquad (J=0,1,2\ldots) \qquad (43.5)$$

The **rotational energy** of a molecule is quantized and depends on the value of the moment of inertia I.

$$E_{\text{rot}} = \frac{\hbar^2}{2I}J(J+1) \quad (J=0,1,2\ldots) \qquad (43.6)$$

The **emission and absorption of photons due to the energy separation between rotational levels** is governed by the selection rule for the rotational quantum number. *J in this formula refers to the larger of the initial and the final rotational quantum number.*

$$\Delta E = \frac{\hbar^2}{I}J = \frac{h^2}{4\pi^2 I}J \qquad (43.7)$$

$$(J=1,2,3,\ldots, \text{ and } \Delta J = \pm 1)$$

A **diatomic molecule modeled as a simple harmonic oscillator**, as shown in the figure, has one vibrational degree of freedom. The atoms have mass values of m_1 and m_2; k is the effective spring constant.

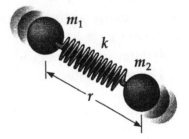

The **vibrational energy for a diatomic molecule** is quantized and is characterized by the vibrational quantum number, v. The selection rule for allowed vibrational transitions is given by $\Delta v = \pm 1$.

$$E_{\text{vib}} = \left(v+\tfrac{1}{2}\right)\frac{h}{2\pi}\sqrt{\frac{k}{\mu}} \qquad (43.10)$$

$$(v=0,1,2,\ldots, \text{ and } \Delta v = \pm 1)$$

The **energy difference between successive vibrational levels** is hf, where f is the frequency of vibration. *The energy separation between successive vibrational levels is much larger than the energy difference between successive rotational levels.*

$$\Delta E_{\text{vib}} = \frac{h}{2\pi}\sqrt{\frac{k}{\mu}} = hf \qquad (43.11)$$

A **molecular absorption spectrum** consists of two sets of lines. When photon absorption by a molecule increases the vibrational quantum number v by one unit, the rotational quantum number J *can either increase or decrease by one unit. See Figure 43.8 in your textbook. J in these formulas refers to the initial rotational quantum number.*

$$\Delta E = hf + \frac{\hbar^2}{I}(J+1) \qquad (43.13)$$

$$(J = 0, 1, 2\ldots, \text{ and } \Delta J = +1)$$

$$\Delta E = hf - \frac{\hbar^2}{I}J \qquad (43.14)$$

$$(J = 1, 2, 3, \ldots, \text{ and } \Delta J = -1)$$

The **ionic cohesive energy** U_0 of a solid is the minimum potential energy per ion pair of the system and occurs when the separation distance between ions (r) equals the equilibrium separation (r_0). U_0 is the energy per ion pair necessary to separate the solid into a collection of positive and negative ions. The Madelung constant, α, has a value which is characteristic of a specific crystal structure. The parameter m is a small integer.

$$U_0 = -\alpha k_e \frac{e^2}{r_0}\left(1 - \frac{1}{m}\right) \qquad (43.18)$$

Ionic crystals:
- form stable, hard crystals
- are poor electrical conductors
- have high melting points
- are transparent to visible radiation
- absorb strongly in the infrared
- are soluble in polar liquids

The **Fermi-Dirac distribution function** $f(E)$ gives the probability of finding an electron in a particular energy state, E. The quantity E_F is called the Fermi energy.

$$f(E) = \frac{1}{e^{(E-E_F)/k_B T} + 1} \qquad (43.19)$$

Electron energy levels in metals are characterized by three quantum numbers. In the ground state, $n_x = n_y = n_z = 1$.

$$E = \frac{\hbar^2 \pi^2}{2m_e L^2}(n_x{}^2 + n_y{}^2 + n_z{}^2) \qquad (43.20)$$

The **density-of-states function**, $g(E)$, multiplied by the energy range dE, gives the number of allowed states per unit volume that have energies between E and $E + dE$.

$$g(E)\,dE = \frac{8\sqrt{2}\pi m_e^{3/2}}{h^3} E^{1/2}\,dE \qquad (43.21)$$

The **number of electrons per unit volume that have energy between E and $E + dE$, $N(E)dE$**, is the product of the number of allowed states per unit volume, $g(E)dE$, and the probability that a state is occupied, $f(E)$.

$$N(E)\,dE = f(E)g(E)\,dE$$

The **total number of electrons per unit volume n_e**, is the integral over all possible states.

$$n_e = \int_0^\infty N(E)\,dE = C\int_0^\infty \frac{E^{1/2}\,dE}{e^{(E-E_F)/k_BT}+1} \qquad (43.23)$$

$$\text{where } C = \frac{8\sqrt{2}\pi m_e^{3/2}}{h^3}$$

The **Fermi energy** (E_F in the Fermi-Dirac distribution function, Equation 43.19) at 0 K is a function of the total number of electrons per unit volume, n_e.

$$E_F(0) = \frac{h^2}{2m_e}\left(\frac{3n_e}{8\pi}\right)^{2/3} \qquad (43.25)$$

The **current-voltage relationship for an ideal diode** depends on the direction of the applied voltage. For forward bias (when the p side of the junction is positive) the current increases exponentially with voltage; a reverse bias (the n side of the junction is positive) results in a small reverse current that reaches saturation value I_0. In the exponent of Equation 43.27, e is the electron charge and T is the absolute temperature. An example I vs ΔV curve for a p-n junction is shown in the figure.

$$I = I_0(e^{e\Delta V/k_BT} - 1) \qquad (43.27)$$

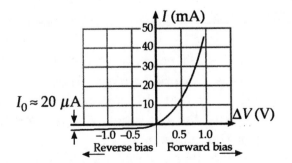

REVIEW CHECKLIST

You should be able to:

▷ Describe the essential bonding mechanisms involved in ionic, covalent, hydrogen, and van der Waals bonding. Determine, given the potential energy function for a molecule, the value of r for which the energy is minimum and the value of E_0. (Section 43.1)

▷ Calculate the reduced mass and moment of inertia of a diatomic molecule about an axis through the center of mass. (Section 43.2)

▷ Calculate allowed values of rotational and vibrational energy levels in a diatomic molecule and determine energy differences between ground and excited states. (Section 43.2)

▷ Use the selection rules for rotational quantum number J and vibrational quantum number v to identify the possible absorptive transitions between $v = 0$ and $v = 1$. (Section 43.2)

▷ Describe the general properties of ionic crystals. Calculate the ionic cohesive energy of a diatomic molecule. (Section 43.3)

▷ Calculate the density of charge carriers, Fermi energy, and energy of conduction electrons in a metal. (Section 43.4)

▷ Use the band theory of solids as a basis for a qualitative discussion of the mechanisms for conduction in metals, insulators, and semiconductors. (Sections 43.5, 43.6 and 43.7)

▷ Describe a p-n junction and the diffusion of electrons and holes through the junction. Discuss the fabrication and function of a junction diode and junction transistor. (Sections 43.5, 43.6 and 43.7)

ANSWERS TO SELECTED QUESTIONS

5. The resistivity of metals increases with increasing temperature, whereas the resistivity of an intrinsic semiconductor decreases with increasing temperature. Explain.

Answer First consider electric conduction in a metal. The number of conduction electrons is essentially fixed. They conduct electricity by having drift motion in an applied electric field superposed on their random thermal motion. At higher temperature, the ion cores vibrate more and scatter more efficiently the conduction electrons flying among them. The mean time between collisions is reduced. The electrons have time to develop only a lower drift speed. The electric current is reduced, so we see the resistivity increasing with temperature.

Now consider an intrinsic semiconductor. At absolute zero its valence band is full and its conduction band is empty. It is an insulator, with very high resistivity. As the temperature increases, more electrons are promoted to the conduction band, leaving holes in the valence band. Then both electrons and holes move in response to an applied electric field. Thus we see the resistivity decreasing as temperature goes up.

□ □ □ □

15. Which is easier to excite in a diatomic molecule, rotational or vibrational motion?

Answer Rotation of a diatomic molecule involves less energy than vibration. Absorption of microwave photons, of frequency $\sim 10^{11}$ Hz, excites rotational motion, while absorption of infrared photons, of frequency $\sim 10^{13}$ Hz, excites vibration in typical simple molecules.

□ □ □ □

SOLUTIONS TO SELECTED PROBLEMS

1. **Review problem.** A K^+ ion and a Cl^- ion are separated by a distance of 5.00×10^{-10} m. Assuming the two ions act like point charges, determine (a) the force each ion exerts on the other and (b) the potential energy of the two-ion system in electron volts.

Solution

The force of each ion is directed towards the other:

(a) $\quad F = \dfrac{k_e |q_1||q_2|}{r^2} = \dfrac{(8.99 \times 10^9 \ \text{N·m}^2/\text{C}^2)(1.60 \times 10^{-19} \ \text{C})^2}{(5.00 \times 10^{-10} \ \text{m})^2} = 9.21 \times 10^{-10} \ \text{N}$ ◊

(b) $\quad U = \dfrac{k_e q_1 q_2}{r} = \dfrac{(8.99 \times 10^9 \ \text{N·m}^2/\text{C}^2)(1.60 \times 10^{-19} \ \text{C})(-1.60 \times 10^{-19} \ \text{C})}{5.00 \times 10^{-10} \ \text{m}}$

$\quad U = -4.60 \times 10^{-19} \ \text{J} = -2.88 \ \text{eV}$ ◊

7. An HCl molecule is excited to its first rotational energy level, corresponding to $J=1$. If the distance between its nuclei is 0.127 5 nm, what is the angular speed of the molecule about its center of mass?

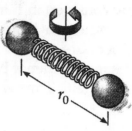

Solution

Conceptualize: For a system as small as a molecule, we can expect the angular speed to be much faster than the few rad/s typical of everyday objects we encounter.

Categorize: The rotational energy is given by the angular momentum quantum number, J. The angular speed can be calculated from this kinetic rotational energy and the moment of inertia of this one-dimensional molecule.

Analyze: For the HCl molecule in the $J=1$ rotational energy level, we are given $r_0=0.127\,5$ nm.

$$E_{rot} = \frac{\hbar^2}{2I} J(J+1)$$

With $J=1$, $\quad E_{rot} = \frac{\hbar^2}{I} = \frac{1}{2}I\omega^2 \quad$ and $\quad \omega = \sqrt{\frac{2\hbar^2}{I^2}} = \sqrt{2}\,\frac{\hbar}{I}$

Common isotopes of chlorine have mass numbers of 35 and 37. Here we will use the average weighted according to abundance. The moment of inertia of the molecule is given by:

$$I = \mu r_0^2 = \left(\frac{m_1 m_2}{m_1 + m_2}\right)r_0^2 = \left[\frac{(1.01\ \text{u})(35.5\ \text{u})}{1.01\ \text{u} + 35.5\ \text{u}}\right]r_0^2 = (0.982\ \text{u})r_0^2$$

$$I = (0.982\ \text{u})(1.66\times10^{-27}\ \text{kg/u})(1.275\times10^{-10}\ \text{m})^2 = 2.65\times10^{-47}\ \text{kg}\cdot\text{m}^2$$

Therefore, $\quad \omega = \sqrt{2}\,\frac{\hbar}{I} = \sqrt{2}\left(\frac{1.055\times10^{-34}\ \text{J}\cdot\text{s}}{2.65\times10^{-47}\ \text{kg}\cdot\text{m}^2}\right) = 5.63\times10^{12}\ \text{rad/s}$ ◊

Finalize: This angular speed is more than a billion times faster than the spin rate of a music CD (200 to 500 revolutions per minute, or $\omega=20$ rad/s to 50 rad/s).

13. Taking the effective force constant of a vibrating HCl molecule as $k = 480 \, \text{N/m}$, find the energy difference between the ground state and the first excited vibrational level.

Solution The reduced mass of the pair of atoms is

$$\mu = \frac{m_1 m_2}{m_1 + m_2} = \frac{(1.01 \, \text{u})(35.5 \, \text{u})}{1.01 \, \text{u} + 35.5 \, \text{u}} (1.66 \times 10^{-27} \, \text{kg/u}) = 1.63 \times 10^{-27} \, \text{kg}$$

The energy difference between adjacent vibration states is

$$\Delta E_{\text{vib}} = \frac{h}{2\pi}\sqrt{\frac{k}{\mu}} = \frac{6.63 \times 10^{-34} \, \text{J·s}}{2\pi}\sqrt{\frac{480 \, \text{N/m}}{1.63 \times 10^{-27} \, \text{kg}}} = 5.73 \times 10^{-20} \, \text{J} = 0.358 \, \text{eV} \qquad \Diamond$$

25. Consider a one-dimensional chain of alternating positive and negative ions. Show that the potential energy associated with one of the ions and its interactions with the rest of this hypothetical crystal is

$$U(r) = -k_e \alpha \frac{e^2}{r}$$

where the Madelung constant is $\alpha = 2\ln 2$ and r is the interionic spacing. [*Suggestion:* Use the series expansion for $\ln(1+x)$.]

Solution

The total potential energy is obtained by summing over all pairs of interactions:

$e \quad -e \quad e \quad -e \quad e \quad -e$

$$U = \sum_{i \ne j} k_e \frac{q_i q_j}{r_{ij}} = -k_e\left[\frac{e^2}{r} + \frac{e^2}{r} - \frac{e^2}{2r} - \frac{e^2}{2r} + \frac{e^2}{3r} + \frac{e^2}{3r} - \frac{e^2}{4r} - \frac{e^2}{4r} + \cdots\right]$$

$$U = -2k_e \frac{e^2}{r}\left[1 - \frac{1}{2} + \frac{1}{3} - \frac{1}{4} + \cdots\right]$$

But

$$\ln(1+x) = x - \frac{x^2}{2} + \frac{x^3}{3} - \frac{x^4}{4} + \cdots$$

Therefore, $x = 1$ for our series, and

$$U = (-2\ln 2)k_e \frac{e^2}{r} = -\alpha k_e \frac{e^2}{r} \qquad \Diamond$$

31. Calculate the energy of a conduction electron in silver at 800 K, assuming the probability of finding an electron in that state is 0.950. The Fermi energy is 5.48 eV at this temperature.

Solution

Conceptualize: Since there is a 95% probability of finding the electron in this state, its energy should be slightly less than the Fermi energy, as indicated by the graph in Figure 43.15.

Categorize: The electron energy can be found from the Fermi-Dirac distribution function.

Analyze: Taking $E_F = 5.48$ eV for silver at 800 K, and given $f(E) = 0.950$, we find

$$f(E) = \frac{1}{e^{(E-E_F)/k_BT} + 1} = 0.950: \quad e^{(E-E_F)/k_BT} = \frac{1}{0.950} - 1 = 0.052\ 63$$

$$\frac{E - E_F}{k_BT} = \ln(0.052\ 63) = -2.944$$

$$E - E_F = -2.944\,k_BT = -2.944(1.38 \times 10^{-23} \text{ J/K})(800 \text{ K})$$

$$E = E_F - 3.25 \times 10^{-20} \text{ J} = 5.48 \text{ eV} - 0.203 \text{ eV} = 5.28 \text{ eV} \quad \lozenge$$

Finalize: As expected, the energy of the electron is slightly less than the Fermi energy, which is about 5 eV for most metals. There is very little probability of finding an electron significantly above the Fermi energy in a metal.

═══════════════════════════════

33. Show that the average kinetic energy of a conduction electron in a metal at 0 K is $E_{av} = \frac{3}{5}E_F$. (*Suggestion:* In general, the average kinetic energy is

$$E_{av} = \frac{1}{n_e}\int EN(E)\,dE$$

where n_e is the density of particles, $N(E)\,dE$ is given by Equation 43.22, and the integral is over all possible values of the energy.)

Solution
$$E_{av} = \frac{1}{n_e} \int_0^\infty E N(E) \, dE \quad \text{where} \quad N(E) = \frac{CE^{1/2}}{e^{(E-E_F)/k_BT}+1} = Cf(E)E^{1/2}$$

But at $T=0$,
$$f(E)=0 \text{ for } E>E_F$$

Also,
$$f(E)=1 \text{ for } E<E_F$$

So we can take $N(E) = CE^{1/2}$:
$$E_{av} = \frac{1}{n_e} \int_0^{E_F} CE^{3/2} \, dE = \frac{2C}{5n_e} E_F^{5/2}$$

But from Eq. 43.24, $\dfrac{C}{n_e} = \dfrac{3}{2} E_F^{-3/2}$:
$$E_{av} = \left(\frac{2}{5}\right)\left(\frac{3}{2}\right)(E_F^{-3/2})E_F^{5/2} = \frac{3}{5}E_F \quad \diamond$$

---

35. (a) Consider a system of electrons confined to a three-dimensional box. Calculate the ratio of the number of allowed energy levels at 8.50 eV to the number at 7.00 eV. (b) **What If?** Copper has a Fermi energy of 7.0 eV at 300 K. Calculate the ratio of the number of occupied levels at an energy of 8.50 eV to the number at the Fermi energy. Compare your answer with that obtained in part (a).

Solution

The density of states at the energy E is $\quad g(E) = CE^{1/2}$

(a) Hence, the required ratio is
$$R = \frac{g(8.50 \text{ eV})}{g(7.00 \text{ eV})} = \frac{C(8.50)^{1/2}}{C(7.00)^{1/2}} = 1.10 \quad \diamond$$

(b) From Equation 43.22 we see that the number of occupied states having energy E is
$$N(E) = \frac{CE^{1/2}}{e^{(E-E_F)/k_BT}+1}$$

Hence, the required ratio is
$$R = \frac{N(8.50 \text{ eV})}{N(7.00 \text{ eV})} = \sqrt{\frac{8.50}{7.00}} \left[\frac{e^{(7.00-7.00)/k_BT}+1}{e^{(8.50-7.00)/k_BT}+1} \right]$$

At $T=300$ K, $k_BT=0.025\,9$ eV:
$$R = 1.10 \left(\frac{2}{e^{1.50/0.0259}+1} \right) = 1.47 \times 10^{-25} \quad \diamond$$

Comparing this result with (a), we conclude that very few states with $E>E_F$ are occupied.

---

39. Most solar radiation has a wavelength of 1 μm or less. What energy gap should the material in a solar cell have in order to absorb this radiation? Is silicon appropriate (see Table 43.5)?

Solution

Conceptualize: Since most photovoltaic solar cells are made of silicon, this semiconductor seems to be an appropriate material for these devices.

Categorize: To absorb the longest-wavelength photons, the energy gap should be no larger than the photon energy.

Analyze: The minimum photon energy is

$$hf = \frac{hc}{\lambda} = \frac{(6.63\times10^{-34}\text{ J}\cdot\text{s})(3.00\times10^8\text{ m}/\text{s})}{10^{-6}\text{ m}}\left(\frac{1\text{ eV}}{1.60\times10^{-19}\text{ J}}\right) = 1.24\text{ eV}$$

Therefore, the energy gap in the absorbing material should be smaller than 1.24 eV. ◊

Finalize: So silicon, with gap of 1.14 eV < 1.24 eV, is an appropriate material for absorbing solar radiation. ◊

47. Determine the current generated in a superconducting ring of niobium metal 2.00 cm in diameter when a 0.020 0-T magnetic field directed perpendicular to the ring is suddenly decreased to zero. The inductance of the ring is 3.10×10^{-8} H.

Solution

Conceptualize: The resistance of a superconductor is zero, so the current is limited only by the change in magnetic flux and self-inductance. Therefore, unusually large currents (greater than 100 A) are possible.

Categorize: The change in magnetic field through the ring will induce an emf according to Faraday's law of induction. Since we do not know how fast the magnetic field is changing, we must use the ring's inductance and the geometry of the ring to calculate the magnetic flux, which can then be used to find the current.

Analyze: From Faraday's law (Eq. 31.1), we have

$$|\varepsilon| = \frac{\Delta\Phi_B}{\Delta t} = A\frac{\Delta B}{\Delta t} = L\frac{\Delta I}{\Delta t} \qquad \text{or} \qquad \Delta I = \frac{A\Delta B}{L} = \frac{\pi(0.010\,0\text{ m})^2(0.020\,0\text{ T})}{3.10\times10^{-8}\text{ H}} = 203\text{ A} \quad \lozenge$$

The current is directed so as to produce its own magnetic field in the direction of the original field.

Finalize: This induced current should remain constant as long as the ring is superconducting. If the ring failed to be a superconductor (e.g. if it warmed above the critical temperature), the metal would have a non-zero resistance, and the current would quickly drop to zero. It is interesting to note that we were able to calculate the current in the ring without knowing the emf. In order to calculate the emf, we would need to know how quickly the magnetic field goes to zero.

---

53. Show that the ionic cohesive energy of an ionically bonded solid is given by Equation 43.18. (*Suggestion:* Start with Equation 43.17, and note that $dU/dr=0$ at $r=r_0$.)

Solution The total potential energy is given by Equation 43.17:

$$U_{\text{total}} = -\alpha k_e \frac{e^2}{r} + \frac{B}{r^m}$$

The potential energy has its minimum value U_0 when $r=r_0$, where r_0 is the equilibrium spacing. At this point, the slope of the curve U versus r is **zero**.

That is, $\qquad \dfrac{dU}{dr}\bigg|_{r=r_0} = 0 \qquad$ with $\qquad \dfrac{dU}{dr} = \dfrac{d}{dr}\left(-\alpha k_e \dfrac{e^2}{r} + \dfrac{B}{r^m}\right) = \alpha k_e \dfrac{e^2}{r^2} - \dfrac{mB}{r^{m+1}}$

Taking $r=r_0$ and setting this equal to zero, $\quad \alpha k_e \dfrac{e^2}{r_0^2} - \dfrac{mB}{r_0^{m+1}} = 0$

Therefore, $\qquad\qquad\qquad\qquad\qquad B = \alpha \dfrac{k_e e^2}{m} r_0^{m-1}$

Substituting this value of B into U_{total} gives

$$U_0 = -\alpha k_e \frac{e^2}{r_0} + \alpha \frac{k_e e^2}{m} r_0^{m-1}\left(\frac{1}{r_0^m}\right) = -\alpha \frac{k_e e^2}{r_0}\left(1 - \frac{1}{m}\right) \qquad \lozenge$$

Chapter 44
NUCLEAR STRUCTURE

EQUATIONS AND CONCEPTS

The **nuclear radius** is proportional to the cube root of the mass number (the total number of nucleons). *Most nuclei are nearly spherical in shape and all nuclei have nearly the same density.* The fermi (fm) is a convenient unit in which to express nuclear dimensions.

$$r = r_0 A^{1/3} \tag{44.1}$$

$$r_0 = 1.2 \text{ fm}$$

$$1 \text{ fm} = 10^{-15} \text{ m}$$

The **binding energy** of a nucleus equals the combined energy of the separated nucleons (protons and neutrons) minus the total energy of the bound system. *The total mass of a composite nucleus is less than the sum of the masses of the individual nucleons.*

$$E_b = (Z m_p + N m_n - M_A)(931.494 \text{ MeV/u})$$

$m_p =$ proton mass

$m_n =$ neutron mass

$M_A =$ nuclear mass

(44.2)

The **semiempirical binding-energy formula** is based on the liquid drop model of the nucleus. C_1, C_2, C_3, and C_4 are constants related to volume, surface, Coulomb repulsion, and symmetry effects; they are adjusted to fit experimental data.

$$E_b = C_1 A - C_2 A^{2/3} - C_3 \frac{Z(Z-1)}{A^{1/3}} - C_4 \frac{(N-Z)^2}{A}$$

(44.3)

The **shell model** of the nucleus accounts for: (i) stability of nuclei with even values of A, and (ii) peaks in the binding energy curve when Z or N equal one of the magic numbers.

Magic numbers of the shell model:

Z or $N = 2, 8, 20, 28, 50, 82$ (44.4)

The **exponential decay law** is illustrated in the graph (N vs t). The decay constant λ is characteristic of a specific isotope and is the probability of decay per nucleus per second. N_0 is the number of radioactive nuclei present at time $t = 0$.

$$N = N_0 e^{-\lambda t} \qquad (44.6)$$

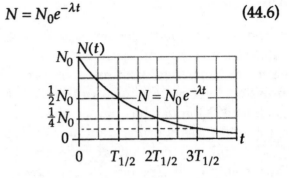

The **decay rate** R or **activity** of a sample of radioactive nuclei is defined as the number of decays per second. R_0 is the decay rate at $t = 0$.

$$R = \left|\frac{dN}{dt}\right| = \lambda N = N_0 \lambda e^{-\lambda t} = R_0 e^{-\lambda t} \quad (44.7)$$

Units of activity are the becquerel (Bq) and the curie (Ci). *The becquerel is the SI unit and the curie is approximately the activity of 1 gram of radium.*

$$1\,\text{Ci} \equiv 3.7 \times 10^{10}\ \text{decays}\,/\,\text{s}$$

$$1\,\text{Bq} \equiv 1\,\text{decay}\,/\,\text{s}$$

The **half-life** of a radioactive substance, $T_{1/2}$, is the time required for half of a given number of radioactive nuclei in a sample of the substance to decay. *After an elapsed time of n half-lives, the number of radioactive nuclei remaining will be $N_0/2^n$.*

$$T_{1/2} = \frac{\ln 2}{\lambda} = \frac{0.693}{\lambda} \qquad (44.8)$$

Spontaneous decay of radioactive nuclei proceeds by one of the following processes: alpha decay, beta decay, electron capture, or gamma decay. The overall decay process can be represented as a decay equation. *The sum of the mass numbers on each side of the equation must be equal.*

$$\underset{\substack{\text{Parent}\\\text{nucleus}}}{X} \rightarrow \underset{\substack{\text{Daughter}\\\text{nucleus}}}{Y} + \overset{\text{Emitted}}{\underset{}{\text{Radiation}}}$$

Alpha Decay

When a nucleus decays by alpha emission, the parent nucleus loses two neutrons and two protons. For alpha emission to occur, the mass of the parent nucleus ($_{Z}^{A}X$) must be greater than the combined masses of the daughter nuclei ($_{Z-2}^{A-4}Y$) and the emitted alpha particle. Two examples of alpha decay are shown.

$$_{Z}^{A}X \rightarrow \, _{Z-2}^{A-4}Y + \, _{2}^{4}He \qquad (44.9)$$

$$_{92}^{238}U \rightarrow \, _{90}^{234}Th + \, _{2}^{4}He \qquad (44.10)$$

$$_{88}^{226}Ra \rightarrow \, _{86}^{222}Rn + \, _{2}^{4}He \qquad (44.11)$$

The **disintegration energy** Q of the system (parent nucleus, daughter nucleus, and alpha particle) is required in order to conserve energy during spontaneous decay of the isolated parent nucleus. Q appears in the form of kinetic energy of the daughter nucleus and the alpha particle.

$$Q = (M_X - M_Y - M_\alpha)(931.494 \, \text{MeV} / \text{u})$$

$$(44.13)$$

Beta decay

When a radioactive nucleus undergoes beta decay, the process is accompanied by a third particle required to conserve energy and momentum: antineutrino ($\bar{v}$) with e^- emission and neutrino (v) with e^+ emission. For each mode of decay, the equations at right show: (i) The equation representing the process, (ii) an example of the decay mode, and (iii) an equation representing the origin of the electron or positron.

Electron Emission

(i) $\quad _{Z}^{A}X \rightarrow \, _{Z+1}^{A}Y + e^- + \bar{v} \qquad (44.18)$

(ii) $\quad _{6}^{14}C \rightarrow \, _{7}^{14}N + e^- + \bar{v} \qquad (44.20)$

(iii) $\quad n \rightarrow p + e^- + \bar{v} \qquad (44.22)$

Positron Emission

(i) $\quad _{Z}^{A}X \rightarrow \, _{Z-1}^{A}Y + e^+ + v \qquad (44.19)$

(ii) $\quad _{7}^{12}N \rightarrow \, _{6}^{12}C + e^+ + v \qquad (44.21)$

(iii) $\quad p \rightarrow n + e^+ + v$

Electron Capture is a decay process that competes with positron beta decay and occurs when a parent nucleus captures an orbital electron and emits a neutrino. *It is typically the K shell electron that is captured by the parent nucleus.*

$$_{Z}^{A}X + \, _{-1}^{0}e \rightarrow \, _{Z-1}^{A}Y + v \qquad (44.23)$$

Gamma Decay

Nuclei which undergo alpha or beta decay are often left in an excited energy state (indicated by the $*$ symbol). The nucleus returns to the ground state by emission of one or more photons. The excited state of $^{12}C^*$ is an example of decay by gamma emission. *Gamma decay results in no change in mass number or atomic number.*

$$^{A}_{Z}X^* \rightarrow {^{A}_{Z}}X + \gamma \qquad (44.24)$$

$$^{12}_{6}C^* \rightarrow {^{12}_{6}}C + \gamma \qquad (44.26)$$

Nuclear reactions can occur when target nuclei (X) are bombarded with energetic particles (a) resulting in a daughter or product nucleus (Y) and an outgoing particle (b). The reaction (Eq. 44.27) can be represented in a compact form as shown. *In these reactions the structure, identity, or properties of the target nuclei are changed.*

$$a + X \rightarrow Y + b \qquad (44.27)$$

$$X(a, b)Y$$

The **reaction energy** associated with a nuclear reaction is the total change in mass-energy resulting from the reaction.

$$Q = (M_a + M_X - M_Y - M_b)c^2 \qquad (44.28)$$

$Q > 0$, exothermic reaction
$Q < 0$, endothermic reaction

The **threshold energy** is the minimum energy required for an endothermic reaction to occur.

The **nuclear magneton**, μ_n, is a unit of magnetic moment in which the spin magnetic moment of a nucleus is measured.

$$\mu_n \equiv \frac{e\hbar}{2m_p} = 5.05 \times 10^{-27} \text{ J/T} \qquad (44.30)$$

SUGGESTIONS, SKILLS, AND STRATEGIES

The rest energy of a particle is given by $E_R = mc^2$. It is therefore often convenient to express the unified mass unit in terms of its equivalent energy, $1\ u = 1.660540 \times 10^{-27}$ kg or $1\ u = 931.494$ MeV$/c^2$. When masses are expressed in units of u, energy values are then $E_R = m(931.494$ MeV$/u)$.

Equation 44.6 can be solved for the particular time t after which the number of remaining nuclei will be some specified fraction of the original number N_0. This can be done by taking the natural log of each side of Equation 44.6 to find

$$t = \frac{1}{\lambda}\ln(N_0/N)$$

Decay	can be written as	Process
Alpha decay	$^A_Z X \rightarrow {}^{A-4}_{Z-2} X + {}^4_2 H$	The parent nucleus emits an alpha particle and loses two protons and two neutrons
Beta decay	$^A_Z X \rightarrow {}^A_{Z+1} Y + e^- + \bar{v}$ (electron emission) $^A_Z X \rightarrow {}^A_{Z-1} Y + e^+ + v$ (positron emission)	A nucleus can undergo beta decay in two ways. The parent nucleus can emit an electron (e^-) and an antineutrino ($\bar{v}$) or emit a positron (e^+) and a neutrino (v).
Gamma decay	$^A_Z X^* \rightarrow {}^A_Z X + \gamma$ * parent nucleus in an excited state	A nucleus in an excited state decays to a lower energy state (often the ground state) and emits a gamma ray.

The table above is a useful summary of the general processes
and characteristics of alpha, beta, and gamma decay.

REVIEW CHECKLIST

You should be able to:

▷   Use appropriate nomenclature in describing the static properties of nuclei. Calculate nuclear radii and densities. (Section 44.1)

▷   Use the binding-energy curve to estimate the energy released by fission and fusion. Calculate values of nuclear binding energy per nucleon. (Section 44.2).

▷   Explain the four major effects which influence the binding energy according to the liquid drop model of the nucleus. (Section 44.3)

▷   Apply to the solution of related problems, the formula which expresses decay rate as a function of the decay constant. Calculate radioactive half-life and the time required for the number of radioactive nuclei in a sample to decrease to a stated fraction of the initial value. (Section 44.4)

▷   Identify each of the components of radiation that are emitted by the nucleus through natural radioactive decay and describe the basic properties of each. Write out typical equations to illustrate the processes of transmutation by alpha decay, beta decay, gamma decay, and electron capture. Explain why the neutrino must be considered in the analysis of beta decay. (Sections 44.5 and 44.6)

▷   Calculate the Q value of given nuclear reactions and determine the threshold energy of endothermic reactions. Balance equations representing specific nuclear reactions. (Section 44.7)

ANSWERS TO SELECTED QUESTIONS

7.   Why do nearly all the naturally occurring isotopes lie above the $N = Z$ line in Figure 44.4?

Answer   As Z increases, extra neutrons are required to overcome the increasing electrostatic repulsion of the protons.

□   □   □   □

13.   Two samples of the same radioactive nuclide are prepared. Sample A has twice the initial activity of sample B. How does the half-life of A compare with the half-life of B? After each has passed through five half-lives, what is the ratio of their activities?

Answer   Since the two samples are of the same radioactive isotope, they have the same half-life; the 2:1 ratio in activity is due to a 2:1 ratio in the number of radioactive nuclei in each sample.. After 5 half lives, each will have decreased in number by a factor of $2^5 = 32$. However, since this simply means that the number of parent nuclei in each is 32 times smaller, the ratio of the number of radioactive nuclei will still be $(2/32):(1/32)$, or 2:1. Therefore, the ratio of their activities will **always** be 2:1.

□   □   □   □

21.   If a nucleus such as ^{226}Ra initially at rest undergoes alpha decay, which has more kinetic energy after the decay, the alpha particle or the daughter nucleus?

Answer   The alpha particle and the daughter nucleus carry momenta of equal magnitudes in opposite directions. Since kinetic energy can be written as $p^2/2m$, the less massive particle has much more of the decay energy than the recoiling nucleus.

□   □   □   □

29.   Suppose it could be shown that the cosmic-ray intensity at the Earth's surface was much greater 10000 years ago. How would this difference affect what we accept as valid carbon-dated values of the age of ancient samples of once-living matter?

Answer   If the cosmic ray intensity at the Earth's surface was much greater 10000 years ago, the fraction of the Earth's carbon dioxide with the heavy nuclide ^{14}C would also be greater at that time, than now. Thus, there would initially be a greater fraction of ^{14}C in the organic artifacts, and we would believe that the artifact was more recent than it actually was.

For example, suppose that the actual ratio of atmospheric ^{14}C to ^{12}C, two half-lives (11460 years) ago was 2.6×10^{-12}. The current ratio of isotopes in a sample from that time would be

$$\left(\tfrac{1}{2}\right)\left(\tfrac{1}{2}\right)(2.6 \times 10^{-12}) = 0.65 \times 10^{-12}$$

We would still measure this ratio today; but, believing the initial ratio to be 1.3×10^{-12}, we would think that the creature had died only one half-life (5730 years) ago.

$$\tfrac{1}{2}(1.3 \times 10^{-12}) = 0.65 \times 10^{-12}$$

□   □   □   □

SOLUTIONS TO SELECTED PROBLEMS

5. (a) Use energy methods to calculate the distance of closest approach for a head-on collision between an alpha particle having an initial energy of 0.500 MeV and a gold nucleus (^{197}Au) at rest. (Assume the gold nucleus remains at rest during the collision.) (b) What minimum initial speed must the alpha particle have to get as close as 300 fm?

Solution

Conceptualize: The positively charged alpha particle ($q=+2e$) will be repelled by the positive gold nucleus ($Q=+79e$), so that the particles probably will not touch each other in this electrostatic "collision." The closest the alpha particle can get to the gold nucleus would be if the two nuclei did touch, in which case the distance between their centers would be about 9 fm (using $r=r_0A^{1/3}$ for the radius of each nucleus). To get this close, or even within 300 fm, the alpha particle must be traveling very fast, probably close to the speed of light (but of course v must be less than c).

Categorize: The initial kinetic energy is equal to the electric potential energy of the alpha-particle-gold-nucleus system at the distance of closest approach, r_{min}.

Analyze:

(a) $K_\alpha = U = k_e \dfrac{qQ}{r_{min}}$

$$r_{min} = k_e \frac{qQ}{K_\alpha} = \frac{(8.99\times10^9 \text{ N·m}^2/\text{C}^2)(2)(79)(1.60\times10^{-19} \text{ C})^2}{(0.500 \text{ MeV})(1.60\times10^{-13} \text{ J/MeV})} = 455 \text{ fm} \qquad \lozenge$$

(b) Since $K_\alpha = \frac{1}{2}mv^2 = k_e \dfrac{qQ}{r_{min}}$, we find that $v = \sqrt{\dfrac{2k_eqQ}{mr_{min}}}$

$$v = \sqrt{\frac{2(8.99\times10^9 \text{ N·m}^2/\text{C}^2)(2)(79)(1.60\times10^{-19} \text{ C})^2}{4(1.66\times10^{-27} \text{ kg})(3.00\times10^{-13} \text{ m})}} = 6.04\times10^6 \text{ m/s} \qquad \lozenge$$

Finalize: The minimum distance in part (a) is on the order of 100 times greater than the combined radii of the particles. For part (b), the alpha particle must have more than 0.5 MeV of energy since it gets closer to the nucleus than the 455 fm found in part (a). Even so, the speed of the alpha particle in part (b) is only about 2% of the speed of light, so we are justified in not using a relativistic approach. In solving this problem, we ignored the effect of the electrons around the gold nucleus that tend to "screen" the nucleus so that the alpha particle sees a reduced positive charge. If this screening effect were considered, the potential energy would be slightly reduced and the alpha particle could get closer to the gold nucleus for the same initial energy.

9. A star ending its life with a mass of two times the mass of the Sun is expected to collapse, combining its protons and electrons to form a neutron star. Such a star could be thought of as a gigantic atomic nucleus. If a star of mass $2 \times 1.99 \times 10^{30}$ kg collapsed into neutrons ($m_n = 1.67 \times 10^{-27}$ kg), what would its radius be? (Assume that $r = r_0 A^{1/3}$.)

Solution

The number of nucleons in the star is

$$A = \frac{2(1.99 \times 10^{30} \text{ kg})}{1.67 \times 10^{-27} \text{ kg}} = 2.38 \times 10^{57}$$

Therefore, $\quad r = r_0 A^{1/3} = (1.20 \times 10^{-15} \text{ m})\left(2.38 \times 10^{57}\right)^{1/3} = 16.0 \text{ km}$ ◊

17. Nuclei having the same mass numbers are called *isobars*. The isotope $^{139}_{57}$La is stable. A radioactive isobar, $^{139}_{59}$Pr, is located below the line of stable nuclei in Figure 44.4 and decays by e^+ emission. Another radioactive isobar of $^{139}_{57}$La, $^{139}_{55}$Cs, decays by e^- emission and is located above the line of stable nuclei in Figure 44.4. (a) Which of these three isobars has the highest neutron-to-proton ratio? (b) Which has the greatest binding energy per nucleon? (c) Which do you expect to be heavier, $^{139}_{59}$Pr or $^{139}_{55}$Cs?

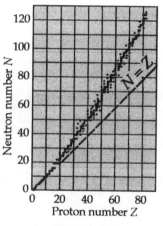

Figure 44.4

Solution

(a) For $^{139}_{59}$Pr the neutron number is $139 - 59 = 80$. For $^{139}_{55}$Cs the neutron number is 84, so the Cs isotope has the greatest neutron-to-proton ratio. ◊

(b) Binding energy per nucleon measures stability so it is greatest for the stable nucleus, the lanthanum isotope. Note also that it has a magic number of neutrons, 82. ◊

(c) Cs-139 has 55 protons and 84 neutrons. Pr-139 has 59 protons and 80 neutrons. When we plot both of them onto Figure 44.4, we see that Cesium is a little farther away from the center of the zone of stable nuclei. Being less stable goes with being able to lose more energy in decay, and with having more mass. We therefore expect Cesium to be heavier. ◊

19. A pair of nuclei for which $Z_1 = N_2$ and $Z_2 = N_1$ are called *mirror isobars* (the atomic and neutron numbers are interchanged). Binding-energy measurements on these nuclei can be used to obtain evidence of the charge independence of nuclear forces (that is, proton-proton, proton-neutron, and neutron-neutron nuclear forces are equal). Calculate the difference in binding energy for the two mirror isobars $^{15}_{8}\text{O}$ and $^{15}_{7}\text{N}$. The electric repulsion among eight protons rather than seven accounts for the difference.

Solution

For $^{15}_{8}\text{O}$ we have (using Equation 44.2)

$$E_b = \left[8(1.007\,825)\,\text{u} + 7(1.008\,665)\,\text{u} - (15.003\,065)\,\text{u}\right](931.494\,\text{MeV}/\text{u}) = 111.96\,\text{MeV}$$

For $^{15}_{7}\text{N}$ we have

$$E_b = \left[7(1.007\,825)\,\text{u} + 8(1.008\,665)\,\text{u} - (15.000\,109)\,\text{u}\right](931.5\,\text{MeV}/\text{u}) = 115.49\,\text{MeV}$$

Therefore, the difference in the two binding energies is

$$\Delta E_b = 3.54\,\text{MeV} \qquad\qquad\qquad \Diamond$$

---

21. Using the graph in Figure 44.5, estimate how much energy is released when a nucleus of mass number 200 fissions into two nuclei each of mass number 100.

Solution

The curve of binding energy shows that a heavy nucleus of mass number $A = 200$ has total binding energy about

$$\left(7.8\,\frac{\text{MeV}}{\text{nucleon}}\right)(200\text{ nucleons}) \approx 1.56\,\text{GeV}$$

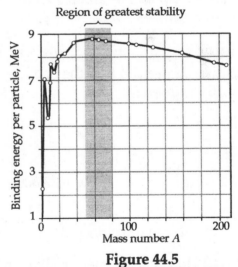

Region of greatest stability

Figure 44.5

Thus, it is less stable than its potential fission products, two middleweight nuclei of $A = 100$, having total binding energy

$$2(8.7\,\text{MeV}/\text{nucleon})(100\text{ nucleons}) \approx 1.74\,\text{GeV}$$

Fission then releases about $\qquad 1.74\,\text{GeV} - 1.56\,\text{GeV} \approx 200\,\text{MeV} \qquad\qquad \Diamond$

This is the energy source of uranium bombs and of nuclear electric-generating plants.

25. A sample of radioactive material contains 1.00×10^{15} atoms and has an activity of 6.00×10^{11} Bq. What is its half-life?

Solution $dN/dt = -\lambda N$

$$\lambda = \frac{1}{N}\left(-\frac{dN}{dt}\right) = (1.00\times10^{15} \text{ atoms})^{-1}(6.00\times10^{11} \text{ s}^{-1}) = 6.00\times10^{-4} \text{ s}^{-1}$$

$$T_{1/2} = \frac{\ln 2}{\lambda} = 1160 \text{ s} \quad \text{(This is also 19.3 minutes)} \qquad \Diamond$$

27. A freshly prepared sample of a certain radioactive isotope has an activity of 10.0 mCi. After 4.00 h, its activity is 8.00 mCi. (a) Find the decay constant and half-life. (b) How many atoms of the isotope were contained in the freshly prepared sample? (c) What is the sample's activity 30.0 h after it is prepared?

Solution

Conceptualize: Over the course of 4 hours, this isotope lost 20% of its activity, so its half-life appears to be around 10 hours, which means that its activity after 30 hours (~3 half-lives) will be about 1 mCi. The decay constant and number of atoms are not so easy to estimate.

Categorize: From the rate equation, $R = R_0 e^{-\lambda t}$, we can find the decay constant λ, which can then be used to find the half life, the original number of atoms, and the activity at any other time, t.

Analyze:

(a) $\lambda = \frac{1}{t}\ln\left(\frac{R_0}{R}\right) = \left(\frac{1}{(4.00 \text{ h})(3\,600 \text{ s/h})}\right)\ln\left(\frac{10.0 \text{ mCi}}{8.00 \text{ mCi}}\right) = 1.55\times10^{-5} \text{ s}^{-1}$ $\qquad \Diamond$

$$T_{1/2} = \frac{\ln 2}{\lambda} = \frac{0.693}{0.055\,8 \text{ h}^{-1}} = 12.4 \text{ h} \qquad \Diamond$$

(b) The number of original atoms can be found if we convert the initial activity from curies into becquerels (decays per second): $1 \text{ Ci} \equiv 3.7\times10^{10}$ Bq

$R_0 = 10.0 \text{ mCi} = (10.0\times10^{-3} \text{ Ci})(3.70\times10^{10} \text{ Bq/Ci}) = 3.70\times10^{8} \text{ Bq}$

Since $R_0 = \lambda N_0$, $N_0 = \frac{R_0}{\lambda} = \frac{3.70\times10^{8} \text{ decays/s}}{1.55\times10^{-5} \text{ s}} = 2.39\times10^{13} \text{ atoms}$ $\qquad \Diamond$

(c) $R = R_0 e^{-\lambda t} = (10.0 \text{ mCi})e^{-(5.58\times10^{-2} \text{ h}^{-1})(30.0 \text{ h})} = 1.88 \text{ mCi}$ $\qquad \Diamond$

Finalize: Our estimate of the half life was about 20% short because we did not account for the non-linearity of the decay rate. Consequently, our estimate of the final activity also fell short, but both of these calculated results are close enough to be reasonable.

The number of atoms is much less than one mole, so this appears to be a very small sample. To get a sense of how small, we can assume that the molar mass is about 100 g/mol, so the sample has a mass of only

$$m \approx \frac{(2.4 \times 10^{13} \text{ atoms})(100 \text{ g}/\text{mol})}{6.02 \times 10^{23} \text{ atoms}/\text{mol}} \approx 0.004 \ \mu g$$

This sample is so small it cannot be measured by a commercial mass balance!

The problem states that this sample was "freshly prepared," from which we assumed that **all** the atoms within the sample are initially radioactive. Generally this is not the case, so that N_0 only accounts for the formerly radioactive atoms, and does not include additional atoms in the sample that were not radioactive. Realistically then, the sample mass should be significantly greater than our above estimate.

---

29. The radioactive isotope ^{198}Au has a half-life of 64.8 h. A sample containing this isotope has an initial activity ($t=0$) of 40.0 μCi. Calculate the number of nuclei that decay in the time interval between $t_1 = 10.0$ h and $t_2 = 12.0$ h.

Solution First, let us find λ and N_0 from the given information:

$$\lambda = \frac{\ln 2}{T_{1/2}} = \frac{0.693}{64.8 \text{ h}} = 0.0107 \text{ h}^{-1} = 2.97 \times 10^{-6} \text{ s}^{-1}$$

$$N_0 = \frac{R_0}{\lambda} = \left(\frac{40.0 \times 10^{-6} \text{ Ci}}{2.97 \times 10^{-6} \text{ s}^{-1}} \right) \left(\frac{3.70 \times 10^{10} \text{ decays}/\text{s}}{1 \text{ Ci}} \right) = 4.98 \times 10^{11} \text{ nuclei}$$

Since $N = N_0 e^{-\lambda t}$ the number of nuclei which decay between times t_1 and t_2 is

$$N_1 - N_2 = N_0 (e^{-\lambda t_1} - e^{-\lambda t_2})$$

$$N_1 - N_2 = (4.98 \times 10^{11}) \left[e^{-(0.0107 \text{ h}^{-1})(10.0 \text{ h})} - e^{-(0.0107 \text{ h}^{-1})(12.0 \text{ h})} \right]$$

$$N_1 - N_2 = 9.47 \times 10^9 \text{ nuclei} \qquad \diamond$$

---

33. Find the energy released in the alpha decay $^{238}_{92}U \rightarrow \,^{234}_{90}Th + \,^4_2He$. You will find Table A.3 useful.

Solution

Table A.3 contains the following values:
$$M(^{238}_{92}U) = 238.050783 \text{ u}$$

$$M(^{234}_{90}Th) = 234.043596 \text{ u}$$

$$M(^4_2He) = 4.002603 \text{ u}$$

We apply to the decay a relativistic energy version of the isolated system model. Some of the rest energy of the parent nucleus is converted into kinetic energy of the decay products according to $Q = \Delta mc^2$.

We calculate $\quad Q = (M_U - M_{Th} - M_{He})(931.5 \text{ MeV} / \text{u})$

$$Q = (238.050783 - 234.043596 - 4.002603)(931.5) = 4.27 \text{ MeV} \qquad \diamond$$

---

39. The nucleus $^{15}_8O$ decays by electron capture. The nuclear reaction is written $^{15}_8O + e^- \rightarrow \,^{15}_7N + \nu$. (a) Write the process going on for a single particle within the nucleus. (b) Write the decay process referring to neutral atoms. (c) Determine the energy of the neutrino. Disregard the daughter's recoil.

Solution

(a) $\qquad\qquad\qquad\qquad e^- + p \rightarrow n + \nu \qquad\qquad\qquad\qquad\qquad \diamond$

(b) Add 7 protons, 7 neutrons, and 7 electrons to each side to give

$$^{15}O \text{ atom} \rightarrow \,^{15}N \text{ atom} + \nu \qquad\qquad \diamond$$

(c) From Table A.3, $\qquad M(^{15}O) = M(^{15}N) + Q/c^2$

$$\Delta M = 15.003065 - 15.000109 = 0.002956 \text{ u}$$

$$Q = (931.5 \text{ MeV} / \text{u})(0.002956 \text{ u}) = 2.75 \text{ MeV} \qquad\qquad \diamond$$

---

41. Enter the correct isotope symbol in each open square in Figure P44.41, which shows the sequences of decays in the natural radioactive series starting with the long-lived isotope uranium-235 and ending with the stable nucleus lead-207.

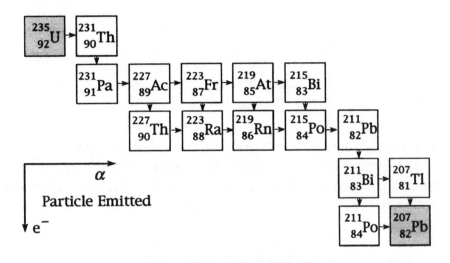

Figure P44.41 (modified)

Solution Whenever an $\alpha = {}^4_2\text{He}$ is emitted, Z drops by 2 and A by 4. Whenever an $e^- = {}^0_{-1}e^-$ is emitted, Z increases by 1 and A is unchanged. We find the chemical name by looking up Z in a periodic table. The values in the shaded boxes (${}^{235}\text{U}$ and ${}^{207}\text{Pb}$) were given; all others have been filled in as part of the solution. ◊

47. Natural gold has only one isotope, ${}^{197}_{79}\text{Au}$. If natural gold is irradiated by a flux of slow neutrons, electrons are emitted. (a) Write the reaction equation. (b) Calculate the maximum energy of the emitted electrons.

Solution

The ${}^{197}_{79}\text{Au}$ will absorb a neutron to become ${}^{198}_{79}\text{Au}$, which emits an e^- to become ${}^{198}_{80}\text{Hg}$.

(a) For nuclei the net reaction is: ${}^{197}_{79}\text{Au nucleus} + {}^1_0\text{n} \rightarrow {}^{198}_{80}\text{Hg nucleus} + {}^0_{-1}e^- + \overline{\nu}$ ◊

(b) Adding 79 electrons to both sides:

$${}^{197}_{79}\text{Au atom} + {}^1_0\text{n} \rightarrow {}^{198}_{80}\text{Hg atom} + \overline{\nu}$$

From Table A.3, $196.966\,552 + 1.008\,665 = 197.966\,752 + 0 + Q/c^2$

$Q = \Delta mc^2$: $Q = (0.008\,465\ \text{u})(931.5\ \text{MeV}/\text{u}) = 7.89\ \text{MeV}$ ◊

49. Using the mass of ^{9}Be from Table A.3 and Q values of appropriate reactions from Table 44.5, calculate the masses of ^{8}Be and ^{10}Be in atomic mass units to four decimal places.

Solution

Conceptualize: The mass of each isotope in atomic mass units will be approximately the number of nucleons (8 or 10), also called the mass number. The electrons are much less massive and contribute only about 0.03% to the total mass.

Categorize: In addition to summing the mass of the subatomic particles, the net mass of the isotopes must account for the binding energy that holds the atom together. Table 44.5 includes the energy released for each nuclear reaction. Precise atomic mass values are found in Table A.3.

Analyze: The notation $\qquad$ ^{9}Be (γ, n) ^{8}Be

with $\qquad$ $Q = -1.665 \text{ MeV}$

means $\qquad$ $^9\text{Be} + \gamma \rightarrow {}^8\text{Be} + \text{n} - 1.665 \text{ MeV}$

Therefore $\qquad$ $m(^8\text{Be}) = m(^9\text{Be}) - m_n + \dfrac{1.665 \text{ MeV}}{931.5 \text{ MeV/u}}$

$\qquad m(^8\text{Be}) = 9.012\,182 - 1.008\,665 + 0.001\,787 = 8.005\,3 \text{ u}$ $\qquad \lozenge$

The notation $\qquad$ ^{9}Be (n, γ) ^{10}Be

with $\qquad$ $Q = 6.812 \text{ MeV}$

means $\qquad$ $^9\text{Be} + \text{n} \rightarrow {}^{10}\text{Be} + \gamma + 6.812 \text{ MeV}$

$\qquad m(^{10}\text{Be}) = m(^9\text{Be}) + m_n - \dfrac{6.812 \text{ MeV}}{931.5 \text{ MeV/u}}$

$\qquad m(^{10}\text{Be}) = 9.012\,182 + 1.008\,665 - 0.007\,313 = 10.0135 \text{ u}$ $\qquad \lozenge$

Finalize: As expected, both isotopes have masses slightly greater than their mass numbers. We were asked to calculate the masses to four decimal places, but with the available data, the results could be reported accurately to as many as six decimal places.

73. Free neutrons have a characteristic half-life of 10.4 min. What fraction of a group of free neutrons with kinetic energy 0.0400 eV will decay before traveling a distance of 10.0 km?

Solution

The fraction that will remain is given by the ratio N/N_0, where $N/N_0 = e^{-\lambda \Delta t}$ and Δt is the time it takes the neutron to travel a distance of $d = 10.0$ km.

We use the particle under constant speed model. The time of flight is given by $\Delta t = d/v$.

Since $K = \frac{1}{2}mv^2$,

$$\Delta t = \frac{d}{\sqrt{\dfrac{2K}{m}}} = \frac{10.0 \times 10^3 \text{ m}}{\sqrt{\dfrac{2(0.0400 \text{ eV})(1.60 \times 10^{-19} \text{ J/eV})}{1.67 \times 10^{-27} \text{ kg}}}} = 3.61 \text{ s}$$

The decay constant is then

$$\lambda = \frac{0.693}{T_{1/2}} = \frac{0.693}{(10.4 \text{ min})(60 \text{ s/min})} = 1.11 \times 10^{-3} \text{ s}^{-1}$$

Therefore,

$$\lambda \Delta t = (1.11 \times 10^{-3} \text{ s}^{-1})(3.61 \text{ s}) = 4.01 \times 10^{-3} = 0.004\,01$$

and

$$\frac{N}{N_0} = e^{-\lambda \Delta t} = e^{-0.004\,01} = 0.9960$$

Hence, the fraction that has decayed in this time is

$$1 - \frac{N}{N_0} = 0.004\,00 \quad \text{or} \quad 0.400\%$$

Chapter 45

APPLICATION OF NUCLEAR PHYSICS

EQUATIONS AND CONCEPTS

The **fission of a uranium nucleus** by bombardment with a low energy neutron results in the production of fission fragments and neutrons. *Equation 45.3 is typical of many possible combinations of fission fragments X and Y. U* is a short-lived intermediate state.*

$$\,^{1}_{0}n + \,^{235}_{92}U \rightarrow \,^{236}_{92}U^{*} \rightarrow X + Y + \text{neutrons} \tag{45.2}$$

$$\,^{1}_{0}n + \,^{235}_{92}U \rightarrow \,^{141}_{56}Ba + \,^{92}_{36}Kr + 3\,^{1}_{0}n \tag{45.3}$$

The **fusion reactions** shown here are those most likely to be used as the basis of the design and operation of a fusion power reactor. The Q values refer to the energy released in each reaction.

$$\,^{2}_{1}H + \,^{2}_{1}H \rightarrow \,^{3}_{2}He + \,^{1}_{0}n \tag{45.4}$$
$$(Q = 3.27 \text{ MeV})$$

$$\,^{2}_{1}H + \,^{2}_{1}H \rightarrow \,^{3}_{1}H + \,^{1}_{1}H$$
$$(Q = 4.03 \text{ MeV})$$

$$\,^{2}_{1}H + \,^{3}_{1}H \rightarrow \,^{4}_{2}He + \,^{1}_{0}n$$
$$(Q = 17.59 \text{ MeV})$$

Lawson's criterion states the conditions under which a net power output of a fusion reactor is possible. In these expressions, n is the plasma density (number of ions per cubic cm) and τ is the plasma confinement time (the time during which the interacting ions are maintained at a temperature equal to or greater than that required for the reaction to proceed).

$$n\tau \geq 10^{14} \text{ s/cm}^3 \tag{45.5}$$
$$\text{(D-T reaction)}$$

$$n\tau \geq 10^{16} \text{ s/cm}^3$$
$$\text{(D-D reaction)}$$

Several different **radiation units** are used to quantify the amount of radiation interacting with a material:

- One **roentgen** (R) is the amount of ionizing radiation that produces, by ionization, an electric charge of 3.33×10^{-10} C in one cm^3 of air at STP.

- One **rad** (radiation absorbed dose) is the amount of radiation that increases the energy of 1 kg of absorbing material by 1×10^{-2} J.

  1 gray (Gy) = 100 rad

- **RBE** (relative biological effectiveness) is a factor defined as the number of rads of x-radiation or gamma radiation that produces the same biological effect as 1 rad of the radiation being used.

- The radiation **dose in rem** (radiation equivalent man) is the product of the dose in rad and the appropriate relative biological effectiveness factor.

  $$\text{Dose in rem} \equiv \text{dose in rad} \times \text{RBE} \qquad (45.6)$$

  1 sievert (Sv) = 100 rem

REVIEW CHECKLIST

You should be able to:

▷ Write several possible equations for the fission of ^{235}U and describe the sequence of events which occurs during the fission process. Calculate the mass of ^{235}U required to release a given quantity of energy by fission. (Section 45.2)

▷ Describe the basic design features and control mechanisms in a fission reactor including the functions of the moderator, control rods, and heat exchange system. Identify some major safety and environmental hazards in the operation of a fission reactor. (Section 45.3)

▷ Describe the basis of energy release in fusion and write several nuclear reactions which might be used in a fusion powered reactor. Explain the three basic requirements of a thermonuclear power reactor and calculate the rate of fuel consumption required to maintain a given power output by the D-T reaction. (Section 45.4)

▷ Define the roentgen, rad, and rem as units of radiation exposure or dose and explain the significance of the RBE factor. Calculate the thickness of shielding required to reduce the intensity of gamma rays to a given fraction of the intensity of the incident beam; see problem 45.26. (Section 45.5)

▷ Describe the basic principle of operation of the cloud chamber, Geiger counter, semiconductor diode detector, scintillation detector, and spark chamber. (Section 45.6)

ANSWERS TO SELECTED QUESTIONS

2. Why is water a better shield against neutrons than lead or steel?

Answer The hydrogen nuclei in water molecules have mass similar to that of a neutron, so that they can efficiently rob a fast-moving neutron of kinetic energy as they scatter it. Once the neutron is slowed down, a hydrogen nucleus can absorb it in the reaction $_0^1n + \,_1^1H \rightarrow \,_1^2H + Q$.

□   □   □   □

9. Discuss the similarities and differences between fusion and fission.

Answer Fusion of light nuclei to a heavier nucleus releases energy. Fission of a heavy nucleus to lighter nuclei releases energy. Both processes are steps towards greater stability on the curve of binding energy, Figure 44.5. The energy release per nucleon is typically greater for fusion, and this process is harder to control.

□   □   □   □

SOLUTIONS TO SELECTED PROBLEMS

5. List the nuclear reactions required to produce ^{233}U from ^{232}Th under fast neutron bombardment.

Solution

First, the Thorium is bombarded: $\qquad$ $_0^1n + \,_{90}^{232}Th \rightarrow \,_{90}^{233}Th$

Then, the Thorium decays by beta emission: $\qquad$ $_{90}^{233}Th \rightarrow \,_{91}^{233}Pa + e^- + \bar{\nu}$

Protactinium-233 has more neutrons than the more stable Protactinium-231, so it too decays by beta emission:

$$_{91}^{233}Pa \rightarrow \,_{92}^{233}U + e^- + \bar{\nu} \qquad \Diamond$$

9.   Review problem. Suppose enriched uranium containing 3.40% of the fissionable isotope $^{235}_{92}\text{U}$ is used as fuel for a ship. The water exerts an average friction force of magnitude 1.00×10^5 N on the ship. How far can the ship travel per kilogram of fuel? Assume that the energy released per fission event is 208 MeV and that the ship's engine has an efficiency of 20.0%.

Solution

Conceptualize: Nuclear fission is much more efficient for converting mass to energy than burning fossil fuels. However, without knowing the rate of diesel fuel consumption for a comparable ship, it is difficult to estimate the nuclear fuel rate. It seems plausible that a ship could cross the Atlantic ocean with only a few kilograms of nuclear fuel, so a reasonable range of uranium fuel consumption might be 10 km/kg to 10000 km/kg.

Categorize: The fuel consumption rate can be found from the energy released by the nuclear fuel and the work required to push the ship through the water. We use the particle in equilibrium model. The thrust force P exerted on the propeller by the water around it must be equal in magnitude to the backward frictional (drag) force on the hull.

Analyze: One kg of enriched uranium contains 3.40% $^{235}_{92}\text{U}$, so

$$m_{235} = 0.034\,0(1\,000\text{ g}) = 34.0\text{ g}$$

In terms of number of nuclei, this is equivalent to

$$N_{235} = (34.0\text{ g})\left(\frac{1}{235\text{ g/mol}}\right)(6.02\times10^{23}\text{ atoms/mol}) = 8.71\times10^{22}\text{ nuclei}$$

If all these nuclei fission, the thermal energy released is equal to

$$(8.71\times10^{22}\text{ nuclei})(208\text{ MeV/nucleus})(1.602\times10^{-19}\text{ J/eV}) = 2.90\times10^{12}\text{ J}$$

Now, for the engine,

$$\text{efficiency} = \frac{\text{work output}}{\text{heat input}} \qquad \text{or} \qquad e = \frac{P\Delta r\cos\theta}{Q_h}$$

So the distance the ship can travel per kilogram of uranium fuel is

$$\Delta r = \frac{eQ_h}{P\cos(0)} = \frac{0.200(2.90\times10^{12}\text{ J})}{1.00\times10^5\text{ N}} = 5.80\times10^6\text{ m} \qquad \lozenge$$

Finalize: The ship can travel 5800 km/kg of uranium fuel, which is on the high end of our prediction range. The distance between New York and Paris is 5851 km, so this ship could cross the Atlantic ocean on just one kilogram of uranium fuel.

---

11. It has been estimated that on the order of 10^9 tons of natural uranium is available at concentrations exceeding 100 parts per million, of which 0.7% is the fissionable isotope ^{235}U. Assume that all the world's energy use $(7\times10^{12}$ J/s) were supplied by ^{235}U fission in conventional nuclear reactors, releasing 208 MeV for each reaction. How long would the supply last? The estimate of uranium supply is taken from K. S. Deffeyes and I. D. MacGregor, "World Uranium Resources," *Scientific American* **242**(1):66, 1980.

Solution   The mass of natural uranium reserves is

$$m_{U,\ total} = (10^9 \text{ metric tons})\left(\frac{10^3 \text{ kg}}{1 \text{ metric ton}}\right) = 1\times10^{12} \text{ kg}$$

For fissionable U-235, $m_{U\text{-}235} = 0.007\,(m_{U,\ total}) = 7\times10^9 \text{ kg} = 7\times10^{12} \text{ g}$

In terms of U-235 nuclei, $N_{235} = (7\times10^{12} \text{ g})\left(\dfrac{6.02\times10^{23} \text{ atoms}}{235 \text{ g}}\right) = 1.79\times10^{34} \text{ nuclei}$

Following the statement of the problem, we take the fission energy as 208 MeV/fission:

$$E = (1.79\times10^{34} \text{ nuclei})(208 \text{ MeV / fission})(1.60\times10^{-19} \text{ J/eV}) = 5.97\times10^{23} \text{ J}$$

Now, $\Delta t = \dfrac{E}{\mathcal{P}} = \dfrac{5.97\times10^{23} \text{ J}}{7\times10^{12} \text{ J/s}} = 8.53\times10^{10} \text{ s} \sim 3000 \text{ yr}$ ◊

17. To understand why plasma containment is necessary, consider the rate at which an unconfined plasma would be lost. (a) Estimate the rms speed of deuterons in a plasma at 4.00×10^8 K. (b) **What If?** Estimate the order of magnitude of the time interval during which such a plasma would remain in a 10-cm cube if no steps were taken to contain it.

Solution   The average kinetic energy per particle $\frac{1}{2}m\overline{v^2}$ must equal the thermal energy $\frac{3}{2}k_BT$. Taking $m\approx 2m_p$ for deuterons,

(a) $\frac{1}{2}m\overline{v^2} = \frac{3}{2}k_BT$

$$v_{rms} = \sqrt{\frac{3k_BT}{2m_p}} = \sqrt{\frac{3(1.38\times10^{-23} \text{ J/K})(4.00\times10^8 \text{ K})}{2(1.67\times10^{-27} \text{ kg})}} = 2.23\times10^6 \text{ m/s}$$ ◊

(b) $t = \dfrac{x}{v} = \dfrac{0.100 \text{ m}}{2.23\times10^6 \text{ m/s}} \sim 10^{-7} \text{ s}$ ◊

23. A building has become accidentally contaminated with radioactivity. The longest-lived material in the building is strontium-90. ($^{90}_{38}$Sr has an atomic mass 89.907 7 u, and its half-life is 29.1 yr. It is particularly dangerous because it substitutes for calcium in bones.) Assume that the building initially contained 5.00 kg of this substance uniformly distributed throughout the building (a very unlikely situation) and that the safe level is defined as less than 10.0 decays/min (to be small in comparison to background radiation). How long will the building be unsafe?

Solution

The number of nuclei in the original sample is

$$N_0 = \frac{\text{mass present}}{\text{mass of nucleus}} = \frac{5.00 \text{ kg}}{(89.9077 \text{ u})(1.66\times10^{-27} \text{ kg / u})} = 3.35\times10^{25} \text{ nuclei}$$

$$\lambda = \frac{\ln 2}{T_{1/2}} = \frac{0.693}{29.1 \text{ yr}} = 2.38\times10^{-2} \text{ yr}^{-1} = 4.53\times10^{-8} \text{ min}^{-1}$$

$$R_0 = \lambda N_0 = (4.53\times10^{-8} \text{ min}^{-1})(3.35\times10^{25} \text{ nuclei}) = 1.52\times10^{18} \text{ counts / min}$$

$$\frac{R}{R_0} = \frac{10.0 \text{ counts / min}}{1.52\times10^{18} \text{ counts / min}} = 6.59\times10^{-18} = e^{-\lambda t}$$

$$t = \frac{-\ln(R/R_0)}{\lambda} = \frac{-\ln(6.59\times10^{-18})}{2.38\times10^{-2} \text{ yr}^{-1}} = 1660 \text{ yr} \qquad \lozenge$$

27. A "clever" technician decides to warm some water for his coffee with an x-ray machine. If the machine produces 10.0 rad/s, how long will it take to raise the temperature of a cup of water by 50.0 °C ?

Solution The energy required to heat the water is $\mathcal{P}\Delta t = Q = mc\Delta T$

Noting that $1 \text{ rad} = 10^{-2} \text{ J/kg}$,

$$\Delta t = \frac{mc\Delta T}{\mathcal{P}} = \frac{m(4186 \text{ J/kg} \cdot °\text{C})(50.0 °\text{C})}{(10.0 \text{ rad / s})(10^{-2} \text{ J/kg} \cdot \text{rad})m} = 2.09\times10^6 \text{ s} \approx 24 \text{ days} \qquad \lozenge$$

(**Note:** the power $\mathcal{P}$ is the product of the dose rate and the mass).

31. In a Geiger tube, the voltage between the electrodes is typically 1.00 kV and the current pulse discharges a 5.00-pF capacitor. (a) What is the energy amplification of this device for a 0.500-MeV electron? (b) How many electrons participate in the avalanche caused by the single initial electron?

Solution

(a) $$\frac{E}{E_0} = \frac{\frac{1}{2}C\Delta V^2}{0.500 \text{ MeV}} = \frac{\frac{1}{2}(5.00\times10^{-12} \text{ F})(1.00\times10^3 \text{ V})^2}{(0.500 \text{ MeV})(1.60\times10^{-13} \text{ J/MeV})} = 3.12\times10^7$$ ◊

(b) $$N = \frac{Q}{e} = \frac{C\Delta V}{e} = \frac{(5.00\times10^{-12} \text{ F})(1.00\times10^3 \text{ V})}{1.60\times10^{-19} \text{ C}} = 3.12\times10^{10} \text{ electrons}$$ ◊

39. Carbon detonations are powerful nuclear reactions that temporarily tear apart the cores inside massive stars late in their lives. These blasts are produced by carbon fusion, which requires a temperature of about 6×10^8 K to overcome the strong Coulomb repulsion between carbon nuclei. (a) Estimate the repulsive energy barrier to fusion, using the temperature required for carbon fusion. (In other words, what is the average kinetic energy of a carbon nucleus at 6×10^8 K?) (b) Calculate the energy (in MeV) released in each of these "carbon-burning" reactions:

$$^{12}C + {}^{12}C \rightarrow {}^{20}Ne + {}^4He \qquad\qquad {}^{12}C + {}^{12}C \rightarrow {}^{24}Mg + \gamma$$

(c) Calculate the energy (in kWh) given off when 2.00 kg of carbon completely fuses according to the first reaction.

Solution

(a) At 6×10^8 K, each carbon nucleus has thermal energy of

$$\tfrac{3}{2}k_BT = 1.5(8.62\times10^{-5} \text{ eV/K})(6\times10^8 \text{ K}) = 8\times10^4 \text{ eV}$$ ◊

(b) The energy released is

$$E = \left[2m(C^{12}) - m(Ne) - m(He^4)\right]c^2$$

$$E = (24.000\,000 - 19.992\,440 - 4.002\,603)(931.5) \text{ MeV} = 4.62 \text{ MeV}$$ ◊

In the 2nd case, the energy released is

$$E = \left[2m(C^{12}) - m(Mg^{24})\right](931.5) \text{ MeV}/u$$

$$E = (24.000\,000 - 23.985\,042)(931.5) \text{ MeV} = 13.9 \text{ MeV} \qquad \Diamond$$

(c) The energy released equals the energy of reaction of the number of carbon nuclei in a 2.00-kg sample, which corresponds to

$$\Delta E = \left(\frac{2000 \text{ g}}{12 \text{ g}/\text{mol C}}\right)\left(\frac{6.02\times10^{23} \text{ atom}}{1 \text{ mol C}}\right)\left(\frac{1 \text{ fusion}}{2 \text{ atom}}\right)\left(\frac{4.62 \text{ MeV}}{\text{fusion}}\right)\left(\frac{1 \text{ kWh}}{2.25\times10^{19} \text{ MeV}}\right)$$

$$\Delta E = 10.3\times10^6 \text{ kWh} \qquad \Diamond$$

---

43. The half-life of tritium is 12.3 yr. If the TFTR fusion reactor contained 50.0 m^3 of tritium at a density equal to 2.00×10^{14} ions/cm^3, how many curies of tritium were in the plasma? Compare this value with a fission inventory (the estimated supply of fissionable material) of 4×10^{10} Ci.

Solution

Conceptualize: It is difficult to estimate the activity of the tritium in the fusion reactor without actually calculating it; however, we might expect it to be a small fraction of the fission (not fusion) inventory.

Categorize: The decay rate (activity) can be found by multiplying the decay constant λ by the number of ${}^3_1\text{H}$ particles. The decay constant can be found from the half-life of tritium, and the number of particles from the density and volume of the plasma.

Analyze: The number of Hydrogen-3 nuclei is

$$N = (50.0 \text{ m}^3)\left(2.00\times10^{14} \frac{\text{particles}}{\text{cm}^3}\right)(100 \text{ cm}/\text{m})^3 = 1.00\times10^{22} \text{ particles}$$

$$\lambda = \frac{\ln 2}{T_{1/2}} = \left(\frac{0.693}{12.3 \text{ yr}}\right)\left(\frac{1 \text{ yr}}{3.16\times10^7 \text{ s}}\right) = 1.78\times10^{-9} \text{ s}^{-1}$$

The activity is then $R = \lambda N = (1.78\times10^{-9} \text{ s}^{-1})(1.00\times10^{22} \text{ nuclei}) = 1.78\times10^{13} \text{ Bq}$

$$R = (1.78\times10^{13} \text{ Bq})\left(\frac{1 \text{ Ci}}{3.70\times10^{10} \text{ Bq}}\right) = 482 \text{ Ci} \qquad \Diamond$$

Finalize: Even though 482 Ci is a large amount of radioactivity, it is smaller than 4.00×10^{10} Ci by about a hundred million. Therefore, loss of containment is a smaller hazard for a fusion power reactor than for a fission reactor. ◊

---

53. Assuming that a deuteron and a triton are at rest when they fuse according to the reaction $^2_1H + ^3_1H \rightarrow ^4_2He + ^1_0n + 17.6$ MeV, determine the kinetic energy acquired by the neutron.

Solution

Conceptualize: The products of this nuclear reaction are an alpha particle and a neutron, with total kinetic energy of 17.6 MeV. In order to conserve momentum, the lighter neutron will have a larger velocity than the more massive alpha particle (which consists of two protons and two neutrons). Since the kinetic energy of the particles is proportional to the square of their velocities but only linearly proportional to their mass, the neutron should have the larger kinetic energy, somewhere between 8.8 and 17.6 MeV.

Categorize: Conservation of linear momentum and energy can be applied to find the kinetic energy of the neutron. We first suppose the particles are moving nonrelativistically.

Analyze: The momentum of the alpha particle and that of the neutron must add to zero, so their velocities must be in opposite directions with magnitudes related by

$$m_n\mathbf{v}_n + m_\alpha\mathbf{v}_\alpha = 0 \qquad \text{or} \qquad (1.008\,7\text{ u})v_n = (4.002\,6\text{ u})v_\alpha$$

At the same time, their kinetic energies must add to 17.6 MeV:

$$E = \tfrac{1}{2}m_n v_n^{\ 2} + \tfrac{1}{2}m_\alpha v_\alpha^{\ 2} = \tfrac{1}{2}(1.008\,7\text{ u})v_n^{\ 2} + \tfrac{1}{2}(4.002\,6\text{ u})v_\alpha^{\ 2} = 17.6\text{ MeV}$$

Substitute $v_\alpha = 0.252\,0 v_n$

to obtain $\quad E = (0.504\,35\text{ u})v_n^{\ 2} + (0.127\,10\text{ u})v_n^{\ 2} = 17.6\text{ MeV}\left(\dfrac{1\text{ u}}{931.494\text{ MeV}/c^2}\right)$

$$v_n = \sqrt{\frac{0.0189\,c^2}{0.631\,45}} = 0.173c = 5.19 \times 10^7\text{ m/s}$$

Since this speed is not too much greater than 0.1c, we can get a reasonable estimate of the kinetic energy of the neutron from the classical equation,

$$K = \tfrac{1}{2}mv^2 = \tfrac{1}{2}(1.008\,7\text{ u})(0.173c)^2\left(\frac{931.494\text{ MeV}/c^2}{\text{u}}\right) = 14.1\text{ MeV} \qquad ◊$$

Finalize: The kinetic energy of the neutron is within the range we predicted. For a more accurate calculation of the kinetic energy, we should use relativistic expressions. Conservation of momentum gives

$$\gamma_n m_n \mathbf{v}_n + \gamma_\alpha m_\alpha \mathbf{v}_\alpha = 0: \qquad 1.0087\frac{v_n}{\sqrt{1-v_n^2/c^2}} = 4.0026\frac{v_\alpha}{\sqrt{1-v_\alpha^2/c^2}}$$

yielding

$$\frac{v_\alpha^2}{c^2} = \frac{v_n^2}{15.746c^2 - 14.746v_n^2}$$

Then

$$(\gamma_n - 1)m_n c^2 + (\gamma_\alpha - 1)m_\alpha c^2 = 17.6 \text{ MeV}$$

and $v_n = 0.171c$, implying that

$$(\gamma_n - 1)m_n c^2 = 14.0 \text{ MeV} \qquad \Diamond$$

55. (a) Calculate the energy (in kilowatt-hours) released if 1.00 kg of ^{239}Pu undergoes complete fission and the energy released per fission event is 200 MeV. (b) Calculate the energy (in electron volts) released in the deuterium-tritium fusion reaction

$$^2_1\text{H} + {}^3_1\text{H} \rightarrow {}^4_2\text{He} + {}^1_0\text{n}$$

(c) Calculate the energy (in kilowatt-hours) released if 1.00 kg of deuterium undergoes fusion according to this reaction. (d) **What If?** Calculate the energy (in kilowatt-hours) released by the combustion of 1.00 kg of coal if each $C + O_2 \rightarrow CO_2$ reaction yields 4.20 eV. (e) List advantages and disadvantages of each of these methods of energy generation.

Solution We obtain the masses of the particles from Appendix A.3.

(a) $\quad E = (1\,000 \text{ g})\left(\dfrac{6.02\times10^{23} \text{ nuclei}}{239 \text{ g}}\right)\left(\dfrac{200\times10^6 \text{ eV}}{1 \text{ nucleus}}\right)\left(\dfrac{1.60\times10^{-19} \text{ J/eV}}{3.60\times10^6 \text{ J/kWh}}\right)$

$\quad E = 2.24\times10^7 \text{ kWh} \qquad \Diamond$

(b) $\quad m_{\text{before}} = 2.014\,102 \text{ u} + 3.016\,049 \text{ u} = 5.030\,151 \text{ u}$

$\quad m_{\text{after}} = 4.002\,603 \text{ u} + 1.008\,665 \text{ u} = 5.011\,268 \text{ u}$

$\quad \Delta m = m_{\text{before}} - m_{\text{after}} = 0.018\,883 \text{ u}$

$\quad \Delta E = \Delta mc^2 = (0.018\,883 \text{ u})(931.494 \text{ MeV/u}) = 17.6 \text{ MeV} \qquad \Diamond$

(c) $\quad E = (1\,000\text{ g }{}_{1}^{2}\text{H})\left(\dfrac{6.02\times10^{23}\text{ deuterons}}{2.014\text{ g }{}_{1}^{2}\text{H}}\right)\left(\dfrac{17.6\text{ MeV}}{{}_{1}^{2}\text{H fusion}}\right)\left(\dfrac{1.60\times10^{-13}\text{ J/MeV}}{3.60\times10^{6}\text{ J/kWh}}\right)$

$E = 2.34\times10^{8}\text{ kWh}$ ◊

(d) Coal is essentially pure carbon. Assuming complete combustion,

$(1\,000\text{ g C})\left(\dfrac{6.02\times10^{23}\text{ C atoms}}{12.0\text{ g C}}\right)\left(\dfrac{4.20\text{ eV}}{\text{C atom}}\right)\left(\dfrac{1.60\times10^{-19}\text{ J/eV}}{3.60\times10^{6}\text{ J/kWh}}\right) = 9.36\text{ kWh}$ ◊

(e) You likely pay the electric company to burn coal, because coal is cheap at this moment in human history. The limit on supply will inevitably drive up the price of fossil fuels. Worldwide, electric energy from fission is an immediate option. Fission may offer advantages of reduced overall pollution, and even a reduction in the radiation released into the atmosphere, compared to mining and burning coal. However, the Chernobyl explosion and other accidents demonstrate that there is a continuing significant risk of catastrophic disaster. For a fair comparison to a coal-fired plant, we should think of a nuclear generating station paying for long-term disposal of its waste and paying for its own insurance, without government subsidies. We hope that fusion reactors will become practical in the future, because fusion might be both safer and less expensive than either coal or fission.

---

57. Consider the two nuclear reactions

$\quad$ (I) $\quad A + B \rightarrow C + E$ $\qquad\qquad$ (II) $\quad C + D \rightarrow F + G$

(a) Show that the net disintegration energy for these two reactions ($Q_{net} = Q_{I} + Q_{II}$) is identical to the disintegration energy for the net reaction

$$A + B + D \rightarrow E + F + G$$

(b) One chain of reactions in the proton-proton cycle in the Sun's core is

$\qquad {}_{1}^{1}\text{H} + {}_{1}^{1}\text{H} \rightarrow {}_{1}^{2}\text{H} + {}_{1}^{0}\text{e} + v \qquad\qquad {}_{1}^{0}\text{e} + {}_{-1}^{0}\text{e} \rightarrow 2\gamma$

$\qquad {}_{1}^{1}\text{H} + {}_{1}^{2}\text{H} \rightarrow {}_{2}^{3}\text{He} + \gamma \qquad\qquad\quad {}_{1}^{1}\text{H} + {}_{2}^{3}\text{He} \rightarrow {}_{2}^{4}\text{He} + {}_{1}^{0}\text{e} + v$

$\qquad {}_{1}^{0}\text{e} + {}_{-1}^{0}\text{e} \rightarrow 2\gamma$

Based on part (a), what is Q_{net} for this sequence?

Solution $Q_I = (M_A + M_B - M_C - M_E)c^2$ $Q_{II} = (M_C + M_D - M_F - M_G)c^2$

$$Q_{net} = (M_A + M_B - M_C - M_E + M_C + M_D - M_F - M_G)c^2$$
$$= (M_A + M_B + M_D - M_E - M_F - M_G)c^2$$

(a)   This value is identical to Q for the reaction $A + B + D \rightarrow E + F + G$. Thus, any product (e.g., "C") that is a reactant in a subsequent reaction disappears from the energy balance.   ◊

(b)   Adding all five reactions, we have $4(_1^1H) + 2(_{-1}^0e) \rightarrow {}_2^4He + 2\nu$

Here the symbol $_1^1H$ represents a proton, the nucleus of a hydrogen-1 atom. So that we may use the tabulated masses of neutral atoms for calculation, we add two electrons to each side of the reaction. The four electrons on the initial side are enough to make four hydrogen atoms and the two electrons on the final side constitute, with the alpha particle, a neutral helium atom. The atomic-electronic binding energies are negligible compared to Q_{net}.

We use the arbitrary symbol "$_1^1H$ atom" to represent a neutral atom. Then we have

$$4(_1^1H \text{ atom}) \rightarrow {}_2^4He \text{ atom} + 2\nu$$

$$4(1.007\ 825\ u) = 4.002\ 603\ u + Q_{net}/c^2$$

$$Q_{net} = [4(1.007\ 825\ u) - 4.002\ 603\ u](931.5\ \text{MeV}/u) = 26.7\ \text{MeV}   ◊$$

---

59.   In addition to the proton-proton cycle described in the chapter text, the carbon cycle, first proposed by Hans Bethe in 1939, is another cycle by which energy is released in stars as hydrogen is converted to helium. The carbon cycle requires higher temperatures than the proton-proton cycle. The series of reactions is

$${}^{12}C + {}^1H \rightarrow {}^{13}N + \gamma \qquad\qquad {}^{13}N \rightarrow {}^{13}C + e^+ + \nu$$

$$e^+ + e^- \rightarrow 2\gamma \qquad\qquad {}^{13}C + {}^1H \rightarrow {}^{14}N + \gamma$$

$${}^{14}N + {}^1H \rightarrow {}^{15}O + \gamma \qquad\qquad {}^{15}O \rightarrow {}^{15}N + e^+ + \nu$$

$$e^+ + e^- \rightarrow 2\gamma \qquad\qquad {}^{15}N + {}^1H \rightarrow {}^{12}C + {}^4He$$

(a) If the proton-proton cycle requires a temperature of 1.5×10^7 K, estimate by proportion the temperature required for the carbon cycle. (b) Calculate the Q value for each step in the carbon cycle and the overall energy released. (c) Do you think the energy carried off by the neutrinos is deposited in the star? Explain.

Solution

(a) The solar-core temperature 1.5×10^7 K gives particles a high enough kinetic energy to get past the Coulomb-repulsion barrier to $_1^1$H $+ _2^3$He $\rightarrow _2^4$He $+ e^+ + \nu$, estimated as $k_e e(2e)/r$. The Coulomb barrier to Bethe's fifth and eighth reactions is like $k_e e(7e)/r$, larger by 7/2, so the temperature should be on the order of

$$\frac{7}{2}(1.5 \times 10^7 \text{ K}) \approx 5.3 \times 10^7 \text{ K} \qquad \Diamond$$

(b) Remembering the conversion factor of $931.494 \dfrac{\text{MeV}/c^2}{\text{u}}$

$$Q_1 = \left[12 \text{ u} + 1.007\,825 \text{ u} - 13.005\,739 \text{ u} \right] c^2 = 1.94 \text{ MeV} \qquad \Diamond$$

The photon carries away this energy, but is omitted from the calculation because the photon has no mass. To calculate the energy released in the second step, add seven electrons to both sides to obtain the reaction for neutral atoms

$$^{13}\text{N atom} \rightarrow {}^{13}\text{C atom} + e^- + e^+ + \nu + Q_2/c^2$$

$$Q_2 = \left[13.005\,739 \text{ u} - 13.003\,355 \text{ u} - 2(0.000\,549 \text{ u}) \right] c^2 = 1.20 \text{ MeV} \qquad \Diamond$$

$$Q_3 = Q_7 = \left[2(0.000\,549 \text{ u}) \right] c^2 = 1.02 \text{ MeV} \qquad \Diamond$$

$$Q_4 = \left[-14.003\,074 \text{ u} + 13.003\,355 \text{ u} + 1.007\,825 \text{ u} \right] c^2 = 7.55 \text{ MeV} \qquad \Diamond$$

$$Q_5 = \left[1.007\,825 \text{ u} + 14.003\,074 \text{ u} - 15.003\,065 \text{ u} \right] c^2 = 7.30 \text{ MeV} \qquad \Diamond$$

$$Q_6 = \left[15.003\,065 \text{ u} - 15.000\,109 \text{ u} - 2(0.000\,549 \text{ u}) \right] c^2 = 1.73 \text{ MeV} \qquad \Diamond$$

$$Q_8 = \left[1.007\,825 \text{ u} + 15.000\,109 \text{ u} - 12 \text{ u} - 4.002\,603 \text{ u} \right] c^2 = 4.97 \text{ MeV} \qquad \Diamond$$

$$Q_{net} = (1.94 + 1.20 + 1.02 + 7.55 + 7.30 + 1.73 + 1.02 + 4.97) \text{ MeV} = 26.7 \text{ MeV} \qquad \Diamond$$

This result is the same as for the proton-proton cycle, because the net reaction is the same as that in Problem 45.57, $4(_1^1$H atom$) \rightarrow _2^4$He atom $+ 2\nu$. The carbon is a catalyst.

(c) Not all of the energy released heats the star. When a neutrino is created it will likely fly out of the star without interacting with any other particle.

Chapter 46

PARTICLE PHYSICS AND COSMOLOGY

EQUATIONS AND CONCEPTS

Pions are in three varieties corresponding to three charge states: π^+, π^-, and π^0. Pions are very unstable; the equations at right show example decays for each type.

$$\pi^+ \to \pi^0 + e^+ + \nu_e$$

$$\pi^- \to \mu^- + \bar{\nu}$$

$$\pi^0 \to \gamma + \gamma$$

Muons are in two varieties: μ^- and μ^+ (the antiparticle). Example decay processes are shown.

$$\mu^+ \to e^+ + \nu_e + \bar{\nu}_\mu$$

$$\mu^- \to e^- + \nu_\mu + \bar{\nu}_e$$

Hubble's law states a linear relationship between the velocity of a galaxy and its distance R from the Earth. The constant H is called the **Hubble constant**.

$$v = HR \qquad (46.7)$$

$$H \approx 17 \times 10^{-3} \, \text{m/s} \cdot \text{ly}$$

We model the universe as composed of field particles and matter particles. The exchange of **field particles** mediates the interactions of the matter particles.

- **Gluons** are the field particles for the strong force.

- **Photons** are exchanged by charged particles in electromagnetic interactions.

- **W$^+$, W$^-$, and Z particles** mediate the weak force.

- **Gravitons** are hypothetical quantum particles of the gravitational field.

The **matter particles** fall into two broad classifications:

Hadrons are particles that interact via all fundamental forces. They are composed of quarks.

Types of hadrons:

- **Mesons** have spin quantum number 0 or 1

- **Baryons** have spin quantum number $\frac{1}{2}$ or $\frac{3}{2}$.

Leptons (e^-, μ^-, τ^-, ν_e, ν_μ, ν_τ and their antiparticles) interact by the weak interaction, the electromagnetic interaction (if charged), and (presumably) the gravitational interaction, but not by the strong force.

- **All leptons** have spin quantum number $\frac{1}{2}$.

All particles have a corresponding antiparticle. A particle and its antiparticle have the same mass and spin, and have opposites of all other "charges" (electric charge, baryon number, strangeness, etc.). For a particle composed of quarks, its antiparticle is composed of the corresponding antiquarks. There are a few neutral mesons which are their own antiparticle.

Conservation laws for elementary particles:

Relativistic total energy is conserved in all reactions, but an apparent fluctuation in energy ΔE can occur in the creation of a virtual particle provided that the particle exists for a time no longer than $\Delta t = \hbar / 2\Delta E$.

Electric charge (a scalar quantity) is conserved in all reactions.

Linear momentum and **angular momentum** (vector quantities) are conserved in all reactions.

Baryon number is conserved whenever a collision or decay occurs. The sum of baryon numbers before the process must equal the sum following the process as illustrated in the example reaction.

Example reaction showing conservation of baryon number:

$$p+n \rightarrow p+p+n+\bar{p}$$
$$(1+1 = 1+1+1-1)$$

Electron lepton number is conserved whenever a reaction or decay occurs. The sum of the electron lepton numbers before the process must equal the sum following the process. See the example reaction to the right.

Example reaction showing conservation of electron lepton number:

$$n \rightarrow p+e^- +\bar{v}_e$$
$$(0 = 0+1 \ - 1)$$

Muon lepton number and **tau lepton number** are also conserved quantities.

Integer quantum numbers for **strangeness, charm, topness, and bottomness** are assigned to baryons. These properties are conserved in interactions involving the strong force, but not necessarily in processes occurring via the weak interaction. These quantum numbers are accounted for in the quark model of hadrons.

Quarks and antiquarks (of six possible types or flavors) make up baryons and mesons; the identity of each particle is determined by the particular combination of quarks. *Each baryon contains three quarks and each meson contains one quark and one antiquark.* The table at right lists the charge and baryon number for each of the six quarks and antiquarks.

Quark, antiquark		Charge	Baryon number
Up:	$u, \bar{u}$	$+\frac{2}{3}e, -\frac{2}{3}e$	$\frac{1}{3}, -\frac{1}{3}$
Down:	$d, \bar{d}$	$-\frac{1}{3}e, +\frac{1}{3}e$	$\frac{1}{3}, -\frac{1}{3}$
Strange:	$s, \bar{s}$	$-\frac{1}{3}e, +\frac{1}{3}e$	$\frac{1}{3}, -\frac{1}{3}$
Charm:	$c, \bar{c}$	$+\frac{2}{3}e, -\frac{2}{3}e$	$\frac{1}{3}, -\frac{1}{3}$
Bottom:	$b, \bar{b}$	$-\frac{1}{3}e, +\frac{1}{3}e$	$\frac{1}{3}, -\frac{1}{3}$
Top:	$t, \bar{t}$	$+\frac{2}{3}e, -\frac{2}{3}e$	$\frac{1}{3}, -\frac{1}{3}$

REVIEW CHECKLIST

You should be able to:

▷ Identify the four fundamental forces in nature and the corresponding field particles or quanta via which these forces are mediated. Calculate energy/mass values involved in pair production and pair annihilation. (Sections 46.1 and 46.2)

▷ Estimate ranges of forces based on the mass of the field particles. (Section 46.3)

▷ Outline the broad classification of particles and the characteristic properties of the several classes (relative mass value, spin, decay mode). Complete a proposed reaction by identifying either a particle in the reactants or products of the reaction. Calculate momentum, and energy involved in pion decay. (Sections 46.3 and 46.4)

▷ Determine whether or not a suggested decay can occur based on the conservation of baryon number and the conservation of electron lepton number. Determine whether or not a predicted reaction/decay will occur based on the conservation of strangeness for the strong and electromagnetic interactions. (Sections 46.5 and 46.6)

▷ Calculate threshold energies for reactions of elementary particles. (Section 46.7)

▷ Show that the charge, baryon number and strangeness of a particle equal the sums of the corresponding numbers/quantities of the constituent quarks. Identify particles corresponding to specified quark combinations. (Section 46.9)

▷ Make calculations based on Hubble's law. (Section 46.12)

ANSWERS TO SELECTED QUESTIONS

5. Describe the properties of baryons and mesons and the important differences between them.

Answer There are two types of hadrons, called baryons and mesons. Hadrons interact primarily through the strong force and are not elementary particles, being composed of either three quarks (baryons), or a quark and an antiquark (mesons). Baryons have a nonzero baryon number with a spin of either 1/2 or 3/2. Mesons have a baryon number of zero, and a spin of either 0 or 1.

□   □   □   □

8. The Ξ^0 particle decays by the weak interaction according to the decay mode $\Xi^0 \rightarrow \Lambda^0 + \pi^0$. Would you expect this decay to be fast or slow? Explain.

Answer This decay should be slow, since decays which occur via the weak interaction typically take 10^{-10} s or longer to occur.

□   □   □   □

10. Identify the particle decays in Table 46.2 that occur by the electromagnetic interaction. Justify your answers.

Answer The decays of the neutral pion, eta, and neutral sigma occur by the electromagnetic interaction. These are three of the shortest lifetimes in Table 46.2. All produce photons, which are the quanta of the electromagnetic force. All conserve strangeness.

□   □   □   □

15. How many quarks are in each of the following: (a) a baryon, (b) an antibaryon, (c) a meson, (d) an antimeson? How do you account for the fact that baryons have half-integral spins while mesons have spins of 0 or 1 ? (*Note*: Quarks have spin 1/2.)

Answer All baryons and antibaryons consist of three quarks. All mesons and antimesons consist of two quarks. When one quantum particle with spin 1/2 combines with another particle to form a new particle, the spin of the new particle must be either 1/2 greater or 1/2 less than that of the original. Since quarks have spins of 1/2, it follows that all baryons (which consist of three quarks) must have half-integral spins, and all mesons (which consist of two quarks) must have spins of 0 or 1.

□   □   □   □

SOLUTIONS TO SELECTED PROBLEMS

3. A photon with an energy $E_\gamma = 2.09$ GeV creates a proton-antiproton pair in which the proton has a kinetic energy of 95.0 MeV. What is the kinetic energy of the antiproton? ($m_p c^2 = 938.3$ MeV.)

Solution

Conceptualize: An antiproton has the same mass as a proton, so it seems reasonable to expect that both particles will have similar kinetic energies.

Categorize: The total energy of each particle is the sum of its rest energy and its kinetic energy. Conservation of energy requires that the total energy before this pair production event equal the total energy after.

Analyze:
$$E_\gamma = (E_{Rp} + K_p) + (E_{R\bar{p}} + K_{\bar{p}})$$

The energy of the photon is given as $\quad E_\gamma = 2.09$ GeV $= 2.09 \times 10^3$ MeV

From Table 46.2 or from the problem statement, we see that the rest energy of both the proton and the antiproton is
$$E_{Rp} = E_{R\bar{p}} = m_p c^2 = 938.3 \text{ MeV}$$

If the kinetic energy of the proton is observed to be 95.0 MeV, the kinetic energy of the antiproton is

$$K_{\bar{p}} = E_\gamma - E_{R\bar{p}} - E_{Rp} - K_p = 2.09 \times 10^3 \text{ MeV} - 2(938.5 \text{ MeV}) - 95.0 \text{ MeV}$$

or $\quad K_{\bar{p}} = 118$ MeV $\qquad\qquad\qquad \Diamond$

Finalize: The kinetic energy of the antiproton is slightly (~20%) greater than the proton. The magnitude of the momentum of each product particle is less than 1/4 of the gamma ray's momentum. Thus, another particle must have been involved to satisfy momentum conservation. It could be a pre-existing heavy nucleus with which the photon collided. This means that the nucleus must have also carried away some of the energy. For a heavy nucleus this could have been small — as little as 3 MeV for a favorable geometry. The actual value cannot be determined without further information. This extra particle also explains why the energy need not be shared equally between the proton and antiproton.

5. One of the mediators of the weak interaction is the Z^0 boson, with mass $93 \text{ GeV}/c^2$. Use this information to find the order of magnitude of the range of the weak interaction.

Solution The rest energy of the Z_0 boson is $E_0 = 93 \text{ GeV}$. The maximum time a virtual Z_0 boson can exist is found from

$$\Delta E \Delta t \geq \frac{1}{2}\hbar \quad \text{or} \quad \Delta t \approx \frac{\hbar}{2\Delta E} = \frac{1.055 \times 10^{-34} \text{ J·s}}{2(93 \text{ GeV})(1.60 \times 10^{-10} \text{ J/GeV})} = 3.55 \times 10^{-27} \text{ s}$$

The maximum distance it can travel in this time is

$$d = c\Delta t = (3.00 \times 10^8 \text{ m/s})(3.55 \times 10^{-27} \text{ s}) \sim 10^{-18} \text{ m} \qquad \Diamond$$

The distance d is an approximate value for the range of the weak interaction.

---

7. A neutral pion at rest decays into two photons according to $\pi^0 \to \gamma + \gamma$. Find the energy, momentum, and frequency of each photon.

Solution We use the energy and momentum versions of the isolated system model. Since the pion is at rest and momentum is conserved in the decay, the two gamma-rays must have equal amounts of momentum in opposite directions. So, they must share equally in the energy of the pion.

$$m_{\pi^0} = 135.0 \text{ MeV}/c^2 \qquad \text{(Table 46.2)}$$

Therefore, $E_\gamma = 67.5 \text{ MeV} = 1.08 \times 10^{-11} \text{ J} \qquad \Diamond$

$$p = \frac{E_\gamma}{c} = \frac{67.5 \text{ MeV}}{3.00 \times 10^8 \text{ m/s}} = 3.60 \times 10^{-20} \text{ kg·m/s} \qquad \Diamond$$

$$f = \frac{E_\gamma}{h} = 1.63 \times 10^{22} \text{ Hz} \qquad \Diamond$$

---

11. Name one possible decay mode (see Table 46.2) for Ω^+, $\overline{K}_S^0$, $\overline{\Lambda}^0$, and $\overline{n}$.

Solution The particles in this problem are the antiparticles of those listed in Table 46.2. Therefore, the decay modes include the antiparticles of those shown in the decay modes in Table 46.2:

$$\Omega^+ \to \overline{\Lambda}^0 + K^+ \qquad\qquad \overline{K}_S^0 \to \pi^+ + \pi^- \text{ (or } \pi^0 + \pi^0)$$

$$\overline{\Lambda}^0 \to \overline{p} + \pi^+ \qquad\qquad \overline{n} \to \overline{p} + e^+ + \nu_e \qquad \Diamond$$

371

15. The following reactions or decays involve one or more neutrinos. In each case, supply the missing neutrino (v_e, v_μ, or v_τ) or antineutrino.

(a) $\pi^- \to \mu^- + ?$

(b) $K^+ \to \mu^+ + ?$

(c) $? + p^+ \to n + e^+$

(d) $? + n \to p^+ + e^-$

(e) $? + n \to p^+ + \mu^-$

(f) $\mu^- \to e^- + ? + ?$

Solution

(a) $\pi^- \to \mu^- + \bar{v}_\mu$ $\qquad$ L_μ: $0 \to 1-1$ $\qquad\qquad$ ◊

(b) $K^+ \to \mu^+ + v_\mu$ $\qquad$ L_μ: $0 \to -1+1$ $\qquad\qquad$ ◊

(c) $\bar{v}_e + p^+ \to n + e^+$ $\qquad$ L_e: $-1+0 \to 0-1$ $\qquad\qquad$ ◊

(d) $v_e + n \to p^+ + e^-$ $\qquad$ L_e: $1+0 \to 0+1$ $\qquad\qquad$ ◊

(e) $v_\mu + n \to p^+ + \mu^-$ $\qquad$ L_μ: $1+0 \to 0+1$ $\qquad\qquad$ ◊

(f) $\mu^- \to e^- + \bar{v}_e + v_\mu$ $\qquad$ L_μ: $1 \to 0+0+1$ $\quad$ and $\quad$ L_e: $0 \to 1-1+0$ ◊

---

17. Determine which of the following reactions can occur. For those that cannot occur, determine the conservation law (or laws) violated:

(a) $p \to \pi^+ + \pi^0$

(b) $p + p \to p + p + \pi^0$

(c) $p + p \to p + \pi^+$

(d) $\pi^+ \to \mu^+ + v_\mu$

(e) $n \to p + e^- + \bar{v}_e$

(f) $\pi^+ \to \mu^+ + n$

Solution

(a) $p \to \pi^+ + \pi^0$ $\qquad$ Baryon number is violated: $\qquad$ $1 \to 0+0$ $\qquad$ ◊

(b) $p + p \to p + p + \pi^0$ $\qquad$ This reaction can occur. $\qquad\qquad\qquad$ ◊

(c) $p + p \to p + \pi^+$ $\qquad$ Baryon number is violated: $\qquad$ $1+1 \to 1+0$ $\qquad$ ◊

(d) $\pi^+ \to \mu^+ + v_\mu$ $\qquad$ This reaction can occur. $\qquad\qquad\qquad$ ◊

(e) $n \to p + e^- + \bar{v}_e$ $\qquad$ This reaction can occur. $\qquad\qquad\qquad$ ◊

(f) $\pi^+ \to \mu^+ + n$ $\qquad$ Violates baryon number: $\qquad\qquad$ $0 \to 0+1$

$\qquad\qquad\qquad\qquad\qquad$ and violates muon-lepton number: $\qquad$ $0 \to -1+0$ $\qquad$ ◊

---

21. Determine whether or not strangeness is conserved in the following decays and reactions:

(a) $\Lambda^0 \rightarrow p + \pi^-$

(b) $\pi^- + p \rightarrow \Lambda^0 + K^0$

(c) $\bar{p} + p \rightarrow \bar{\Lambda}^0 + \Lambda^0$

(d) $\pi^- + p \rightarrow \pi^- + \Sigma^+$

(e) $\Xi^- \rightarrow \Lambda^0 + \pi^-$

(f) $\Xi^0 \rightarrow p + \pi^-$

Solution   We look up the strangeness quantum numbers in Table 46.2.

(a) $\Lambda^0 \rightarrow p + \pi^-$ Strangeness: $-1 \rightarrow 0 + 0$
    (−1 does not equal 0: strangeness is not conserved)   ◊

(b) $\pi^- + p \rightarrow \Lambda^0 + K^0$ Strangeness: $0 + 0 \rightarrow -1 + 1$
    (0 = 0: strangeness is conserved)   ◊

(c) $\bar{p} + p \rightarrow \bar{\Lambda}^0 + \Lambda^0$ Strangeness: $0 + 0 \rightarrow +1 - 1$
    (0 = 0: strangeness is conserved)   ◊

(d) $\pi^- + p \rightarrow \pi^- + \Sigma^+$ Strangeness: $0 + 0 \rightarrow 0 - 1$
    (0 does not equal −1: strangeness is not conserved)   ◊

(e) $\Xi^- \rightarrow \Lambda^0 + \pi^-$ Strangeness: $-2 \rightarrow -1 + 0$
    (−2 does not equal −1: strangeness is not conserved)   ◊

(f) $\Xi^0 \rightarrow p + \pi^-$ Strangeness: $-2 \rightarrow 0 + 0$
    (−2 does not equal 0: strangeness is not conserved)   ◊

27.   If a K_S^0 meson at rest decays in 0.900×10^{-10} s, how far will a K_S^0 meson travel if it is moving at $0.960c$?

Solution   The motion of the K_S^0 particle is relativistic. Just like the astronaut who leaves for a distant star, and returns to find his family long gone, the kaon appears to us to have a longer lifetime.

That time-dilated lifetime is:

$$t = \gamma t_0 = \frac{0.900 \times 10^{-10} \text{ s}}{\sqrt{1 - v^2/c^2}} = \frac{0.900 \times 10^{-10} \text{ s}}{\sqrt{1 - (0.960)^2}} = 3.21 \times 10^{-10} \text{ s}$$

During this time, we see the kaon travel at $0.960c$. It travels for a distance of

$$d = vt = \left[0.960(3.00 \times 10^8 \text{ m/s})\right](3.21 \times 10^{-10} \text{ s}) = 0.0926 \text{ m} = 9.26 \text{ cm}$$ ◊

33. Analyze each reaction in terms of constituent quarks:

(a) $\pi^- + p \rightarrow K^0 + \Lambda^0$ (b) $\pi^+ + p \rightarrow K^+ + \Sigma^+$

(c) $K^- + p \rightarrow K^+ + K^0 + \Omega^-$ (d) $p + p \rightarrow K^0 + p + \pi^+ + ?$

In the last reaction, identify the mystery particle.

Solution

We look up the quark constituents of the particles in Tables 46.4 and 46.5.

(a) $d\bar{u} + uud \rightarrow d\bar{s} + uds$ nets 1u, 2d, 0s before and after   ◊

(b) $\bar{d}u + uud \rightarrow u\bar{s} + uus$ nets 3u, 0d, 0s before and after   ◊

(c) $\bar{u}s + uud \rightarrow u\bar{s} + d\bar{s} + sss$ nets 1u, 1d, 1s before and after   ◊

(d) $uud + uud \rightarrow d\bar{s} + uud + u\bar{d} + uds$ nets 4u, 2d, 0s before and after   ◊

A uds is either a Λ^0 or a Σ^0 ◊

39. A distant quasar is moving away from Earth at such high speed that the blue 434-nm H_γ line of hydrogen is observed at 510 nm, in the green portion of the spectrum (Fig. P46.39). (a) How fast is the quasar receding? You may use the result of Problem 38. (b) Edwin Hubble discovered that all objects outside the local group of galaxies are moving away from us, with speeds proportional to their distances. Hubble's law is expressed as $v = HR$, where Hubble's constant has the approximate value $H = 17 \times 10^{-3}$ m/s·ly. Determine the distance from Earth to this quasar.

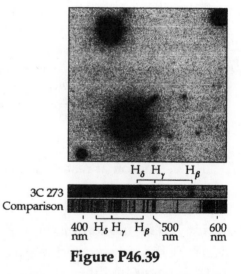

Figure P46.39

Solution

Conceptualize: The problem states that the quasar is moving very fast, and since there is a significant red shift of the light, the quasar must be moving away from Earth at a relativistic speed ($v > 0.1c$). Quasars are very distant astronomical objects, and since our universe is estimated to be about 15 billion years old, we should expect this quasar to be ~10^9 light-years away.

Categorize: As suggested, we can use the equation in Problem 38 to find the speed of the quasar from the Doppler red shift, and this speed can then be used to find the distance using Hubble's law.

Analyze:

(a)
$$\frac{\lambda'}{\lambda} = \frac{510 \text{ nm}}{434 \text{ nm}} = 1.18 = \sqrt{\frac{1+v/c}{1-v/c}}$$

Squared this becomes $\quad \frac{1+v/c}{1-v/c} = 1.38 \quad$ or $\quad 2.38\frac{v}{c} = 0.381$

Therefore, $\quad\quad\quad\quad\quad v = 0.160c \quad\quad\quad\quad$ (or 16.0% of the speed of light) ◊

(b) Hubble's law asserts that the universe is expanding at a constant rate so that the speeds of galaxies are proportional to their distance R from Earth, $v = HR$.

So, $\quad\quad\quad\quad R = \frac{v}{H} = \frac{0.160(3.00\times10^8 \text{ m/s})}{1.70\times10^{-2} \text{ m/s·ly}} = 2.82\times10^9 \text{ ly}$ ◊

Finalize: The speed and distance of this quasar are consistent with our predictions. It appears that this quasar is quite far from Earth but not the most distant object in the visible universe.

═══════════════════

53. The energy flux carried by neutrinos from the Sun is estimated to be on the order of 0.4 W/m² at Earth's surface. Estimate the fractional mass loss of the Sun over 10^9 years due to the emission of neutrinos. (The mass of the Sun is 2×10^{30} kg. The Earth-Sun distance is 1.5×10^{11} m.)

Solution

Conceptualize: Our Sun is estimated to have a life span of about 10 billion years, so in this problem, we are examining the radiation of neutrinos over a considerable fraction of the Sun's life. However, the mass carried away by the neutrinos is a very small fraction of the total mass involved in the Sun's nuclear fusion process, so even over this long time, the mass of the Sun may not change significantly (probably less than 1%).

Categorize: The change in mass of the Sun can be found from the energy flux received by the Earth and Einstein's famous equation, $E = mc^2$.

Analyze: Since the neutrino flux from the Sun reaching the Earth is 0.4 W/m², the total energy emitted per second by the Sun in neutrinos in all directions is

$$(0.4 \text{ W/m}^2)(4\pi r^2) = (0.4 \text{ W/m}^2)\left[4\pi(1.5\times10^{11} \text{ m})^2\right] = 1.13\times10^{23} \text{ W}$$

In a period of 10^9 yr, the Sun emits a total energy of

$$(1.13 \times 10^{23} \text{ J/s})(10^9 \text{ yr})(3.156 \times 10^7 \text{ s/yr}) = 3.57 \times 10^{39} \text{ J}$$

in the form of neutrinos. This energy corresponds to an annihilated mass of

$$E = m_\nu c^2 = 3.57 \times 10^{39} \text{ J} \qquad \text{so} \qquad m_\nu = 3.97 \times 10^{22} \text{ kg}$$

Since the Sun has a mass of about 2×10^{30} kg, this corresponds to a loss of only about 1 part in 50 000 000 of the Sun's mass over 10^9 yr in the form of neutrinos. ◊

Finalize: It appears that the neutrino flux changes the mass of the Sun by so little that it would be difficult to measure the difference in mass, even over its lifetime!

57. Determine the kinetic energies of the proton and pion resulting from the decay of a Λ^0 at rest: $\Lambda^0 \rightarrow p + \pi^-$.

Solution

We first look up the rest energy of each particle:

$$m_\Lambda c^2 = 1115.6 \text{ MeV}; \qquad m_p c^2 = 938.3 \text{ MeV}; \qquad m_\pi c^2 = 139.6 \text{ MeV}$$

The difference between starting rest energy and final rest energy is the kinetic energy of the products:

$$K_p + K_\pi = (1115.6 - 938.3 - 139.6) \text{ MeV} = 37.7 \text{ MeV}$$

In addition, since momentum is conserved in the decay, $\left| p_p \right| = \left| p_\pi \right| = p$. We use one symbol for the magnitude of the momentum of each product particle. Applying conservation of relativistic energy:

$$\left[\sqrt{(938.3)^2 + p^2 c^2} - 938.3 \right] + \left[\sqrt{(139.6)^2 + p^2 c^2} - 139.6 \right] = 37.7 \text{ MeV}$$

Solving the algebra yields $p_\pi c = p_p c = 100.4 \text{ MeV}$

Thus,
$$K_p = \sqrt{(m_p c^2)^2 + (100.4)^2} - m_p c^2 = 5.35 \text{ MeV} \qquad ◊$$

and
$$K_\pi = \sqrt{(139.6)^2 + (100.4)^2} - 139.6 = 32.3 \text{ MeV} \qquad ◊$$